A Science Conspiracy to Conceal and to Cover Up

ISBN-13: 978-1533361691
ISBN-10: 153336169X

Part of

www.sirnewtonsfraud.com

WRITTEN BY Peet (P.S.J.) SCHUTTE

ISBN 978-1-920430-07-8

©KOSMOLOGIESE EN ASTRONOMIESE TEGNIKA

Module One The Prologue

Years ago I was reading of a remark Einstein made about his realisation whiles being a patent clerk. Einstein realised that had Einstein fell from the window of the patent office Einstein would feel as if he was as weightless and as weightless as the chair and a pen falling alongside Einstein down the building would be. The only principle Einstein therefore could not accommodate in his theory on Relativity was Newton's gravitational pulling by the value of mass. Einstein saw this "feeling" as a psychological experience more than it was physics. This moment reading this gave me the breakthrough that I waited for and it (then) took me twenty-three years to make the breakthrough. It was not Einstein or the chair or the pen that fell but it was the space the three components occupied that descended.

Then I realised Einstein felt weightless because he was falling and part of falling was feeling what was happening to him. He was not pretending to fall whereby he then would feel as if…he was really falling and with that there is no as ifs. What he experienced came by means of what he was experiencing in as much as undergoing. If Einstein was experiencing weightless ness, it would be because he was weightless while falling. Weightlessness is not having mass and not having weight is feeling being without mass. If he felt being without weight he then was without mass and then mass has nothing to do with falling or the process of going towards the earth. Einstein would not imagine the weightless ness because Einstein was truly falling. He was at that moment truly weightless. He saw himself falling alongside (not faster as he should if he had weight or mass) than the chair and the pen that dropped at the same pace and at the same sped as he was falling. Einstein, the pen, and the chair had the same weight since they were all weighing the same because they were descending at the same rate. All three items would be equally weightless during the falling…that was what Galileo found because objects of different size and different mass travel equal while descending. This is what Galileo found and that is what Newtonian science can't reject but have to accept in spite of Newton contradicting this founded his pendulum theorem on. Newton contradicting this is because Newton claims things falls by applying mass. Galileo says all things fall equal whereby mass plays no role and Newton says it is only mass that plays a part. Reality TV shows that the bigger objects do not fall quicker than a smaller object and that can only be attributed to one fact; it can only be true if they weighed the same while falling.

From this one can deduct that gravity is motion or the intent to commit motion and mass is one the motion of gravity is frustrated by blocking the continuing of the motion. Gravity is motion of space and mass is the restricting of the motion of space. Having mass does not bring about gravity but it does restrict gravity's motion. Gravity produces mass but mass does not produce gravity. Mass is the restraining motion and gravity is material moving about. Mass only comes into the application when two objects filled with space moves into a position where both want to claim space the other occupy. In essence it still is the frustration of motion and the commitment to move once the blocking of space is relinquished.

I then after reading this realised that gravity is not mass orientated, but gravity is motion differentiation between objects. While falling, The object moves less or slower in the direction that the Earth rotates and will fall in the direction of the Earth centre until such a time as the movement of the object is in synchronising with the speed that the Earth spins or if not the object will and on the Earth surface at the edge of the Earth and that will bring about having mass. The gravity applies as speed that is putting time in relation to the distance travelled and distance travelled is space. While the object is in a process of falling, the motion confirms gravity, both by getting the object's distance or band in which the object travels in harmony with the Earth that conducts all the spinning taking place at that point. That will reduce the height in which the object spins until it lands on the Earth and then can't reduce such reducing of a travelling band any further. It has to do with specific density. If the specific density is increased by filling the object with helium we will find there arrives a point where the conducted speed is at a level that the Earth no longer will claim the body into having mass. When motion downward ends and the Earth disallow any further movement to secure a better specific density in relation to rotating movement, then mass sets in and becomes what is than point holding mass where the constraining of the object takes place to secure frustration of further movement and the Earth's motion annexes the object's freedom. While experiencing mass the motion is still there but now incarcerated by mass and locked onto the Earth by the rotation of the Earth and the superior or equal specific density of the Earth. By connecting to the Earth the motion that the object is experiencing is what nails to object to the Earth by the force of mass and the object is then experiencing mass and not falling further through the loss of downward movement and now only conducts with the Earth rotating side-on movement. In this the downward movement is not lost altogether but remains as detectable movement is the form of having a tendency to move although the object in mass is applying by forcing the downward motion to stand still. While the object is in mass and seems to be as if it is resting the tendency to move downward remains applying but that tendency to continue to move downwards is the tendency he named mass. However mass then restricts motion and becomes motion tendency. While falling, gravity applies as equal motion to all objects relying to place all objects in relation to specific density and because of this motion counteracts any size, mass or weight by making everything able to fall equal in specific density. When falling, the object is either equal to what might be in the air according to allowed specific density, or has more than the specific minimum required density that is what is allowed to serve as the minimum required specific density and therefore will spiral down to the Earth. When the Earth restrains further downward motion of the object that comes as the result of finding an allocated position of motion according to the specific density of the falling object, this readjusting of allocated position is stopped from conducting further downward or readjusting movement and all such further movement of gravity is hindering in the form we call mass. The falling object remains individual and still tends to move while Earth individuality resists movement. Further movement is disallowed as other material fill space. While the bonding of the atoms forming the object will secure any further deforming the object will remain to be independent but it is this bonding that is the value of the specific density of the object applying. By securing a [lace on the Earth, the falling object will finally rest and from that motion resistance comes mass.

While falling, the object is experiencing gravity because the object is in gravity but when on the soil the object experience mass which is the restricting of gravity or motion   of the space filled with material.

 Moreover, I came to another conclusion of equal importance. When any person is standing on any place anywhere, while viewing the Universe, that person is filling the centre of the Universe. Let's get more personal. When you, the person that is reading this, are standing at night and are looking at the Universe you are seeing the Universe from the centre of the Universe. All the light, every single beam that ever left any destiny at any time acknowledges this fact. You are the most important person in the Universe because you are holding the most important position in the Universe. All the light that comes across all of space runs directly in a straight line towards you filling the centre of the Universe. Not excluding the effort of one photon, all light is heading to meet you where you are in that centre spot and not one photon will pass you by. Not one photon dare miss you because if they do they miss the effort that all light has to accomplish and that is to locate you as the person filling the centre of the Universe. If you find this funny, or laughable you are in for a shock because this is what gravity is and this principle dictates gravity. It is the most complex issue one can imagine and expanding on this thought takes thousands of pages. It forms the crux to all cosmic principles and embraces every successful and meaningfully theory ever used to explain the Universe. Without taking this aspect in to account, there is no valid explanation available to understand the cosmos. Al the light coming from wherever meets the point you fill in time and in space. For al the light travelling you hold the spot it was on route to.

Should you decide to shift your position to any other place in the Universe you will shift the centre of the Universe to that location as well. If you install a camera on Mars, the light is obliged to acknowledge your relocating the centre of the Universe at your will to reposition you're being that centre of the Universe. All the light that ever left its destination crossing the vast spaces of the Universe, excluding no particular light, travelled all the way just to find you filling the centre of the Universe, right where you are. By you're standing anywhere, you fill the centre of the Universe, and the entire Universe admits to that because all the light comes to meet you there. If you shift from the North Pole to the South Pole you will shift the centre of the Universe because all the light travelling throughout the Universe will find you where you then moved the centre of the Universe. The light left its destination billion years ago as it travelled through space at the speed of light anxious to acknowledge you're being in the very centre of the Universe. No photon will pass you by where you are in the centre of the Universe. No wonder every person born has the idea they were born to fill the centre of the Universe, which we do fill. The Universe is spinning around you or I, which is filling a centre where all motion is connected. That is the Coanda effect on the uttermost grandest scale imaginable; nevertheless it is only a manifestation of the Coanda effect. It implicates gravity as wide as can be…

Then I reviewed the Universe. If gravity is motion, what causes motion? What stops motion? That answer is in the Black Hole. If a star is about fusing atoms thereby growing, what happen when all the atoms fused into one all collective atom? What is the gravity if the star has one all-inclusive atom providing all the gravity that the star had when the star still had massive volumetric space? If all that space that once filled an entire giant star fused into one enormous gravity applying atom and that enormous force has been secures in the space that one atom holds, the atom would then show a force that would pull the surrounding Universe flat. Where does the gravity of the star end when all the atoms in the star became one giant atom? Gravity is smallest where space is least. Where space of an entire massive star is left in the size of one atom the gravity coming from that will pull the Universe flat at that point.

Coming to the conclusion about gravity being motion and mass being the restriction of motion was the easy part. What produced the motion and what prevented the restriction from overcoming the motion was the tough part. Figuring out why was everything on the move and where did the motion stop that was the part that took some figuring and some explaining. What made gravity move and why does gravity move…the answers are in the four phenomena never yet explained to satisfaction but now turns out to be the cradle of gravity.

Gravity is The Roche limit,
> **Gravity is The Lagrangian system**
>> **Gravity is The Titius Bode law**
>>> **Gravity is The Coanda affect**

And gravity as the Roche limit forms the principle in producing the sound barrier. Read the book and find out why this is the case. I explain these in the next chapter or the chapter following the next chapter.

Newton's claims about the principles that he declared is responsible for guiding physics carries no validated proof and only after I realised that, was I able to start forming another line of thought on gravity. This had the purpose of confronting the corner stone of modern physics and at first I tried desperately to do just that. At first I was not confrontational towards Academics in physics and avoided any indication about disagreeing with Newton, although avoiding to show my disagreements was also totally impossible too but every time I approached academics with my new concept the academics always threw Newton at me . Facing Newton or facing defeat became a two-sided blade and I had to start to confront them by confronting Newton, with which I was in disagreement from the beginning. At first I was reluctant to voice any opinion about the matter of how far I was prepared to challenge Newton because Newton was and is an icon. But slowly it dawned on me that if I had any serious plans to introduce my ideas I had to dispute Newton's gravity principles and do it head. When the slight confrontation did not bring results I finally decided to go all the way and show the inconsistencies that were prevailing in Newtonian science. That worked neither and it brought me the same results as before whereby I decided to go public and straight to John and Jane Dow avoid arrogance academics have with only one motto they serve and that is their autocracy and in particular their megalomania especially to my case as well as me in person. I wrote them (nine in total) letters in which I warned them that I was going public to show the extent of their dishonesty in their Newtonian's approach and lacking of substance and proof their physics has. The lack of honesty and furthermore the absolute dishonest on their part is there whether I avoid it or attack it; the inconsistencies are part of forming the basis for modern accepted science.

This process I now described is explained in a paragraph or less and it seems I got that far in a breath or two, but getting this far took me the best part of seven years to get to I tried my best not to attack them or Newton but left with the option to leave the project and lose thirty years of work and then fail after I concluded an answer on every aspect they never even thought of or take them on and dish out what they should have received years ago made me decide on the latter. After being avoided and taunted by their powerful positions and arrogance vested in their mentality they show in regard to their positions as well as the disregard they show in the mentality of others I slowly concluded that only and after I can get people forming the general public and the opinion of those that holds their disregard just as I do to see what they hide will I get a response from the Mater's of fraud. First I had to show the general public the true colours of the academics in physics and get every one to see how incorrect Newton is, and only then do I stand any chance to introduce my line of thought. I am so sure of the ideas that I propose of being correct that I dare any one to disprove any part or the entirety that my concepts about cosmology forms! But that can only come about when I can get an audience to see how I expose Newton for what Newton was and it is in that where I find no luck. I can't find one academic with influence that is brave enough to stand up and face my attack on Newton and argue me down or prove me wrong in a sound debate. Now I see frowning coming from everywhere because it is madness on my part to think the world is wrong and only I am correct!

I realise that it shows signs of madness on my part and in my thinking to even regard any possibility that I am the only person on Earth that is correct and all others that ever studied physics are wrong but mad as it seems, if that is what I have to say to find an audience to listen and to judge my case, then that is what I say. I don't say this lightly or without understanding the enormity of what I suggest is going on, but be that as it may seem, it is the truth without question that Newton went on for three hundred and fifty years defrauding science with no one testing his claims. Argue me down or prove me wrong but don't discount me before hearing me out and only after considerable consideration while studying my arguments then form an opinion that disputes what I say but when disputing what I say, do it while confronting me in a sound argument when proving me incorrect! This not one academic could achieve and I challenge the lot to do so. But do it after studying all my work and being in a position to account for all the details I propose. Don't just dismiss me because I dismiss Newton because following that road is the way of the coward and the mentally impaired. Read my challenge about the correctness of Newton's proposals when he brought no more than suggestions into science and when I dispute Newton, then take me on by proving Newton correct... do it just once... prove Newton correct just once...prove that his formula is working and that his principles apply on the grounds he principled his ideas.

Detecting Newton's misconduct is possible because I saw a way to break away from the invalid concepts Mainstream physics hold. I went about and tried to prove Newton and when that was not happening I tried to apply Newton's ideas into the greater fields of cosmology. That also wasn't possible. I tried to amalgamate the four cosmic principles applying in cosmology with what Newton said was happening in the cosmos with mass and with gravity and in light of what the cosmos showed was happening Newton just wasn't happening! Notwithstanding the pose Mainstream physics try to uphold, the entirety of physics still use the idea of magical forces intervening in nature and they still base concepts on unexplained novelties. Think of finding four unexplained forces going around and influencing persons in an unexplainable manner except that the magic of gravity keeps people attracted to the Earth. To say the least, the concepts physics use in terms of Newton would not even be acceptable to children in the modern informed era we live in, I challenge any person to prove Newton, not to accept Newton but to undoubtedly prove Newton correct! Prove how Newton's formula of mass forming the force of gravity can apply as Newton said it does! I recognised the impossible double standards Mainstream physics apply to promote their much shady explaining. In short I tested Newton's principles and found the principles to be wanting.

The inconsistencies Newton introduced brought science double vision and to compensate for these bogus truths supporting their incredible theories, they simplify issues to such a level where what they embark on, is the meaningless acceptance of the unproven and they proclaim to understand what are meaningless inconsistencies and to achieve this they create scenarios which uses the entanglement of deception. Prove the attraction Newton said was enforcing gravity that is pulling by mass and is gathering plants by contracting the diameter between planets. Show how much the Moon came closer to the Earth since the time of Kepler. Show proven distances taken by radar tracking and indicate just how accurate Newton was. Show how much the Moon came closer to the Earth since the time of the Moon walk in sixty nine. The figures are available but are kept in a grave of silence where no one ever speaks about what science found applies and how much the distance between the Earth and the Moon is shrinking as Newton said is happening or then how much is the is expanding which will contradict the very principles Newton brought about! What they declare as unwavering facts can't even be supported in some form when tested by a silly test as to show that the distance between the Earth and the Moon is shrinking. Even the least degree of verification of correctness is absent when trying to find support of Newton and Newton lacks all evidence of authentication in any investigation of even the simplest terms. It is as if they never read with interest that which they explain when they embark on explaining Newton and they never scrutinise that which they advocate when they teach Newton's principles applying. They give values that are senseless and the very values they use make that which they say meaningless.

In this book I am going to investigate how much truth there is in mass pulling by the force of gravity. To most if not to all of the persons reading this such a venture of investigating Newton is time wasted and just the thought about me embarking on the investigation of the issue is totally senseless to investigate. It is senseless because the concept it carries became accepted as household practise and life science from where it proceeded to become everyday culture in every person's mind. The worst part is that the group of people normally considered as the wisest bunch there is, never did prudent testing on Newtonian presumptions, while to test the presumptions is most easy to do. I will not believe that a lot that lives up to the veneer of being the best mathematical intellectuals on Earth, never though of testing Newton's very simple formula and in that disregard the formula because of the incorrectness the formula holds.

Do you think of astrophysics as being the department that is run by the wise and the level minded, the honest and pure at heart, the nobility of well-to-do academics and the sober thinking standing in front of the world as the absolute trustworthy? If you are a student, there is no other choice you have but to trust them while they feed you absolute hogwash! If you would so much as dare to doubt any thing they say they will banish you from the institution they rule so absolutely. The banishing process is dome under the blanket of examination. They teach you what to think and to make sure you think what they wish you to think, they tell you to confirm their teachings on a blank piece of paper. You write what they prescribe and you supply the answers they demand in the words (sometimes) of what they demand. Should you in any way say anything different from what they tell you to think, your presence will not be tolerated any further as they abolish you from their institution of academic tutoring. After reading this book I invite you to…no I dare you to challenge their statements with evidence gained from this book and see them wilfully further their culture of deceit by bringing unfounded arguments just in order to silence you and prevent you from getting behind the truth. If you think those in charge of astrophysics are the pillars of trust, then get wise by reading the following facts and arguments this book presents. What you are about to read is simply mystifyingly simple and yet to this day I have not had the privilege to challenged one academic any where that had the honesty to admit to the fact of Newton being wrong. After you have considered the following you might agree with me that even small Children can reach a higher level of clear-minded logic and find more sensibility than what those scientists promoting astrophysics have because science lives in a make believe fool's paradise.

The manner of regard to life that the Academic Physicist holds and the outlook on life that the followers of Newton physics have (I call them plainly Newtonians and to me they are sheepish because they resemble to the image that to me seems the same as sheep running after their leader without having the ability to think for one second any thought spawned out of personal intellect) is quite the opposite of what I think of them. They keep their forming the establishment of the order the Academic Physicist in high regard and consider their order to be the top thinkers in society. This religion that they practise of self promotion and sublimely self regarding their status being next to God has them so high that we down on Earth forming the waste of human garbage can be told anything and we will believe what they say just because they with their supreme intellect tell us to think what they wish us to think. This they do because we human waste living way down below their supremacy have not the ability to think and therefore they must think on our behalf. In their view and so far very correctly judged on their part, they, the persons being in the group that forms the Academic Physicists, believe very correctly that can dish up whatever they wish and we, those forming the group in the gutter, those that are mindless in their eyes, we will have to accept what they say without being allowed to form an opinion other than having the opinion they give us to have because in their view we are unable to have a mind other than what they are able to control. This attitude they have is the result of a relationship that worked for so long and thee fact hat it worked that long is what confirmed their opinion that we, the public, are fools to believe anything and everything because of blind stupidity.

But in spite of their aggravating conduct and mischief towards us, it is not because of a lack of insight and inability of controlling a mind that we have our childlike belief and blind trust in their opinions and which there was. It is the faith we shown that they misused for their scandalous cheating. Our faith is what we have shown towards them and is that, which became used as the reason why we accepted what they said blindly. We didn't accept their word on the grounds of us being utterly stupid as they perceive us to be but our trust depended on our good nature and believing in their trustworthiness. This trust we have is brought on by a culture of trusting the King to do the people well and somewhere in every person's cultural past there was Kings that did us well in leadership. But their underestimating of our abilities is the testimony of their poor understanding and their weak insight ability, which results from their arrogance and stupidity. You are about to see just how stupid they really are in the thinking aspect of science. It will become clear as you page along while reading! They didn't fool us half as much as they fooled themselves and you are about to read all about it. The fact that they could fool us for centuries didn't run on their intelligence being so much superior but served their purpose as it stemmed from the trust we had in them resulting from good intentions on our part. This betraying on their part and misusing the public's good nature to be used in schemes to get the public conned must end and I pray that this book form the first step in resisting the arrogance of the Academic Physicist.

Any one not in their group of the Academic Physicist is part of the lowest order of mindless being and to become part of their order and those that have minds with an ability to think, students have to accept what they say when they say whatever they wish to say without having to prove the correctness of what should back their saying so and as a result of this students may never question what they say. Only when and after proving that a student has totally lost all ability to think for him or her self may a student be promoted into the ranks of their sublime intellectual group. The sifting process they named examinations. You write on paper what they told you and never question their opinion and after passing that examination will you ever enter their sphere of intellectual brotherhood. Does this sound far fetched? Then you better read on and I will remove your blindfold and show you what a world of deception the Academic Physicist force on us into.

Read the following and see how they, the high and the mighty, those that think they can replace God and those who think they can think on our behalf and think what to tell us to think, how much they are clowns and the jokers in society. Read how little are they, the Academic Physicists, able to understand concepts about Creation while they think they are able to replace God in their superior intellect.

If you are a student in the science of physics, then ask your Educated Masters to please explain the following abnormalities you are about to read in this book and insist on a clear explanation about the inconsistencies they promote while tutoring physics as if the physics they present are the most flawless and accurate institution there has ever been. Ask those academics supporting Newton about the following flaws that no one mentions …ever… except me in this book you are about to read and get them to explain the inconsistencies never talked about, which I present in this book and then after confronting those charged with tutoring physics and seeing who should be believed, then get wise instead of brainwashed. Let them mathematically show how one would go about and use Newton's visionary formula $F = G\dfrac{M_1 M_2}{r^2}$ to calculate the force of gravity by replacing the symbols with the actual values in mass that the items referred to have. Put in the Earth's mass in place where it belongs and put in your mass in place where it should be and then divide that with the distance between your soles and the Earth measured in micro millimetres by the square thereof!

In the book named an ***Open Letter on Gravity Part 1 and Part 2,*** I bring the solution to the mystery behind gravity. I tried in vane to introduce the principles I find valid to the academics in charge of astrophysics. Facts that Science present as being the uttermost explicit and unwavering truth, fails to bring any logic answers to so many questions that it should address. It fails to have substance in addressing the most basic and simple questions about gravity and physics. Yet to every question science can't answer my approach does bring many solutions. The presentation and the delivery of my answers that I reach are understandable and simple where it serves both logical science and the truth. Since my answers do not match Newton and his misconception about gravity and that mass generates gravity, those in charge of science don't even bother to read my work. With their affixation to the corruption they portrait I can do little to the giants where they are in the mighty positions they have and just because of that they can go about to sideline and ignore my work and this is notwithstanding the correctness that my work delivers compared to the utter failing that Newton's work shows. When confronted with my evidence and they have to match my work with the hypocrisy and misleading nature of Newtonian cosmology their defence in substantiating their claims is to ignore me. Since I do not applaud mainstream science and the clear fraud they embrace and fraud it is that they embrace, I am silenced. Why is it that my work is going unrecognised or even in the least goes never debated and never commented on…it is because it will then trash every article anyone has ever written about astrophysics and cosmology. They show little integrity when academics with such supposed high standing or then such as they should have, play a dishonesty game where those in commanding positions will rather protect fraud and save their skins. They would rather protect the corruption they have than seek the truth and find honesty in physics. Those academics in charge would much rather protect their un defendable ethos they maintain as forming the back bone in science and what gives their personal position legality although it is corrupt than admit to the truth they find when they begin reading my work and in agreement they then have to back the truth my work brings. Doing that (accepting the truth in my work) will trash all work in cosmology delivered thus far and condemn it to the waste paper basket and render all work invalid and void. It will put all the Newtonian's bias and fraud into the place where it belongs. Considering that such acting will lose them money, those academics in controlling positions then will rather rape the truth in order to benefit from continuing to corrupt student's minds further. If they wish to justify their inconstancies they have to attack my work and disprove the accuracy of my work. That they can't do. They then ignore my work because they can't attack my work. In that sense they also place their work beyond my approach, as they can simply ignore me as if I represent the plague while they carry on with little consequence to bother them. I challenge them to prove Newton correct and not just declare Newton being beyond reproach after all has seen the evidence I bring. After reading this all students must challenge them to defend what they can't or get honest.

$$F = G \frac{M_1 M_2}{r^2} \qquad F = \frac{r^2}{M_1 M_2}$$

This is the basis that Mainstream science uses as the foundation of all physics anywhere. If this is wrong then everything they have got to work with goes out the window. They put mass and the distance that parts objects in a relevancy, in other words the one is a ratio to the other. The one factor brings a measure to the other factor's value. The one cannot be without the other. The increase in one becomes the reducing of the other and the other way round also applies. When the distance is large, the influence of mass will be small and when the distance is small, the influence of mass will be overwhelming. Then they state we are in a Big Bang expanding of the entirety. Why then, when considering that if it is mass that produces an inclining force of contraction as Newton says there is going on then…why didn't the expanding stop before it started when the Universe was small. Today using hindsight after the fact of the exploding Universe became apparent by the studies Hubble brought to light did the lot of everything that is not implode as Newton would have us believe whereas, instead it did expand just as Hubble proved. The radius at the time of the first instant back then was no factor, which makes the gravity at the time a totality of unrivalled force. The radius being that insignificant leaves the mass unchallenged in asserting power in relation to the non-existing radius it had.

I dare any physicist to show me where they apply Newton's formula just and exactly as Sir Isaac Newton suggested gravity applies. Show me just once where the mass of the Earth is multiplied with the mss of the object in normal physics. Show me just once how $F = \frac{r^2}{M_1 M_2}$ or $F \, \alpha \, \frac{M_1 M_2}{r^2}$ where one M represents the mass of the Earth while the other M represents the mass of the object and in this formula the end result will have a value of 9.81 Nm/s^2 … show just once one example… where the use of the mass of the Earth comes into play. If multiplying the mass of the Earth with the mass of an object and dividing that with the distance parting the two mass factors does not deliver 9.81 Nm/s2, and then any claim by Newton indicating that $F \, \alpha \, \frac{M_1 M_2}{r^2}$ is equal to gravity, such claiming constitutes to deliberate fraud…even if Sir Isaac Newton said this. Prove that the mass of the Earth with the mass of an object and dividing that with the distance parting the two mass factors delivers 9.81 Nm/s2 or admit physics is conducting fraud to protect Newton!

To whom it may concern:

My introduction as well as introducing the readers to general cosmology in a very brief and compressed manner but first, I have to give the emphatic warning to all prospective contemplating readers.

Please take note of a conscientious warning about the gravity of the misgiving there is on the part of the most respected Academics in physics about a much concerning matter. I state it emphatically that science accuses me to be not schooled to the point where I am able to have any form of an opinion on any matter concerning Sir Isaac Newton. Notwithstanding that my research proves I did my private studies and through which I skipped the indoctrination and mind control academics place on students goes unrecognised by their standards and so too my ability to have any insight on matters regarding physics. However my skipping their methodical and systematic brainwashing enabled me to see and allowed me to be able to express the incorrectness in Newton's teachings and allowed me to show in clarity what destructive force Sir Isaac Newton used to corrupt the laws of mathematics, corrupting to science along the way and mostly raping to the work of a great man, Johannes Kepler and what Sir Isaac Newton did can only be expressed as being blatant criminal fraud. What his deeds amount to is to corrupt the laws of mathematics, to render the laws of cosmology useless and to rubbish all of science. Should you find this to be unbelievable, then I am glad to announce that this book is more for you than any other person, so go on and read what academics guarding science never wanted published. I challenge any one that disputes any claim I make to prove me wrong by proving me wrong and not merely suggesting claims in that direction.

PROVE ME TO BE INCORRECT IN ANYTHING I SAY!

To whom it may concern and all others reading this document:
This is my introduction and this is my prologue:
But before I can commence with that task I have another duty administer: I AM ABOUT TO WARN EVERY PERSON IN SIGHT OF MY WORK ABOUT MY SLENDER ABILITIES.

Therefore in the light of what the most respected academic group on Earth accuses me of, I therefore have to issue a most serious warning to any person with the intention of making some kind of inquiry to the content this book holds, then the most concerning matter involving any content within the pages of this book you hold are that you must please seriously consider that where the stating declares the possibility that the content in this book has been (written by...) then don't take the announcing Written By Peet Schutte (Petrus S. J. Schutte) very seriously for there are grievous doubts leaving considerable dispute about the possibility, which underwrites the authenticity of Peet Schutte achieving the (written by...Peet Schutte) status. Please take note of the following dehortation. In the light of the reference to me serving in the capacity as being responsible for authoring, (written by...) in line of keeping fairness and justice to members of society, where all civil beings should carry reputed honesty, then: Please be warned before any reader starts reading about the following extremely serious admonition: I am bound by my conscience to warn all intended readers that I am placed under caution by the Academics in Physics. Those most esteemed members responsible for the guardianship and maintaining the ethos in physics are of the opinion that I, Peet Schutte, am unable to write any book on the science of Physics as well as Astrophysics. Therefore I, Peet Schutte, must declare that I should be considered as not very able to write anything, because I am incapable thereof. I suppose, I merely generate new information, which I establish as thoughts and then gather as concepts. I further collect the result as words, which I put on paper using alphabetic symbols. I then compile that in a format that others may confuse with a book, but a book it cannot be, since the Masters in science found me unable to write a book. But before you go further and follow my arguments, I first have to level with you about how academics view me in the position I hold. Please do not allow me to fool you, for this then cannot be, or represent a book. Now I have done my duty in warning everyone and in that, I denounce further participating with any purposive intention to wilfully bring down the crux of civilization by acting unacceptable and irresponsible.

I didn't write a book since I am not schooled to do so. It is my guess that I merely generated uninformed thoughts, which I collected as alphabetic symbols and plotted that in ink on paper. This effort I achieved from harbouring my delusional ideas spawned by a dehumanised brain. It only proves my weak and under developed mentality, due to my lack of an informed insight that is a typical symptom that all those have that is suffering from a disadvantaged past that one can only have when the person obviously lacks formal education. While you are reading the letter deciding to regard or dismiss my work, then also please keep in mind when reading my language used and also please give credit where it belongs...if you do find linguistically improper use of words or misspelling, then remember that I am a feeble minded motor mechanic and not a literal giant. I do find much pride in my status as being Afrikaner and would like to have my names used by pronouncing it in the manner Afrikaans dictates...therefore I would sincerely appreciate the courtesy when readers will take note that my name and last name are pronounced in Afrikaans, which is originally from Dutch and must be pronounced that way. Peet one would pronounce "here" which is the closest English to the pronouncing of the "ee". The "Sch" in Schutte is pronounced exactly as school is where both actually are pronounced Skutte or "skool". By pronouncing my name in Afrikaans you do me the utmost courtesy any one can. Being an Afrikaner is what I am most proud of.

Should any person challenge me about the legitimacy of the statements and content of this book, please do so at any time after you have familiarised yourself with what it is I say. However do not do it on the grounds of only the information provided in this book. For such persons believing totally in the accuracy of science, the believability of Newton, and the uncorrupted nature of Physics academics first get to know the truth by going to and reading www.sirnewtonsfraud.com, www.questionablescience.net as well as http://www.singularityrelavancy.com/ and see what facts you ace. There are so much facts pointing the truth and that much more detail when you read the six part theses called THESIS"

As you read the title of the book
www.SIRNEWTONSFRUAD.com

I know and realise that you are disgusted by my attitude when I degrade the name on which physics are founded. In this introduction part I am going to show you just some of the deceptions all students are forced to believe since all physics students are forced to believe in Newton, **Sir Isaac Newton** that is.

In the following am giving you a choice. You can say I am going to commit fraud by aligning the planets' positions according to mass but then Newton has committed the fraud because I only follow his lead. If I am judged to be the culprit that is guilty of deception then it is because Newton misled me. You can choose.

You are expected to believe the following: Newton stated under the nametag of Kepler that there is so called Conversions for "Unknown" factors.

Tell me, can you find any credence in the "Conversions for "Unknown""

$$4\pi^2 a^3 = P^2 G(M + m)$$

In this comes the fraudulent part because there is no evidence of mass playing a part or forming an actual presence in the solar system.

If the cosmos supported Newton's claims of $P = \left(\dfrac{4\pi^2 a^3}{G(M + m)}\right)^{0.5}$ then the planet arrangement would have been much more likely as I show above, but the picture indicates the mass as well as the planet formation.

You must judge; it is either the cosmos that is incompetently wrong or it is Newton that is incompetently wrong because what the cosmos has in place Newton knows nothing about and what Newton claims the Universe uses, the cosmos knows nothing about. Who would you say knows more about the cosmos' method of workings, Newton or the cosmos? If Newton is correct then the planet layout must be as I show with Jupiter very close to the sun. It seem the cosmos is just as unaware of Newton's ideas as Newton is of what is happening in the cosmos. Who would be correct about cosmic principles applying, the cosmos or Newton?

$P = \left(\dfrac{4\pi^2 a^3}{G(M + m)}\right)^{0.5}$ What hogwash does the factor $\dfrac{}{G(M + m)}$ indicate?

The same can be said in the formula $M = \left(\dfrac{4\pi^2 a^3}{GP^2}\right) - m$ when $P = \left(\dfrac{4\pi^2 a^3}{G(M + m)}\right)^{0.5}$ that the

factor $\dfrac{P^2}{}$ is senseless and $\left(\dfrac{P}{2\pi}\right)^2 = \dfrac{a^3}{G(M + m)},$ has no foundation other than fraud.

$$M = \left(\frac{4\pi^2 a^3}{GP^2}\right) - m$$

is complete fraud. The Cosmos does not support the Newtonian formula even in one place where it could apply.

Position as a function of time

$$P = \left(\frac{4\pi^2 a^3}{G(M+m)}\right)^{0.5}$$

This is what Newton said is in place and with no evidence ever founding this ridiculous proposition, all Newtonians that ever come after Newton. This is what Newton and his Newtonian followers tell the solar system it has in place and tell the cosmos it uses to operate. I have indicated that mass has no place or use in the solar system according to what the solar system puts in place.

Visit www.singularityrelevancy.com to obtain more information on the subject free of charge.
Visit LULU.com and request The Absolute Relevancy of :
By going to LULU.com the following books are available in e-book format as individual books wherein I share with you the newly discovered information about Go to Lulu.com and download
Questionable Science
http://www.lulu.com/content/e-book/questionable-science/7742175]
Questioning Newton's Mythology
http://www.lulu.com/content/e-book/questioning-newtons-mythology/7570956]
QUESTIONING NEWTON'S PHYSICS PRINCIPLES Part 1 – 5
http://www.lulu.com/content/e-book/questioning-newtons-physics-principles-part-1---5/7763188]
www.sirnewtonsfraud.com
http://www.lulu.com/content/e-book/wwwsirnewtonsfraudcom/7182451]
www.questionablescience.net
http://www.lulu.com/content/e-book/wwwquestionablesciencenet/7607865]
singularityrelevancy.com The Comprehensive Article
http://www.lulu.com/content/e-book/singularityrelevancycom-the-comprehensive-website/7319058]

These are the closest because these are the massive giant gas plants and having the most mass must put them the closest to the Sun.

$$P = \left(\frac{4\pi^2 a^3}{G(M+m)}\right)^{0.5}$$

Get your professor to prove Newton correct in the face of and if he can't let him admit he has been conducting in a fraudulent practise all the time he was teaching.

1. ALL THERE IS... ABOUT NOTHING?

The single most tedious problem I faced in writing this book was where to start. This book is everything about the **Universe** in time and space, which on their own is eternal and put together, remains eternal. Wherever I decided to begin there were always some factors that lead to that event. Firstly, I had to explain those factors that the readers were to understand, before I could explain the events that took place wherever I decided to begin. Unfortunately, these factors did not stand alone, but were supporting other events that I first had to explain. If there were not a well founded comprehension about the factors that led to the events which supported the factors in explaining the events what occurred before the start of whatever point I decided to start off with, it seemed as senseless as this sentence which I just wrote. Writing one sentence, while sounding stupid is one matter, but to write a whole book and coming over sounding like an idiot is quite a different kettle of fish to fry.

The obvious point to start with, should be the Big Bang, with one minor problem. In the Big Bang, the Universe was not a **"NOTHING"** that came from "nowhere". The Universe is forever "SOMETHING" which is flowing from "somewhere" to another destiny which seems to be right in the middle of wherever the Universe is heading. That was what forced me to start with the "NOTHING", and only by starting with "NOTHING", I stood a chance in not achieving "NOTHING".

In order to understand the Universe you had to understand Einstein's theory. Einstein said there are three known substances in the Universe that we know. One is matter, two is time and three is space. Matter is one substance, which we consider everything should consist of; time is a **"NOTHING"** that man created and space is a "NOTHING" that God created.

However, have you, as the reader, ever considered what **"NOTHING"** is? What is **"NOTHING"?** It is a notion used by all and understood by none. No person can define the exact meaning of this word. In every day language, it applies to an understanding of not understanding. This seems to apply to the world of cosmology as well. The Universe takes its birth from this very word. However, there must be some explanation to the concept of the meaning of the word, if it has any right to exist.

Only when one put into context with its contrary meaning, which is **"SOMETHING"** there comes validity to the meaning of **"NOTHING"** and only then can one visualize the concept of "NOTHING". Let us use an imaginary scenario. Let us pretend it is nighttime and there is a dog outside that starts barking. The owner of the house gets uneasy and goes out to investigate. According to the dogs, observation there was "SOMETHING" to investigate. However, to the person's observations there was "NOTHING" outside. Who is right?

Let us take this same scene a little farther. Now the dog enters the house. In the living room, he sees humans staring at a light that flows from a square apparatus. The people's faces seem taut and hypnotized. Because of the dog's inability to see the picture in the T.V., the dog will find the human behaviour confusing. To him, they focus on, "NOTHING" but to their vision's ability, "SOMETHING" excites them.

With this in mind, that the concept of **"NOTHING"** can only be defined by the use of the word **"SOMETHING"** and in this doing, **"SOMETHING"** validates the meaning of **"NOTHING".** With the next verbal sketch that might seem to be a little tedious, I would like to prove my point. I use myself as an example as not to gossip about the facts of another person.

There are parties in my family who thinks of me as **"NOTHING".** According to those persons, my outlook on life is flat, boorish, racist, mindless and rude because of my personal standings and viewpoints.

According to me, I consider myself to be to the point and without frills. I am the owner of a B.M.W. 735 I but the model is a 1986, thus almost 16 years old. It has little value and nobody is interested in buying it, though it was an expensive German car.

I live in an ordinary farmhouse that is far from good looking. I own a farm with an enormous debt. My five tractors, two centre pivots, three trucks and all my implements are between 10 and 30 years old. According to my standards, these antiques that the bank regards, as **"NOTHING"** is all I have. That means I am financially worth **"SOMETHING"** because it is better than **"NOTHING".**

My relatives, seen from my point of view are gaudy overindulging with liberal notions because they benefit financially from being "liberated". To my mind, they hide their shortcomings behind a curtain of money and therefore their outlook on life is **"NOTHING".** They consider me as **"NOTHING".** I consider them as **"NOTHING".**

We cannot both be **"NOTHING"** because then "NOTHING" must be sharing in a common and specific value. If this was true, we then must have had **"SOMETHING"** in common though it will be **"NOTHING"**, but we have **"NOTHING"** in common therefore we share **"NOTHING".** We have **"NOTHING"** in common. With all this personal gossip, my only intention is to give meaning to the **"NOTHING"** that I share with my relatives and not even that **"NOTHING"** is **"SOMETHING".**

This brings about that **"NOTHING"** only has meaning when compared to **"SOMETHING"** which then can put a certain value to "something "and rob **"NOTHING"** of every value it might have.

This misconception divides families political convictions, religious believes, and cuts through society to the bone. In addition, all of this comes about because of **"NOTHING".**

Let us take this **"NOTHING"** even farther. As a concept in the universe, we are able to take this **"NOTHING"** much farther. The atom compromises a nucleus in the middle and electrons orbiting the nucleus. In a sketch on paper this two dimensional drawing seems exceptional and the atom seems to be logic. However, if you put the atom in the correct perspective, your comprehension of **"NOTHING"** becomes mesmerized, stripped of all logic even to a point of absolute inconceivability. To prove this point I will place the atom to a more lifelike scale.

Put an object the size of a sugar grain in the middle of a rugby field in the very centre. Consider this the nucleus of a hydrogen atom. Now put a grain of salt right in the middle of the upright post, fifty yards away from the grain of sugar. You now have constructed your hydrogen atom. The orbit of the grain of salt (electron) will now move in a circular motion around the grain of sugar at the speed of light from the one post to the other post.

Now go and stand on the pavilion at any given angle and admire your hydrogen atom. Your atom would be unsighted to yourself or any other person with normal eyesight. Remember the two grains are representing the **"SOMETHING"** and the grass is representing the **"NOTHING"** part. The grain of sugar and the grain of salt is the portion that one can regard as the matter part and the grass represents the "NOTHING" in the atom. There seems to be a very little of **"SOMETHING"** and a tremendous amount of **"NOTHING"** in this picture.

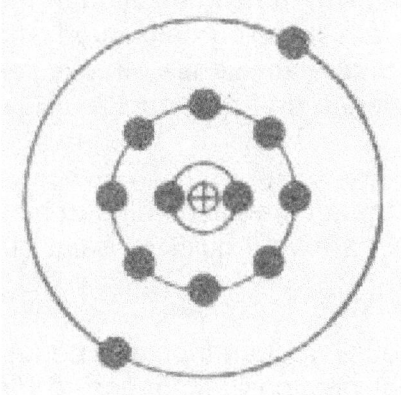

Now, let us construct the iron atom, which is the atom that keeps everything known to man in our Universe in a structural form by means of so called gravity. (Later in this book, I shall qualify this statement.)

For this experiment, you would have to go out to a vast expansion like the Namib dessert. Choose any point and put down a golf ball. Now put two grape pips opposite each side of the golf ball at a radius of 10km. Now in a radius of 20 km from the golf ball 40 km in diameter put down eight grape pips evenly apart. Now place another 16 grape pips in a diametric circle of 80 km (radius 40 km) evenly spread over the circle.

If you want to construct an even denser atom, you would have to go another 20 km and construct a circle with 32 pips in this circle. However, let us stick to the iron atom. Now you would have to fly in an air balloon at a height of about 500 km to picture your construction of one of the densest atoms in the universe, which make up stars. Even if you had all the imagination of a space scientist, (and do they have over inflated imaginations!) you still would only be able to see sand. Now this represents an atom of enormous density! The sand in fact represents the **"NOTHING"** that this enormous dense atom is made of. This boils down to an enormous misconception that overruns any sort of sane logic and it is no wonder that none of the "SUPER- EDUCATED- MASRER- OF- FACT" and most learned cosmologist experts ever tries to explain the workings of this **"SOMETHING"** to **"NOTHING"** ratio.

From this stance the **"NOTHING"** is so overwhelming that the **"NOTHING"** to the **"SOMETHING"** ratio becomes unexplainably unreal. If the **"NOTHING"** in this argument is so overwhelming, the only conclusion I can come to is that that **"NOTHING"** must be **"SOMETHING"**.

In a question I once put to a "SUPER- EDUCATED- MASTER- OF- FACT" Genius, how the construction of this atom can bring about a dense and solid structure as in the case of iron, he started blabbering about the electrons moving so fast that it forces the atom to become impenetrable. The worst part was that I had to sit and listen to this rubbish. What happens at 3 000 degrees when iron becomes a gas. Does the electrons then become slower as it orbits the nucleus?

This means that I have to conclude that the speed, at which the electrons travel, will bring about whether matter is solid, a liquid or a gas. This means the negative particles determine the density of the atom. This is not the end of the argument.

Now, picture the scene where the electrons orbit the nucleus of the atom, in an orbit that place them right next to the nucleus. The electrons are now rubbing against the nucleus in such a way that **"NOTHING"** separates the two. Will there be more **"NOTHING"** between the electron and the nucleus in the last mentioned relation or will there be more **"NOTHING"** in the previously mentioned example. In which of these two examples are there more **"NOTHING"** that separates the electron and the nucleus of the atom? Whatever your answer may be, then why? I do not care how brilliant you may think you are, but to this question there can be no answer as to the value of **"NOTHING"**.

I believe predominant in my Creator. That includes my belief that His creation is perfect. If the creation were perfect, there would obviously be logical, calculated and obvious solutions in the composition of the universal creation. With this construction of the atom, science obviously fails in being calculated and logic. There is far too much **"NOTHING"** and far to little **"SOMETHING"** to make any sense. In this lies my main motivation why I conducted my studies, because I believe that science and religion should be the very same thing if the same Creator creates it.

The atheists believe in **"NOTHING"** after death. When he breathes his last breath then according to him, there is **"NOTHING"** that follows this departure of life. Those same atheists are very quick to point out how valued life is and how dear we must consider it to be.

No atheist, up to now could explain to me the reason why, if life is so precious, would it then just be lost after death to **"NOTHING"**? Why then would **"SOMETHING"** which is that precious (according to the humanists), just vanish into **"NOTHING"**?

Personally, I think that all atheists are on a self-sublimation crusade. Their way of thinking is that if man, including the atheists are without a god, man then is put right on top of the priority list and becomes a god! As scientists, they become even more of a **"SOMETHING"**, because the scientist becomes one of only a few that can understand and explain to a very trifling point how creation actually works.

However, if there is a Supreme Being, then this blasphemer and self appointed god becomes just another **"NOTHING"** as all the other "nothings" that call themselves humans. The atheist tumbles down to the same level that we, the "nothings" find ourselves to be in the order of importance. Then the atheist is not a **"SOMETHING"** any longer and he could not consider himself one. By recognizing a Supreme Being, the atheist will become the same **"NOTHING"** that he regards the rest of us to be and that would be the end of his sublimation in regarding himself to be **"SOMETHING".**

3. IS THERE DEATH AFTER LIFE?

Science believes in science with no reservation. Therefore scientists believe in science with a belief more that Theologians could believe in God. To Physicists everything there is must be science. They formulate a mathematical equation and when doing so the explain everything in man's field of knowledge. Those in science propagate that we will live up to a thousand years in a very short time in the future. However ask them to explain aging and they have no idea what they are talking about. Science puts life down to some acids and a jolt of electricity and with that life could be programmed. According to science life is mostly electricity stored in the brain and by electricity the body moves life around through the body. What a lot of hogwash and this attitude of simplification is so Newtonian junk as Darwin's simplified ideas are about the origins of life. To any Newtonian any suggestion in a simplified form regardless of proofs everything science requires to become accepted as fact. The overwhelming majority of physicists with Doctoral degrees in physics will not have the ability to read this article and too understand the arguments. That is because there is a shortfall in their argumentative ability. This inability runs very deep in physics.

When there are aspects in nature that physics can't address then we have to look for the shortfall in physics. Physics holds the opinion that God cannot be proven or be substantiated. The fact that physics can't accommodate a certain fact or feature does not exclude the fact, but it underlines the failure in physics. In example when we look at the fact that I can think and my thinking is a fact beyond proof, yet physics can't prove how I think or why I think, except put it down to a flow of electricity in my brain and body. Other than that they have no capacity to know what life is or to understand the concept of life. If they shock a frog leg with electricity and the muscles show spasm they then conclude life is electricity. It never dawns on them that life is the ability to generate electricity whereby muscles are controlled and life is a lot more than just the flow of electricity. This shows the absolute fallibility of physics and not the absence of my thoughts. The Newtonian mindset is to keep physics above reproach and beyond suspicion while it is desperately poor in all senses.

If physics tries to put my thought down to some brain activity somewhere in my mind it shows how incompetent the reasoning is behind the argument. The activity is electrically induced and could be harvested by jolting on nerves from some exterior source and that does not prove that my thinking comes from the brain being jolted by electricity, it proves I am some thought with the ability to jolt the brain into action by supplying the electricity. As the medium could arouse action by stimulating the brain with electricity, so the true "I" has the ability to stimulate the mind with a jolt of electricity to provide electricity. If it was the electricity that did the job, we then could jolt a cadaver's mind and get the cadaver filled with life. The brain is not what stores life but it is life that keeps the brain with life. If life was in the brain we could generate the brain back into "life" by shocking the brain with electricity hours after death.

Once life is lost, no electricity can restore the factor we call life. Life we know is in thought and thought is the presence that physics never can accommodate and yet the entirety thought to be life in whatever form is held by thought and with thought the body is motorised by life. Yet, when physics can't accommodate thought we do not discard thought as an absence that is unproven. My thoughts are with me until I die. If physics fail to accommodate God then it is not God not being a reality but it is physics being an utter failure. I can and I do prove God as the absolute factor in the cosmos, but before one can get there a lot of Newtonian garbage has to be discarded and a lot of misleading Newtonian disbelief has to be dismissed. If you look at the night sky you will see many specks of light. In that understanding but more advanced to understand than any person can realise we find the proof of God.

It is not in the light but in understanding the principle forming light and the understanding behind the concept of light that we can mathematically prove God by using mathematics. I have done just that but it takes a book of 700 plus pages just to explain the reality forming the concept. Atheism is not proof of intelligence but it is proof of the lack thereof. My dog is the biggest atheist walking this earth but he is that because he is stupid and has a lack of mental capacity. One must star to understand light and understand the light is the Universe. The Universe is not material and darkness but it forms by light. One doesn't see a galactica, one se light that formed a galactica a very long time ago. That what I see is the Universe and the Universe forms by light left behind by time as space. I am not going into that because I wish to try and keep it simple.

Have you as you sit reading this part at this minute sat back and gave a thought about the light enabling you to read? Such a thought brings to mind the most simplistic answer one can imagine. The light hits the page bounces from the page and contacts the lens of my eye where the lens conveys the photons becoming electricity to a part of the brain that translate the electricity to an understandable message and that makes one read. It is as simple as that! Ever gave a deeper thought about light streaming across the night sky, coming from the visible limits we think of as ends of the Universe we do not even realise it is there? How does the photons manage to convey one complete picture coming from as far apart and as wide an area as it does? With a few photons connecting the eye or lens no one ever noticed the wonder of light. The photons reflect a view that seems as if coming from all the billions upon billions of stars. But most is coming from darkness covering an area no man can measure. Yet how many photons can actually connect to the lens of the camera or to the eye considering the size the eye allows light to pass through? We see by using a few photons.

We see with a few photons going past our lens a Universe representing immeasurably many photons. Still a few photons coming from a single direction directly ahead eventually tell the entire storey of a Universe larger than we could ever understand. What we see is bigger than any person can comprehend and what we apply to see that much is smaller than any human mind can comprehend to understand. It is very simple to take the process of seeing by means of photon conducting very lightly and I have never heard one of the Brainy Bunch really in sincerity uncover the process to its utter and full potential. Moreover let those Mathematical-Masters put this notion into an equation and conduct the understanding they have by proving the concept I just explained. It is impossible that light from such an array of assorted sources can simply come together at the eye lens and show a picture of objects spanning across a Universe as wide as our mind can receive where the objects they reflect is beyond human measurement and the quantity we can apply to receive a vision is inconceivable many. With that small space within our eye how can we see the space that is large as the Universe that we see? I am never going to try and Simplify the idea behind understanding God in terms of physics, but very deep inside this understanding we can begin to fathom the presence of God in terms of physics, not religion according to a Bible, but in terms of Physics. But understanding that is way ahead and light-years advanced from what I explain in this book.

Light is much more than the medium science takes it to be. Light connects the Universe in a way we cannot contemplate. Light being far apart originating from regions not in the same time or Universal space connects in a way that present us with a picture holding the Universe in an understandable content. From the point we stand and we watch the Universe the significance of what we see surpasses the sense of understanding of what we are experiencing and more so surpasses what we are able to understand. How can the few photons that our lenses catch coming from such an area as the night sky cover transmit the complete picture of what we see. Take a few seconds and study the picture of the night sky then rethink the picture applying the full content in the picture to what the size of you eyes is. Think how big the picture is that your eyes take in and translate that area to the size of your eyeball in an effort to determine a ratio.

One will be forgiven if one thinks of the ratio as eternal to nothing. Yet a few pages back I showed that according to mathematics there couldn't be anything as nothing. Consider the path the light followed from the source connecting to light from all other sources where all particles of the other light may come from and bringing a full picture to the lens one use to look through. In your mind connect a line from every atom producing light and connect the lines to your eyeball and see how you can manage to fit all the lines, as small as the lines may be. Understanding this can only come when we understand singularity and I still have to find a person educated or otherwise that would be able to understand this idea. **The Absolute Relevancy of Singularity The Website** http://www.lulu.com/content/e-book/the-absolute-relevancy-of-singularity-the-website/7517996] **starts to introduce you to how singularity apples.**

If it is lenses that enable us to see what we can't see in outer space it also means we cannot see the light, which is outer space because we haven't got the lens to match the curb of outer space. Newtonians think of outer space as geodesic zero, with nothing in outer space but space. Geodesic zero means the light travels in a straight line from where it originates unhindered all across space to where the light connects the eye. Such an idea by itself is outrages because the stream of photons reduce in space to such a minute quantity that taken the area the photons travel and the space in vastness it covers, the chances of one photon coming across many hundreds of light years through billions upon trillions of cubic kilometres of space and selecting my eye to convey the electricity is less than infinite. Yet such conveying takes place every second of every minute. The position of the location of the second singularity, which is the precise duplication of the first singularity but in a diminished capacity, is obvious to miss when one is not applying a detective mentality, as one should in scrutinizing the cosmos. Culture will have us believe that when one sees a colour shining from an object the colour is associated with the object. Logic tells a different storey. A yellow dot is all the colours in the spectrum but yellow because it is disassociating with the yellow. That goes for red blue and all other colours we may visualise. I think the norm accepts this as scientific fact with very little argument or substantiating proof about that required.

 If light came as individual streams of photon flurries, then our visage would translate that as such shown in the fragmented as telescopes enlarge images. If the light only held what we think of as light, we would be unable to see the dark parts because only the light parts would contain light. The total picture we could see would be a picture unconnected bringing across some photons in the manner where every object stands apart not being related in any way and that will be what we see, if it is anything that we see. That we know is not the case but that means geodesic zero is as much rubbish as anything Newtonians regard with simplicity and with careless thought. Geodesic zero means nothing and how can I see nothing as darkness because "nothing" is not darkness, nothing is "nothing" and the darkness I see is darkness showing the darkness as something. The darkness we see is as much "light" as the "light" we see because we see Light" and in that we see "darkness" as another form of light.

What then about colours that are technically not colours as is the case with black and white? White is simple. By spinning all the colours in the spectrum the colour white shines through. Black is quite another matter. A friend of mine whom is one of the best painters I have ever come across told me that one couldn't paint black but have to make black a dark blue to show shade on the canvass. That apparently is his success in achieving the realism. He also went on to explain how many variations of dark blue form the shadows in one simple tree. This remark set my mind in motion. One cannot see black because black has no colour to show, but black is the colour most prevalent in the universe. One can see only by colour and since black is not a colour we should not see black, but we do.

The fact that we see light means that the dark next to the light cannot be "nothing", If the darkness was the representation of "nothing", then that should be exactly what we must see, nothing but the stars. Taken from the top picture some stars and leaving the rest to nothing is what we see in the picture below. A blind person sees nothing but when we look at space, we see something that we think nothing of as we see as space. One cannot have the ability of sight and see nothing. It is light that we see and it is light that we use, which enable us to see.

That proves the darkness that we see in outer space is light that we see without recognizing it as such. If the darkness was the representation of "nothing", then that should be exactly what we must see, nothing but the stars. Taken from the top picture some stars and leaving the rest to nothing is what we see in the picture below. A blind person sees nothing but when we look at space, we see something that we think nothing of as we see as space. One cannot have the ability of sight and see nothing. It is light that we see and it is light that we use, which enable us to see. That proves the darkness that we see in outer space is light that we see without recognizing it as such.

What puts us humans in a category one higher than animals (or so we like to think) is our ability to think about that what we can see. The less develop an animal is the more it has the attitude of eat or be eaten. The higher developed animals are the more the animal find reason to argue. One may teach a crocodile not to eat you if you start feeding the animal. That is a mindless reptile and yet it can think above eat or be eaten. What we see is not merely the truth and it requires reasoning to see the truth and substantiate between culture motivated observations and thought through decisions.

To all the **Super-Educated-Mathematically-Superior-Intellectuals** physicists believing in mathematics and those trying to replace God with mathematics, prove the ability to live by calculating thought as the sole factor of life where it is life that controls the body and not the other way around. Should any of them insist that the mind is responsible for life, then revive a cadaver by filling it with whatever acids you claim produces life or shock the corps until it roasts or force movement onto the body. If it can't revive, then go and calculate by mathematics the precise ingredient it is that has left the body and therefore has filled the body with death. Mathematics is as unaccommodating to reality as Newton is to cosmology. All those Super-Superior Newtonian mathematicians, use you Newtonian inclined mathematics to explain my previous argument about how all the light that fits and fills the entire Universe can bring one picture of the entirety to fit into my eye.

The fact that **your** Newtonian physics will not allow you calculate this does not remove my ability to see the entire Universe in as large as it is through one tiny hole at the back of my eye. When you understand this entire concept you will have the ability to understand God's presence in the Universe and until then your mathematics removes your ability to understand physics and Newton promotes you blind stupidity about real cosmic physics. Moreover, I prove all of this ability mathematically but only after removing the falseness of the factor of mass and from the myth presented as Newtonian corrupt science.

This article is as much about proving what energy is as it is about knowing the difference to the state in which alive person is and in which a dead person is. Newton considered all forms of energy to be the same, and oh boy, was he mistaken. It is not surprising he formulated gravity the way he did. There is a worldwide fashion amongst the very well educated that in order to be regarded by those with the know how as a supremely informed person, one must at least be an atheist. The key to science is apparently to be completely atheistic. Atheists do not believe in the life after death, a Creator or a Force that does not exist outside the technological criteria of mathematical science.

Everything that does exist only exists because it exists in the perceptibility. Any force that might lie outside this norm is quite unthinkable and that thought could never present itself as to be present in the material universe. The ironic of this fact is that those well-educated scientists have only one source of information and that is light waves. Still they permit themselves to be atheists in their blind state of ignorance.

I do not condemn them, because they apparently know more than I will ever know. However, because I am not that knowledgeable, I must feel my way through the tunnel of ignorant darkness like a blind person. However, in doing that, I stumbled across a heap of questions that has no answer, even by those who carry the flame of knowledge. That forced me to form my own theories, think and come up with sensible conclusions, which answers all those questions their light of knowledge could not answer.

I declare to be of average intelligence and like millions of others on earth, all these millions are believers, like me, and are confronted by the same questions these super intellectuals are seemingly incapable or unwilling to answer. Then I realized the super intellectuals only have one source that lead them and that is measured light.

Let us look at the definition of energy. Energy is, as I understand it, indestructible, which means it cannot be destroyed. Energy can only be transferred from one form to another form. Let us look at the example, which is used to teach scholars at school. We take a rock and move it from a ditch up a hill. On top of the hill, we have a lot of potential energy that was transformed from static energy by means of kinetic energy. In the transformation, other losses occurred, like heat, sweat vapour and friction losses.

The science apparently does not take into account energy losses brought about by anger, fighting and frustration brought about by incumbency. These are also energy losses. After all the sweat and wrestling, the rock is on the top and we have a situation with potential energy from which we can derive kinetic energy when the rock is rolled down hill. I do not agree with any of the above mentioned, and will later state my point of view. I will however declare at this point that Newton's statement of energy and work being the same thing is utter nonsense.

This is the simplest example we teach children in school. I too had to teach the children this nonsense, in the period when I too was a teacher.

Life starts of being in a sperm that has to couple with an egg. The sperm only carries life but does not even represent life. If life leaves the sperm you can do to the sperm whatever you wish and it would not represent life. Therefore the sperm is a vehicle for life and life forms the sperm sell. If life abandons the sperm sell the sell goes back to atoms. If it were the other way around, the sperm cell would remain intact and start hunting for a new life form to hold. The same argument applies to the egg. Thee egg carries life and life supply the other half of the life that will become human. If the egg does not hold life any longer, then the egg will disintegrate into billions of atoms once more.

Brain waves monitored

ADAM.

This is extremely important to realise that from the first second of life forming life collects tissue that will become a human body. It is not like your halfwit Newtonian professor believes that the human body represents life. From the first moment life forms the body and it is not the body that forms life. Therefore you with your life forms your human body and it is not your human body that takes the responsibility for life.

It is by the thought process that life collects material to form the human body and the human body does not collect life as it goes along. Every one in modern science think it is the brain that controls the human body but are they so completely wrong. You use your thought process to control your body and in this thought process you form your body to be as strong as you wish it to be. Can I make you strong; no I cant because you are already as strong as a giant. What I can do is help you realise you potential strength by helping you learn to control your muscles. Before you complete any action of movement a thought first have to apply. It is the thought that command the muscle and not the brain that commands the Muscle. The thought takes charge of a cell in the mind and then takes information stored in the cell of the brain, which a thought directs to a channel that by electricity which the thought is also responsible, inflicts current in the muscle to pull the muscle. The thought generates the electricity that collects the information and the thought sends the electricity carrying the information to the muscle that has to do the job. It is the thought and not the brain and there science falls flat in their hogwash they use as information.

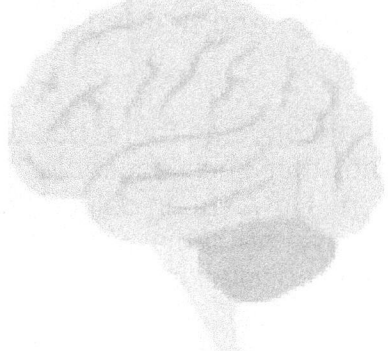

Life which is what you are, not a decomposable body, started accumulating material by thought when you were sperm and egg, and after the Unification you started accumulating useful building material. It is done by mind controlling the body. Don't allow the atheistic senseless stupidity tell you different. If your body was what is in charge which is you, then when you are dead someone with life can pump some oxygen into you and shock you with electricity until you bounce around like a ping-pong ball and you will begin life again. That is total rubbish. If life leaves the body there is no structural formation left to control and maintain the structural I integrity of the body.

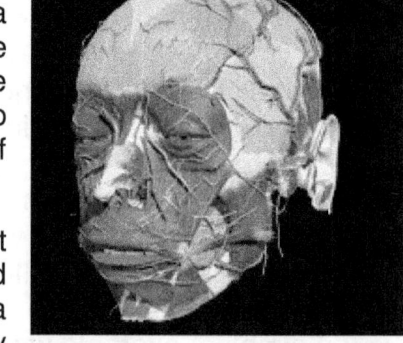

Even in the very beginning life formed the sperm and the sperm did not represent life. You can't have a tube filled with sperm and when you find the lot are dead you revive the sperm with an electric jolt. You can't have a jar filled with D.N.A and by regrouping the composition you build the body of the person once more. You build your body through thinking with your mind. You construct your body cell be cell by using your mind to do so. How can I prove this? The instant your life vacates the body; the body's ability to restructure the structure leaves that very second. The moment life vacates the body, the body degenerates by fragmenting the structure until the entire construction disassembles into forming atoms again. It is life that keeps the body into form and without life the body de-fragments into atoms once more. Your body doe not hold life but your mind by thought controls your body.

Your medical doctor will tell you your body represents your life and when you die your body dies.
He will be of the opinion that when your human body dies you have died. The problem with this attitude is that while your human body is still intact, one should be able to resurrect the body by supplying heat and electricity. That is not possible. That is the way one goes about killing people. By electrocuting people on a chair the state removes life from the body ands so giving the body electricity does not bring life to the body. Therefore the body does not use electricity to instate life but by duplicating the transmitting of electricity the body becomes confused and the body relinquishes life. It allows life to depart.

Your physics teacher / professor will tell you that you are what you are because of the body you have. Hogwash I say that concept is and I prove it is rubbish. You build your body with your mind but you control your mind through thought and without though you will not even move a muscle. It is by thought that you tell a muscle to move and it is by thought you tell the muscle to get active and by the same thought you form the muscle which puts the ability in the muscle to form the strength. I prove that this is how gravity forms and it does not form by the pulling force of mass. Newtonians such as your physics teacher or professor wouldn't even read my books in which I prove they (those teaching physics) are all brainwashing students to believe that physics is what Newton said it is. I prove they are submitting all students to mind control in order to force you to believe physics is what they teach it is. If you want more facts to see if I am correct about your teachers brainwashing you and submitting you to cruel mind control you are welcome to go to **www.sirnewtonsfraud.com or another slightly more complicated version I call www.singularityrelevancy.com** or another slightly less complicated version I call **www.questionablescience.net** , where the websites will show you how much those you trust deceive you with science they can never prove.

Those that think they are experts in physics has no idea about physics and I challenge all of them to prove Newton is correct, not to surmise that Newton is correct or to force students to admit and confess that Newtonian physics is correct but to prove it is correct. Prove the formula $F = G\ \dfrac{M_1 M_2}{r^2}$ does form gravity by forming a force by the value of mass or prove that the formula $P = \left(\dfrac{4\pi^2 a^3}{G(M+m)}\right)^{0.5}$ does put planets in positions allocated according to mass.

It is life that allows the semen to swim and it is the life within that allows the egg to be receptive of the semen. When either the semen or the egg holds no life and the egg or the semen is still intact. There is no life forming possibility. The semen does not swim it is life that allows the semen to swim. This proves that it is life that allows movement from the beginning of where we think life starts.

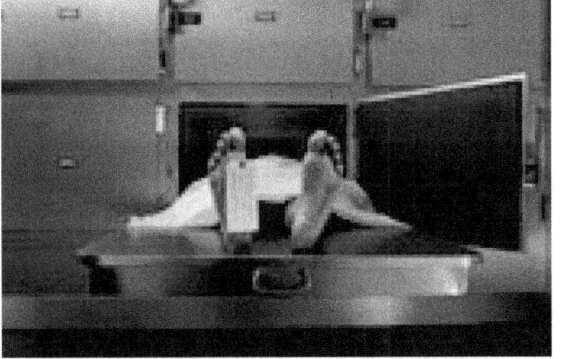

This is a picture of a cadaver. It is not something to be scared of because it is a body NOT containing life, as anyone of us will be someday. So it is the same as you being scared of you as you are going to be somewhere in the future and that is pretty silly. On the condition that you were born the only thing you will be someday is dead. If you are alive then you will face death. What we have to answer is what is the difference between this cadaver in the mortuary and me. One is that the cadaver has no life and I show vital signs filling me with life.

The biggest factor is movement and that movement is linked to thought. Considering the implication of this is vital if you wish to enhance your physical strength and build your body. There are persons in hospital in a coma for years and they apparently show no thought because their muscles don't move and therefore they wither away. The thought gives control over the body and the thought form the muscle and the thought form the size of the muscle.

This cadaver or dead person can't get up and walk as I can. Why can't this dead body get up and walk, it is because the dead has no thoughts. If you think the Newtonian idea is correct that life is part of the body then rethink. I dare you to conduct some tests. If life is electricity as they say it is, then why can you shock that cadaver until it hums like an electric transformer and life will not return? If life is as they say it is electric convulsions then try and shock the brain with electricity and you will find no response. The fact that you can manipulate muscle spasm with electric convulsion shows that life controls the brain by charging electricity and that process is done by thought in life. Life generates electricity that life then implements to control the body life extends for the purpose of serving life. Life is in charge of the body and of thought and not the body being in charge of life. By electrocuting a body with life you merely short circuit life's actions with a stronger jolt of electricity but the electricity is just a modem through which life controls muscles and growth in the body. Then you burn the electricity conducting connection that life has as life controls the functions of the body and do that long enough and life may not find a manner to form conduction of electricity whereby the organ control will become suspended.

Your mind charges electricity and that electricity are created by thought and thought is life. By creating thought you form your body and by forming your body through thought you establish you level of strength. That is why when you are in shock you are able to perform in a manner not even you ever thought you are capable of. Around your head there are electricity flowing which science named "brainwaves". These brainwaves are just a form of electricity and that current is the same as what flows around every electrical motor or any planet charging gravity or any star forming a gravitational field.

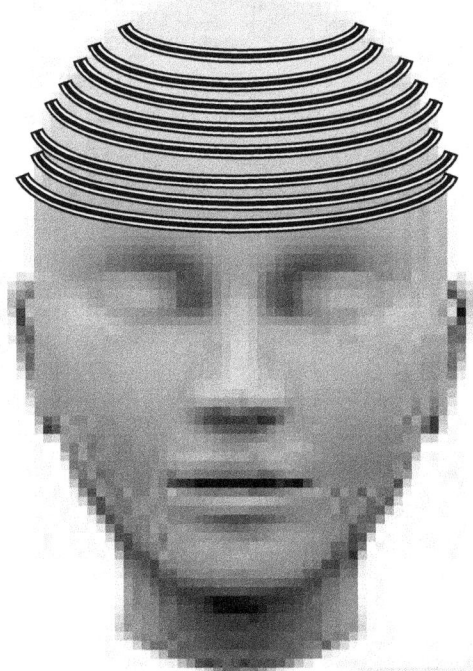

A human is not the body you have but the mind that forms the body

Now we take the scenario of a person's life. When that person is born, and after that momentous occasion of the birth episode, he or she continuous to live on this planet for the best part of the next sixty or seventy years. In this, period a great deal of energy is used to walk, run, laugh, cry, think, produce and reproduce. By doing that, he would from time to time state that he feels tired or without energy. What energy is the man referring to. I have once heard a scientist that made such a fool of him. That scientist declared that if God was energy, God could be coal, because coal is energy as well. Now I would love to invite him to a meal and see him devour a plate of coal. If coal is energy, he can make a meal of it, and then live very cheaply. What he does not seem to grasp, is that there are many forms of energy, which differ totally as we distract the heat and in doing so one can tap the energy. However, coal cannot walk, run, jump and laugh. I cannot even begin to imagine one brick crying and moaning because his friend was thrown into a fire. Coal cannot have sexual intercourse producing an offspring and then caring for it afterwards. Life on the other hand does have that energy quality. This means that there are different values and forms of energy, of which life is one. If life is no different to other forms of energy, God could be another total different concept of energy. This is the problem that I have with these "SUPER-

EDUCATED- MASTERS- OF- FACT" geniuses. They can make the most bizarre statements and could be away with it unchallenged.

When circumstances starve the body of food, life occupying the body would begin to devour the body in order to sustain life's ability to occupy the body. To life the body is only a vehicle to serve its purpose. When the body starving, the body does not suspend life until conditions are favourable to have the body reinstall life. It is so very typical of the "SUPER- EDUCATED- MASTERS- OF- FACT" to uncomplicated issues to serve their insight. They make something such as life so simple as to pretend they are completely in control of the knowledge that subject has to offer. The body does not turn life off, life eats up the body until the body is so feeble it can't host life any longer after which when life then rejects the body. The body does not maintain structure but as soon as life evacuates the body, the body breaks down the structure it held when it hosted life. Life maintains the body and will even devour the body and consume the fibre until the body becomes useless to life and until it cant serve life any longer. After life abandons the body, the body returns to a state of atoms with no resemblance of what it were when life formed the body. It is life that constructs the body, maintains the body, and controls the body and consume either the body or as food that is some other life form that had a body.

The world contains a wide spectrum of different occupations that people earn their livelihood from. Seen from my personal occupation, there are two types. Those that farm and produce wheat, corn, barley, nuts, sugar cane, vegetables and many other produce. These are potential energy producers. They produce food, for the other group of the human population that uses this energy product to maintain their strength to apply it to other methods of occupation. Cattle and sheep farmers produce meat that is used by others to convert into energy for their personal use

All people have one thing in common. They devour one form of energy, which is known as food. That is needed to maintain a life cycle, and the consuming of food must be done on a regular basis, to enable a human to live and reproduce for a lifespan of seventy of eighty years. The only precondition is that life would sponge on other carbon-based forms of life, whether it is plants or animals.

This person maintains his way and means of life, thus transferring energy from a form of food to a form of work. Then one day he collapses and becomes still. That person becomes unable to move. We call this state that the person is in, being dead. Even if I take a shovel of food and force it down his throat, he still would lack the ability to transform that energy to movement. But why would this then not bring back life?

Simply because he does not breathe any more. And why isn't he able to breathe? The reason for the person's inability to function, as a human should is because the cadaver is dead. When a person is considered dead, he lacks energy to such an extent that he cannot bring his own body to the grave. Others like me, and I have to carry him to his grave. We, that are alive, and maintain the process of translating food into life, have to carry the dead (he who is without life) to his grave.

The only difference between him and me is the energy form known as life. However, life is not the same form of energy as food, oxygen, heat and electricity. Even if I force all the food down his throat, and pump his lungs with air, while I heat his body with a blowtorch and shock him with electricity, he would still find himself unable to walk himself to his grave. That means the one form of energy is not the same as the other form of energy.

It is widely accepted that there seems to be a generator in the brain that generates electrons which enables the body to function. We know the flow of electrons is due to the process called electricity. On the other hand, do we? In a later chapter, I shall point the difference out between this flow of electricity. However, for the mean time I would stick to this accepted fact that electricity is conducted by the flow of electrons. Now, you can shock the cadaver with electricity until it hops about like a ping-pong ball, if life has gone absent, conducting a flow of electricity would not reinstate life.

You could put the cadaver on life support, with a heart machine a lung machine and all kinds of other machines. This method has nothing to do with life being precious, but fare more with the money paid by his medical aid, being precious. Once the cadaver's line of financial support dries up, his life instantaneously becomes worthless.

Then the cadaver finds the problem that it seems unable to live which means it is dead. Death means the brain is unable to send electronic signals by means of amino acids to the muscles, which would enable those muscles to continue with its normal function. The cadaver finds itself without the energy called life.

At this stage, I think that I pointed out to the difference between a body filled with energy called life, and a body that lacks energy and is called death. However, the energy that I pointed out called life, is miles apart from the energy that consists of food, air, the burning of it and the destruction of it. There is a broad difference between the food process and the actual form of energy called life.

Now I would like to ask those Super Intelligent Atheists and consumers of food and air to explain where the energy form that is called life has gone. Energy cannot be destroyed, but can merely be transformed from one form to another. This is scientific gospel. Life as I pointed out, has a different value to heat. Life cannot be destroyed, that means it has to be transferred from one form to another form, and life itself is not heat, electricity, or food, because applying all those other forms of energy cannot raise the dead.

The fact that energy must be transformed and cannot be destroyed is proved by science to be unquestionable. The life energy started assembling a body albeit sperm or an egg before conception or procreation took place. If the sperm was dead the sperm would not swing and all those that swam in vane died. They did not hang around as lifeless sperm to be vitalised with life as soon as the next opportunity arrived. The very second life left the sperm or the egg without fore filling the process of fertilisation extending the ability of life to assemble more material in order to form a body filled with life, life left the sperm or egg and in that the body holding the sperm or egg destructed. It is life that captures material to form a body notwithstanding how small and in that no one can remove God from physics. The idea that life sprang from somewhere as soon as a sperm was there is as mad as having mass being able to form gravity. With my physics I can prove God being responsible for the flow of time within the Universe and if Newtonian science are not able to accomplish that, it is Newtonian science that is dismally inadequate, but then again that is what Newtonian science is in almost every sense.

The only answer I can conclude is that science is ignoring their own findings to prove their own religion fashions. With life being an undisputable form of energy and energy cannot be destroyed, it seems very unscientific to propagate atheism as a fact.

From these facts, one has to conclude that there does exist another form of life after death.

5.   INTELLIGENCE VERSUS   EXTELLIGENCE

In this book, I shall introduce you, as the reader, to a completely new line of thought about the science of cosmology. Some of these scientific facts date back to the time when man became aware of a lifestyle that just started to include a civilized order. According to some discoveries by archaeologists, it seems that mankind had its survival mostly due to the way it accomplished knowledge about primitive science, this enabled man to survive in a total hostile environment.

Man's first awareness about forces that he could not control, was explained as forces unleashed by pagan gods. In that is seated mankind's belief and mankind's desire to be in total control of these godly forces. This desire therefore became one of the biggest incentives that drove mankind to a civil obedience and law-abiding standard of living.

We can even today go as far as to except that the role that intelligence played in the development of our specie was far bigger than the role was of the more brutally and physical force. In the animal world, the strongest in specie would ultimately be the leader of the pack. With his brutal power and brute force no one in the pack would dare to challenge the leaderships hierarchy and in that the leader himself. If one challenger should dare to do so, the challenger would pay with his life. It is a well known fact that male baboons not only kill the previous leader, but he will wipe out all the siblings, no matter how much the female baboons might protest against it.

This is even more so in species that has much closer links with mankind. The chimpanzee male just simply murders all possible male challengers until the day he himself is also murdered by his successor. The orangutan male is another example of a male that would not even tolerate any male in a smelling distance. This confrontation will definitely lead to the death of the weaker one of the two.

There may be a distinct possibility that fear for the unknown was the only reason mankind's development lead us to a higher norm in development than our close relatives. In case of other species the generational development benefits the physical strongest and do not favour the more intelligent of the species. These animals are still much stronger than man is, although man tamed all animals, at one point or another. Therefore, all animals submit to man.

In this, one must define the difference between intelligence and the idea, which I refer to as extelligence

In the understanding of the meaning of intelligence brings to mind how the animal socializes with its own species to guarantee the social survival of the species. That means that all animals have intelligence. Dogs has been with man as long as we can trace back human development, so in doing that he forced the dog into acknowledging mans intelligence. However, the dog still communicates with his own species in the way its intelligence dictates. He sees man as the leader of the pack, rather than a completely different species. When it communicates with man, it will wag its tail or show submission by lying on its back. The dogs intelligence never allowed him to try and communicate with man by using mankind's standards, although his intelligence placed him in a certain advantage point to share to some extend mankind's intelligence. However, the only reason it did so, was to farther his own needs in surviving in the pack with man then becoming leader of the pack.

As a farmer, I often watch the manner in which cows interact. At one predestined time during the morning, the mothers leave in a group to have a drink of water. I admit there is nothing strange about that. The strange part is in the procedure, when taking into account that we regard these animals to be thoughtless beasts. One dry cow gathers the entire suckling calves, takes them to a safe, and secured area, where the calves would play and enjoy one another's company. I refer to this as the "kindertiun" which is the kindergarten. After the water drinking, the mother's would gather in a shady spot, and ruminate for about two hours or so. Then they would get up, and stroll in the direction of the kindergarten. Only when they come to a certain point will the calves leave the seclusion, and run to meat their mothers to feast in the generous supply of milk The biggest amazing part, is that the kindergarten hostess is never the same cow. Everyday another dry cow takes on the responsibility of playing stepmother to the calves. Not once is there an incident of one of the calves being disobedient or not under standing what is expected. They always seem to know which cow to follow, and are never fearful of leaving their mothers. They never are obstinate and wonder off in search of their mothers or run to their mothers before someone gives the signal, whatever the signal may be. If cattle are that mindless who decides who's turn it will be to play stepmother, who and how are the calves informed about who to follow, and why are they acting that responsible and disciplined. After all, they are only young mindless beasts. I concluded that in our self-righteousness we under estimate our fellow living species. However, in all fairness to my own species I have to admit, mankind disposes of intelligence as well as extelligence. Extelligence is the acquired knowledge to deal with matters not relating to its survival. A part of this development was due to the need of extelligence to eat. Mankind's progress in becoming a forceful species lagged behind because of his awareness to the fact that he could manipulate certain forces in nature to his advantage. A part of this manipulating process was accomplishing the skill to control and use fire. However, man also noticed that fire came from the heavens and clouds. These clouds formed part of the sky where the sun, moon and other stars are located.

With this argument, astronomy had to play a huge role in the development of mans culture, especially the religious aspect. His health, happiness, belief, future and wealth all derived from the gods that was found in the stars. This fascination and even religious fears was derived from the stars that even today is still apart of the science of cosmology. That is why even today, people are still motivated by the stars. Ironically enough, the other big motivation lies in man's lust for power and his war games to commit murder, to demolish other's property, and to dominate other members of its species to the point beyond that of slavery. The role of slave owners and slave drivers today is in the hands of the Mammonites. They use John and Jane Dow and all mankind that belongs to social grouping lower than they do. Mammonites are those that control every facet of the man on the street's life, should it be by job supply, political law enforcement, food and house supply, by dictating to the politicians in what manner and which laws should be applied and enforced. Mammonites are the bankers, the Wall street brokers, the Insurance firms, the drug lords, those that Motorcar manufacturing belongs to, the Shopping chain owners, the Oil barons and the Petrol companies but to name a few and all those and all other evil proprietors of wealth and monetary fortunes that force the law of the merchant onto the public. The merchant never has scruples, never has a conscience, adhere to one God which is Mammon has love for wealth alone and only has greed as a driving force. They that ask in all sincerity "What is wrong with having greed" but that question is on the lips of all other criminal elements in society. Mammonists on the other hand is the smaller and lesser counterpart that would pass his hungry brother and not help him, although he has more money than he needs, but he shares in the greed of the Mammonites.

To them, their love for money is far greater than their love for their fellow brothers and sisters and they will force millions to go hungry just so that they could have billions. This can be found in any social structure, be it capitalists, socialists, communists or kingdoms. They are the ones society looks up to while they are the one buying the politician to change law that bleed the population dry to enrich the those that already have everything anyway. Al Capone was nobody. Al Capone had one house in a neighbourhood of wealth. The others were more crooked and bigger gangsters that he was because they owned more than he was. They only pretended to be legal while they branded him as illegal. Who were the crooks that owned the other houses in the suburb that houses the rich? They are as big Mafioso because they pretend to be cream of the social structure while they would bleed those below him dry without mercy and always to the cream's advantage.

The fear that man experienced about forces, which was, according to him, inexplicable and therefore stronger than him, was considered by him to be of a godly nature and therefore only the wise amongst the wise could explain and philosophise about the nature of these forces. Common man never questioned the correctness of these layouts. However, man still prevailed in explaining to the best of his ability, the logic about his viewpoints and in so doing to guide the incredible forces that torched his fear. Today in retrospect, we consider these arguments laughable. Think how ridiculous it seems to regard the sun as a god on a chariot of fire that patrols the sky on a daily basis. Today it seems ridiculous to regard the earth as being flat, or to see a face of a goddess on the surface of the moon.

However, the tendency to except these super intellectual's arguments remains a custom until this day, no matter how silly it seems to be. Man is still upholding the ancient culture that in a case where it cannot be understood, the idea must be correct and therefore the ordinary man would inevitably be too stupid to follow. In addition, to this day, the "SUPER- EDUCATED- MASTER- OF- FACT" among us still misuse this phenomenon in common practice. Think about the silliness of Einstein's single dimension theory. How ridiculous is such a notion. Once we accept the earth to be round, Einstein came along and invented a flat universe!

In a hundred or two hundred years from now, it would be our generation's turn to be considered short sited and backwards because we accepted these ridiculous arguments. Thus, no matter how dynamic our visions of the cosmos seem to be, we shall still be regarded as non-intellectual and stupid by generations to come.

In writing this book, I too attempt to deliver a contribution to clarify a certain line of thought that is unclear and to give an explanatory value to it. You, the reader, will evaluate the acceptability of my reasoning and you will remain the only evaluator that will approve or reject my work.

My viewpoints are not the consequence of a big literacy, but rather due to a lack thereof. Because I shall entrust you as the reader with my thoughts, I shall have to introduce myself in a brief manner. Relatively spoken I can be considered as stupid. I do not try to sell myself short, but in accepting this fact, I was able to use it to my great advantage.

First, I have to qualify my statement that I am stupid. All people know what they know. However, they do not know what they do not know. For us mortals, the sum total of what we know, is enormously big. That comprises the total amount of our total human existence and our accumulated knowledge gained over a lifetime of labour. On the other hand, we regard the part that we do not know, as so insignificant small, that it bears a value of nothing. Because we do not know how much we do not know, we cannot evaluate the sum total of that.

People always concentrate on the part that they know and therefore realize how intelligent they are. In this lies the accumulation of their absolute arrogance. With this arrogance the part that they do not know, becomes even more insignificant. The normal procedure of man is that he will concentrate on the part that is of value, disregarding the worthless part that is of no value.

In my case, I had to concentrate on the part that I did not know, because of the lack of formal education and therefore not knowing how much I knew. This brought about that I always had to regard the part that I knew as being insignificant small. I had no formal examinations in testing how much I knew and thereby evaluating my field of knowledge. In the absence of examinations, I had to disregard the amount that I know and always had to consider myself as being stupid. This brought about that I had to remain humble, because I was untested and stupid. This book is the consequence not of my enormous intellect, but rather as the result of my stupidity. I always had to fight my ignorance and had to seek answers to my own questions because I was too stupid to accept the official answers given by the educated..

At school, I was a rebel. Because I was so stupid, I refused to accept facts without testing the reasons behind the answers and just because the teacher said so. An intellectual person would have accepted those given facts without causing him self all the inconveniences. I always insisted on outlined explanations by my teachers. I bluntly refused to "learn" anything. My point of view was that if I understood a subject brought about by a good explanation, studying was unnecessary. If the teacher could not explain the subject in depth, then I disregarded him and his subject and treated him with disrespect. I reasoned that the blind could not lead the blind.

This of course enraged the teachers and they tried to break my resistance with corporal punishment, which I deserved. This went into a spiral, where I got more rebellious and they had even more reason to apply the cane. In the end, I was the one that got the short side of the stick due to my rebellious stupidity, and brought about that I had to teach myself all that I know. I tell you this to point out that since my earliest days as a child I could never conform only because my superior said a certain thing and I was supposed to accept it. Sometimes (I guess) I was wrong and they were correct but only convincing me with intellectual arguments could allow me to see the other side. I was ultimately the one that paid the penalty for my behaviour because the road I took was tough. Some things I had to go through because of my stubbornness I do net even wish on the devil himself.

As I said, I am plainly stupid. But in saying that, I believe I am just your everyday person and if I could write this book, being as stupid as I am, any high school pupil can read and appreciate this book.

Because I do not have any noticeable academic background, I believe that these "SUPER- EDUCATED- MASTERS- OF- FACT" Academics will try to shoot the information down in flames, (if there are any that even would read it). The hostility I received so far from these Super Intellectuals did not surprise me in the least. However, I spent 21 years of my life to come to the conclusion that I share with you in this book and like every human being, I would want to defend my work against the onslaught of these sublime intellectuals.

The first few chapters are everyday common sense, but by regarding it, it will enable you to grasp my line of reasoning, which you will need in understanding the last few chapters. If one does not read the simple chapters, the terminology used in the last few chapters might seem somewhat incomprehensible but that is only because of new terminology that compelled me to introduce new terminology is brought in, in order allow he reader miss conceptions used and the new definitions I introduce.

This book must comply with a commercial value, to pay for the publishing costs and to introduce my work to the broadest range of reader's possible. That is the only way I believe I can force the academics to take note of my work. People associate cosmology with Einstein and his complicated brilliance and brainpower. That complication is only because Einstein told half the story. When I tell the other half in this book, you will see that it is not half as complicated as it seems. Every person with a normal mind will find all those unbelievable complicated statements that Einstein made, is in fact rather simple, when told in its full content.

On the other hand, this book must comply with certain technical facts to stop those "SUPER- EDUCATED- MASTER- OF- FACT" super intellectuals from blowing the statements that I make away with a few words. This work comprises of hundreds of new thoughts on cosmology, laws and processes in nature that was never noticed before and arguments that is now seeing the first light of day. Many of these arguments might be old statements that is purposely withheld by scientists, to the general public, because in admitting to the follies that exist in modern science, they then have to admit that their scientific layouts are in the least, foolish. As I said, the rejection received up to now, will be but a drop in the ocean in comparison to what I expect when this book is published in English. In the past, the only rejection on their part was because of the lack of my education, and therefore they refused to listen to my arguments. Therefore, I belabour this point concerning my education because if any person feels a need to reject my work based on the lack in education on my part, do it from the start, but if you do so, it will be to your own disadvantage.

I realize I do not beat around the bush, when confronting certain statements by some ingenious jokers that cannot be taken seriously. For that reason many of the academics would be sensitive to my work, but if they feel the need to make foolish arguments on international T.V. and in books, they must prepare themselves to be made the fools they are. These same intellectual gurus go to extreme detail in their own publications how the Roman Catholic Church denied science freedom of speech and prosecuted scientists in the dark Middle Ages. In that time, such people were prosecuted.

The powers that control the media today is far more powerful and much more methodical than what the Roman Catholic Church was then, and the modern day media's inquisitions are far less merciful and more subtle. If there is a certain school of thought, whom the modern media does not want to propagate, it will be killed by silencing its publication. If not for a medium like the Internet, this would have been the lot of this book as well.

This book will serve as a modern test in press freedom to see how science currently will respond to a new school of thought when their beliefs are ostracized. This book will put the shoe that was 500 years ago on the foot of the Roman Catholic Church, right back on the foot of science and the Newton apostles.

It may seem that I have a hate campaign against the intellectuals of the day. It is not true. However, I refuse to believe that with all their geniuses combined, they are unable to see the facts I have seen. I may be wrong, but there seems to be a sinister motive in the published work of modern science in that they promote atheism and use every chance to degenerate Christian belief. If a "mister nobody" as myself can see these facts so clearly, surely they have to see it too. Why do they then keep silent about it? They are the ones with almost unlimited IQ's, not me!

You might find the first four chapters to be simplistic and uncomplicated but if you do not get a well grasped understanding of the introduction of my theory, the complexity in the last four chapters will then seem impossible to follow.

The contents of this book is not aimed to relax and entertain, but the possible enrichment of your comprehension to the layout of God's creation will richly compensate for the effort, especially in the last four chapters. Thus, if you may find the first few chapters below your mental capacity and development, I ask you to bare with me, it would be worth your while when you reach the last four chapters and the truth starts to dawn on you. However, the last four chapters would seem complicated if the golden thread were not drawn right through to the end.

The main contents of this book, is as far as my knowledge goes, never been written or spoken by any person dead or alive and should be fresh to all.

If you do not approve of the very simplistic mathematical calculations, please feel free to ignore it. As I have shown in the prologue, it is merely put in to prove a point. If the simplest calculations on gravity are ridiculous, there is no applicable scientific applicability on reality. I did use it to point out how illogic certain scientific arguments are, but is will not enhance the explanation of this theme in any way what so ever. It is merely placed in this book to silence some of the "SUPER- EDUCATED- MASTER- OF- FACT" intellectuals.

As a background sketch to how I was motivated in writing this book, I must share the incident with you, being my reader. Due to my interest in the science of cosmology, I get asked certain questions from time to time to explain these known and accepted theories, principals and definitions propagated by the "SUPER- EDUCATED- MASTER- OF- FACT" academics.

As my personal studies progressed it became increasingly more difficult to explain certain accepted theories and concepts promoted by scientists and more even, to agree with these hypothetical mumble jumble and fairy tales. How does one defend certain accepted ideas that science promotes, but is faced on contra dictionary of how you relate to these facts? How does one explain your own concepts when it differs completely or does even vaguely been accommodated in the science of the day. I would have been able to ignore these conflicting feelings, if I did not know there are millions of John Dows out there who, as I do, did not appreciate (like me) or understand (and therefore do not accept) this misleading information.

To consider yourself part of the intellectual cosmology know how elite, certain recommended directions for use should be meticulously followed when confronted with the unexplainable concepts that is being promoted by the intellectual of the day. Applying these evading methods is very unacceptable to me. However, these directions are being used by the utmost intelligential on ignorant persons. When being confronted by a question you do not know the answer to, throw a mind boggler and complete mesmerizing question back at him, with the knowledge that nobody on earth knows the answer to that question. The questioner would find himself so bewildered that by the time, he recovers his senses, he would find himself still without an answer but he then would be out of a chance to insist on an answer. That will keep him ignorant and well in his place. On other occasions, other methods can be used with similar results.

First, if being asked an unanswerable question, congratulate the questioner on his brilliance and well thought question. Share the brilliance and well thought question. Share the brilliance of the question with the whole audience. Let the audience comprehend how brilliant the question really is and let the audience applaud his brilliance. Then abruptly ignore the question by going on to the next question. The first questioner will be so pleased with his own brilliance, he would never insist on an answer! However, beware; these methods can only be used in extreme cases and definitely not too often. The actual rudimentary way of dealing with the problem should be as follows.

(i) When nobody understands a certain concept give the questioner an impressed but unmistakable superior smile.

(ii) With a tone dripping with sympathy, you should use a stance of high and mighty superiority and let the questioner realize that you sympathize with his intellectual shortcomings.

(iii) Let the person very well realize (still with dripping sympathy) that if he was blessed with your intellectual insight and capacity, these facts would seem trivial, as it does in our own case, and you do understand his mental shortcoming, but he has to accept his inevitable weak minded position.

(iv) When the questioner's reason becomes far to logic suppress with your own ignorance of these illogic matters, like gravitation and electromagnetism, time and the black hole, immediately and without any further hesitation and with all the haste you can muster, refer to rule number one as stated above. Do not delay another instant. I shall damage your personal reputation to a point of no return.

I know all these methods of question evading by those super blessed intellectuals because it was used on me so many times by some of the most renowned geniuses. It was because of this that I started my personal search some 33 years ago. I had no intentions ever to put it down in published writing, because the motivation lacked on my part. For the past twelve years I kept myself occupied by trying to introduce my findings while not finding an audience very interested in what I have to say.

One night the owner of the local, Wimpy, Johan, asked me what space was. After trying to explain by starting with the atom, I saw that I was making very little to no progress. Then I decided to put these explanations down on paper so that he would be able to read it in his own time. Before long I realized that I could only explain this by means of a book, because in order to understand one explanation, I had to explain the facts that leads and follows that explanation. Now I find myself more than a decade years later, still trying to bring across my point where I wrote the book in Afrikaans, but has to translate it to English because of the weak market for such books in Afrikaans and for what it is worth here it is. *"Johan hier is jou antwoord in Engels!"*

The problem is those that should be interested in what I write are not interested because they are those that think they know more than God and that is not meant to be blasphemous because I show later on that those "SUPER- EDUCATED- MASTERS- OF- FACT" would rather have the cosmos change and start to contract in stead of expand as it does than admit Newton had everything wrong all along. Then we have the other lot that think physics are above them and they don't understand physics all the while it is that the physics they can't understand is so crooked not even Newton understood his own physics.

Between these two options of prospective readers I have not yet found that big understanding about the message I try to convey. I hope this effort will be simple enough so that everyone not connected to physics will show interest and understand what is wrong with the physics they don't understand and those that do understand physics will see how big fools they are too understand the physics Newton couldn't understand.

Those that do not understand physics will see why they were brilliant enough not to understand physics because it is one big hoax and those that do understand physics I hope will feel as stupid as they are because they think they understand the physics that is a complete joke.

9.   SUBLIMATION; The Newtonian Mythology

The purpose of this book is not to echo the **Newtonian** n version of Greek Mythology.
The Greek Mythology had aimed to bring, what they considered religion to be, in line with their perspective on what they considered science, cosmology and astronomy had to be. In Newton, a change a change came about in the contents of the mixture, but not the ingredients as such.   The Anglo American Mythology now preaches a religion called atheism, which, as were in the case of the Greeks, based on what they perceive science and scientific facts to be. If you find a desire to run along with these myths of star travel, speeding through the Universe at the speed of light, encountering alien societies and indulge yourself in such modern mythology, please do not read this book. In this book, the magic spells that Newton named gravity, and which Einstein took to a single dimension fantasy, is discarded in dismay once the Anglo American Mythology is replaced by factual truth, the creation according to the Bible becomes a detailed analysis of the actual creation.   When Newton's lies are replaced by the real functioning of the universe, the author of the Bible's firs book has such a precise recollection of events, that it put all scientific facts, being broadcasted at present to pitiful shame. That includes the ideas of modern atheistic Anglo American science cult. This book does not repeat all the traditional nonsense, but explain in detail, how the cosmic year structure works, in such detail that children can understand and accept it.

Civilization throughout the ages always used Cosmic Science, but especially, by the Roman Catholic Church, to prove the importance of the earth as the centre of the universe. That brought about that the sun shone on the earth, which was the centre of the universe. Since the earth was made for man and was the centre of the universe, God made everything with mankind in mind. Whereas the Roman Catholic Church was the only representation of God on earth saw they represent God and all His Powers on earth. It meant that the Pope was God on earth (being the head of the church of God on earth) and the Catholic hierarchy was the most elite and privileged on earth. That was precisely how the church considered them and how they could conduct their teachings.

Right at the top was the Pope who ruled the earth and as the earth was the centre of the universe, the Pope for that matter, ruled the whole universe. God and the saints ruled the heavens and the Pope with the Cardinals ruled the creation of God. This brought about the sublime picture that suited every person, which was considered anybody. He whom the Pope blessed was blessed and he, whom the Pope damned, was doomed.   Everybody op importance could buy the Pope's blessing and could die in reinsurance that his life was ultimately saved.

By the middle of the 15th century, some unimportant persons with no real social standing came along and disagreed with this ultimate and universal accepted hierarchy. They stated that the earth was not the centre of the universe, which meant that the Pope and his cardinals were not in control of the universe. This new perspective on the Universe and the chain of command was very unacceptable to anybody of social standing. This (then considered) blasphemy was to be killed in its very infancy, even before birth. However, as with everything else, the truth eventually prevailed and certain intellectuals took notice of statements by Galileo and Kepler. The reaction that followed can be regarded as one of the most important historical events in the route that man's civilization took in forming modern man.

In this book, we put the work of the two giants, Galileo and Kepler, under the magnifying glass. In contras to popular belief, there is an astonishing difference between the work of Galileo and Kepler and modern science. Modern science is based on the findings of the father of modern science, which I consider Anglo-American mythology. The father of Anglo-American mythology is none other than Isaac Newton in person. Galileo and Kepler had the truth of the Universe unlocked when Isaac Newton came along and raped their findings.

First, let us consider Galileo's work.

When a pendulum swings, the pendulum's rhythm remains while the stroke tarnish. All big clocks are based on this working principle and can therefore keep time mechanically. Time has been measured by means of this method for about half a millennium.

When two structures of different mass is dropped at an equal distance and time, the two structures will hit the earth at precisely the same time, as long as the wind resistance is equal to the two bodies. This experiment was the first that was done on the surface of them moon on July 1969 and billions of T.V. spectators bared witness to the outcome of this experiment. A hammer and a feather were dropped and they hit the surface of the moon on precisely the same instant.

Kepler, on the other hand proved that the earth and all other planets rotate around the sun in an elliptic orbit. All three these findings never mentioned any force called gravity.

At this stage, science used precisely the correct argument. The findings were noted and it could afterwards be checked for the same results repeatedly.

Afterwards an English genius by the name of Isaac Newton noticed an apple falling from a tree. This prompted him to calculate a force that existed between the apple and the earth, in which the matter of the apple and the matter of the earth was drawn by a force he named gravity. He never took into consideration any of the findings of the previous two giants, although he praised them for their work. Newton went and calculated the existence of a force that existed between the above-mentioned bodies. At this very point, science took a wrong direction. The force that Newton calculated is a secondary function to the primary condition that holds matter in place throughout the universe.

By using the findings of an even bigger genius, Tycho Brahe, Johannes Kepler proved that the radius of the planet, taken from the sun to the planet and is measured in astronomical units, is equal to the square of the rotation period. Kepler said that there exists a relation between two bodies where $T^2 = R^3$. He never mentioned a force. Galileo's findings proved the same as Kepler's, that there is a ratio between two bodies, not a force.

The second statement of Galileo can compared to driftwood on water. If two pieces of wood which is different in size and weight floats on water, and both pieces of wood is subjected to the same force value in the stream, both pieces of wood float at equal force, no matter what the difference in size and weight is that comes into play. This would be caused by the difference in the drag resistance on the different surface area of the two wooden bodies. However, the difference in weight that comes into play, at this point is only due to the drag that the water experiences, not the actual weight.

If this were compared to, the findings of Galileo one would find that this is the precise method how matter moves towards the earth. Galileo made no mention of a force between the two bodies. Then Newton came along and published his mathematical findings, which is totally out of line with Galileo's findings and ever since then no person ever gave the actual findings of Galileo and Kepler a second thought. There can be no force such as gravity, electromagnetism, strong and weak forces, or nuclear energy. These so called forces are part of precisely the same value that is in relation and exist between space and time.

You, the reader may ask yourself why is there such importance in these findings as to know the correct way that gravity actually works? All calculations have already been made; mankind already possessed the knowledge, expertise and willpower to visit out of this world's Tara novas and even to colonize them.

All knowledge about physics and astrophysics has been studied, formulated and tested! I will reply to this question by asking two other questions.

A certain man drives his car down a lonely road. After a while, the car comes to a standstill. He knows that a car needs fuel to run. He takes a 25-liter can and walks 10 km to a filling station to buy fuel for the car. He carries the 25-liter fuel 10 km back to the car and puts it in. After trying for 10 minutes, the car still would not start. He takes the 25-liter can and walks 15 km to another fuel station to buy some more fuel. After walking all the way back and filling the tank with the petrol, the car still refuses to start. He then takes the 25-liter can and walks 25 km to yet another fuel station for fuel. After returning, he filled the tank yet again. Do you, as the reader think for one minute the car would start?

The American scientists spends 1 000 million dollars to pressurize four hydrogen atoms into the same space of one helium atom. The experiment seems unsuccessful because the helium unfolds back to the original four hydrogen atoms after four seconds. After that they use three 000 million dollars to pressurize four hydrogen atoms into the space that one helium atom occupies. After seven seconds the four hydrogen atoms depressurizes back to its original state. Now the American nuclear scientists use 10 000 million dollars to pressurize the four hydrogen atoms into the space of the one helium atom where the experiment lasts for 12 seconds. After 12 seconds the hydrogen atoms moves back to their original condition. Question 2 now is this: "Can you see any connection between the two examples I have put to you? "No pressure in the world can fuse four hydrogen atoms to one helium atom, even if Einstein said so! Those super brains and academics are blind with their own mathematical genius of mathematics and physics that they fail to see the most basic and elementary principals in science. What is tragic is that it does with the taxpayer's hard earned money!

Another rudimentary example is the so-called "falling star". The conventional theory that is propagated is that the dust speck burns to ashes because of the friction the particle has with the air it collides with in the atmosphere. This is the biggest mindless rubbish one can imagine. Just because the person that tells me this have six doctoral degrees, does not make it the gospel truth. Far from it… What actually happens is that when the grain of dust enters the earth's atmosphere, the time aspect changes and forces the dust particle into a different space occupation which then changes the heat value of the grain of dust. The dust grain is forced into such a smaller volume of space-time occupation that the heat it generates just burns it to ashes. Therefore the whole structure glows itself into nothing.

These "SUPER- EDUCATED- MASTER- OF- FACT" giants might use the most breathtaking mathematical formulae that can humanly be dreamt up, but if the principle, on which they have based their calculations on, is wrong, the whole exercise is fruitless.

In my book, I discard the "conventional" standpoints by following the unconventional principals, based, on the work of Galileo and Kepler. The basis for my theory has never before been propagated except for me, and therefore all arguments will be new and fresh.

I am no writer and even less a scientist. Furthermore, I do not pretend to be regarded as any of the above mentioned. I simply came to certain conclusions. When I shared my conclusions with other people, I had the very same reaction repeatedly. The reaction followed spontaneously without exception. Those who were prepared to listen to me knew even less about cosmology than myself and never understood a word I said. Those who know more than I do ignored me immediately when I said Newton and Einstein are wrong in their views about gravity.

Partly the writing of this book is to prove that there is little difference to science and the Bible, even if scientists and theologians try to make it their lifetime task to prove the other opposing side wrong. However, all their arguments are based on their individual agnostic belief in their self-righteousness and have nothing to do with the truth.

Free thought has always been a fact that all that is in influential positions proclaim to strive to but the minute the free thought differs from their concepts, it is the very first thing they crush as hard as they might. In this book, I strive to accomplish the very essence of free thought and try to lead the reader away from the brainwashing that all intellectuals try to force onto the public. Every person that is in any position is busy with their own sublimation that it renders them blind to the truth that is out there.

RE-(W)RIGHTING COSMOLOGY

I have realized the extension of the Einstein hypothesis, and concluded the other part of the relativity theory. In short, the part of "gravity" that Newton saw, and the part of "gravity" that Einstein saw is not the same identical thing, but two different values to the same "gravity". That is why Einstein concluded his theory on a single dimensional Universe. The part of "gravity", which Newton related too, Einstein did not see. "Gravity" consist of a single dimension value (Newton's gravity), which is part of (not the same thing) the "gravity", that Einstein saw. These two values are contained in a balance in space-time. In this, the Universe is in a three-dimensional state of space (three dimensions) in time. Therefore, space is not flat, as EINSTEIN SAW IT, but has three dimensions, locked in singularity where singularity is single dimensioned but by directional relativity forms the fourth dimension of time. I am not going into space-time in this book and I am steering clear by a country mile of any arguments about space –time in this book but Newtonians knows less about space-time than what my dog knows about communism.

In this, there is no force, but only an ever-altering balance, which forms the time component. Only the energy value we regard, as "life" is a force, because only "life" can inflict a change in the balance of space-time and this forms a force. That is why science has been unable to recognize the value of "life" as a separate energy form. If you regard the Universe driven by a force, the value of life has to disappear. The Universe is in a state of balance, and the scale of the balance is determined by time. In this aspect lies the reason why time has never been detected. Only life can upset the balance, but then life will run into upsetting time. When time is upset, we see it as an explosion. The more time is upset, the bigger the explosion will be. I have put everything into dimensions:

1. Newton's "GRAVITY" which is the single dimension. It is a single value, starting nowhere and stopping nowhere.
2. Is Einstein's "gravity" which is a wave that is forever, circling forever by reducing and expanding.
3. These two values above are combined in the third dimension of space, the third dimension we, and everything in the universe, find ourselves in. In this, the first and second dimensions form an inseparable combination.
4. Everything in the Universe is moving. Even the part we relate to as "NOTHING" moves. This movement forms the fourth dimension of time. Time contains the third dimension, which is formed by the combination of the first two dimensions, and dictates the tempo, or balance in the first two dimensions, and which forms the third dimension. As all things the Universe are part of the balance in movement all things stand equally and evenly effected in the third dimension every thing is connected by movement (time). Therefore, everything connects to time and by time, from the smallest part of matter to the Universe as a whole. When space becomes less, the movement becomes more therefore time becomes more.
5. As life are connected to the movement of time, but can also move independently from the balance of time, life is a separate energy value, and is in the fifth dimension, but part of the first four dimensions. Being part of the fifth dimension, life is subjected to the criteria dictated by the balance of the fourth dimension, and therefore has to abide by the laws of the fourth dimension, while it is part of the fourth dimension. In all of this is the outstanding factor that nothing happens by chance, except for the intervention of "life".

Everything I am about to show Newtonians say is nothing new. Newtonians say they are aware of everything that I show and then they turn around and simplify physics by putting the entirety of their physics down to the pulling power of mass that forms gravity. The sound barrier is gravity and to understand gravity is to understand the sound barrier. They say that I bring nothing new to the table and they say they apply every aspect I indicate that forms gravity. When I ask them to explain the sound barrier they use the Mach principle and Doppler's effect, which both dates from a time when no one was aware anything man-made could fly let alone go faster than sound can. Doppler made his contribution to science at a time when the train he measured was slower than a horse. The statistics he left to science applies to going as fast as a man on a horse could ride. That is a far cry from a jet breaking the cosmic boom. They can claim what they like because they never have to prove anything and they never have to listen since they know everything while their shortcomings makes science more the jesting of a buffoon. That annoys me to my guts because all of that comes down to the conspiracy whereby they whitewash their stupidity from the brainwashing they inherited from their predecessors and pass the brainwashing on to their students. Their ignominious stupidity makes me want to shout to the mountains in agony and unbelievable frustration. When a cloak of arrogant self-righteousness hides their incomprehensible stupidity under a cloud filled by their belief in self-importance it becomes hideously loathing.

"GRAVITY IS DIVIDED IN TWO FACTORS, BEING LINEAR DISPLACEMENT (Π / Π⁰) WHICH IS WHAT NEWTON'S GRAVITY IS AND,

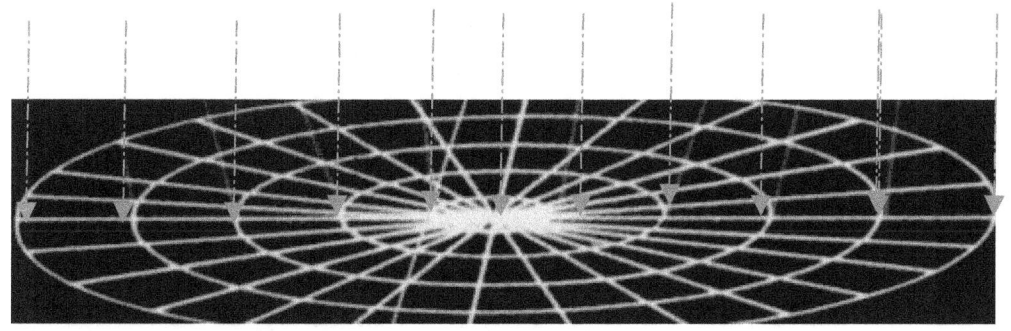

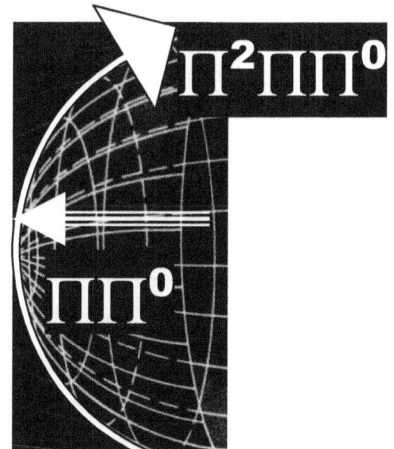

CIRCULAR DISPLACEMENT (Π²/ Π)

WHICH IS THE "GRAVITY" EINSTEIN RECOGNIZED

Those that are of the opinion that it is mass that "pulls' the object towards the earth might be a little wiser when reading the next part carefully as I explain the following:

Anything entering the earth from space it travels in a straight line. The mathematical formulating of this I do not provide at this stage but it is another "some simple algebraic relations" as Prof Friedrich W. Hehl from Annalen Der Physics put it (and I also explain this a in the following chapter). But using the law of Pythagoras and putting 7 in relation to 10 provides the value of Π. The value of Π has two different dimensional values where the one is Π and the other is $21.991°/7°$

Any object entering from space and that does not have direct contact with the surface of the earth, encounters gravity in terms of being directionally diverted from going straight to turning towards the earth by $7°$. This then is what provides the "pulling of mass" and the other is having contact with the surface of the earth, which then puts Π in terms with the centre of the earth at Π^0.

What becomes clear from these two illustrations is the following: Where there is a direct entry by the spacecraft, the time factor is not sufficient in duration.

This will not allow the time-duration needed for this structure of the aircraft to revaluate its space occupation an therefore the space factor of the spacecraft cannot adopt to its new position in space-time occupation as it has to relate to a new value in accordance with the value that is determined by the earth's concentration of space-time.

In the case of the second entry illustration, a lot more time lapse is allowed for the spacecraft to revalue its structural position in accordance to its new value of space-time occupation. This is the very reason why objects like "falling stars" burn out when they enter the earth's atmosphere.

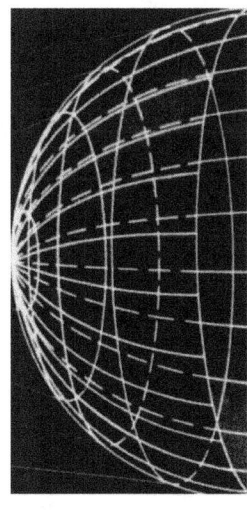

Forget about mass forming gravity it is a fool's rhetoric. When an object holds a steady position to the earth while making a sound, the sound will go in all directions evenly. It will go left and right equally fast. The concept of gravity connects to roundness and to Π. Gravity is Π. Gravity is the movement in terms of Π by duplicating the position of Π per specific time units applying. Then the object making the sound moves left it will hasten the flow of sound in the direction it moves by moving towards the sound while it will increase the distance the sound has to travel by also moving away from the departing sound going to the right side. This means that gravity has two values one being linear and the other being circular and the linear affects the circular in movement as much as the circular affects the linear. This is not only connected to sound but is connected to everything applying to gravity as gravity. When an atom moves the size of the atom must shrink to compensate for the directional change of the orbit that decreases the electron's orbit and therefore decreases the size of the moving atom where atoms form an object.

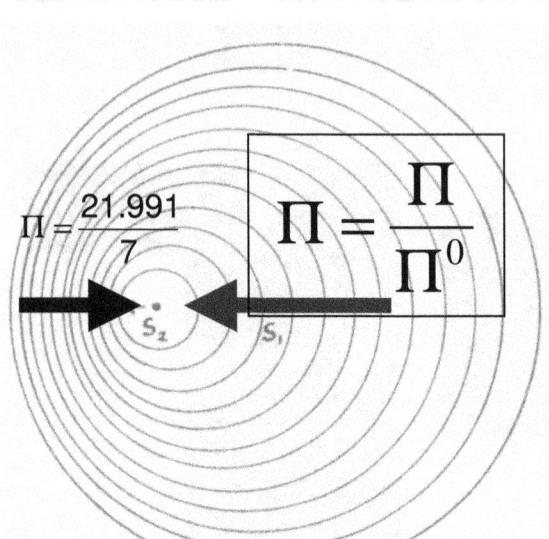

Coming toward the earth from space the object travels straight but the spin of the earth by 7° re-directs the object to follow an inclining line of 21.991 /7. At such a point the object moves towards the earth but the object cannot have weight yet, since the object does not connect directly with the earth. When the object touches the earth it relates to $\Pi\Pi^0$ and at such a point the object receives a value of mass by which it becomes a unit within the earth. By connecting to singularity in terms of $\Pi\Pi^0$ or in terms of $\Pi=21.991$ /7 puts the object in liquid or in solid.

CURCULAR ATMOSPHERIC DISPLACEMENT $\Pi^2 \setminus \Pi$

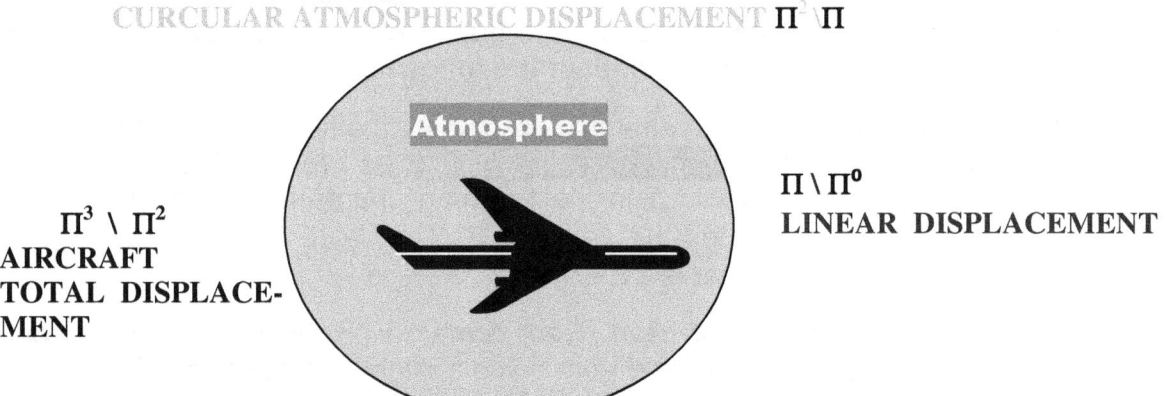

$\Pi^3 \setminus \Pi^2$
AIRCRAFT TOTAL DISPLACE-MENT

$\Pi \setminus \Pi^0$
LINEAR DISPLACEMENT

This principle applies to the sound barrier but also to the object moving. As the object moves faster the effect the movement has on sound would transform the structure of the aircraft as well as the movement alters the atoms forming the aircraft. The structure shrinks because the movement shrinks the molecules as the atoms shrink that make up the molecules that make up the structure.

BETWEEN MACH ONE AND THREE Π^3 / Π^2

ATMOSPHERIC CIRCULER DISPLACEMENT $\Pi^2 \setminus \Pi^0$

ADAPTED   CIRCULER DISPLACEMENT $\Pi^2 \setminus \Pi^0$

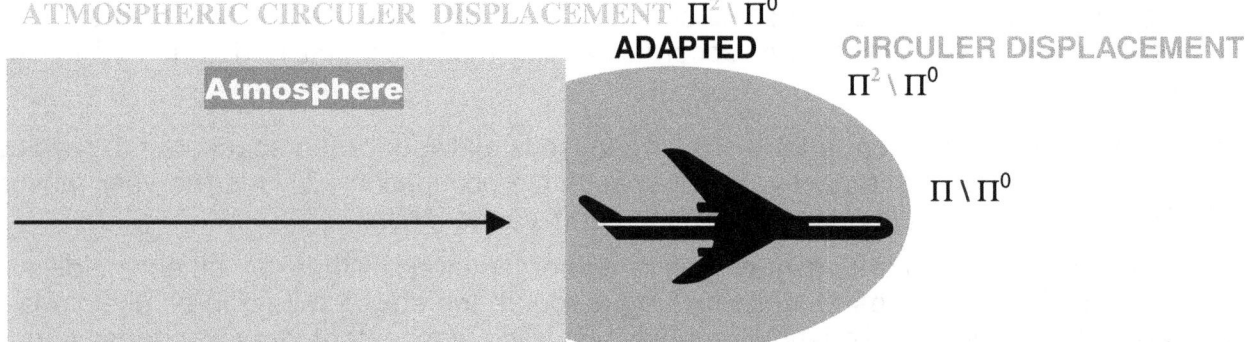

$\Pi \setminus \Pi^0$

LINEAR DISPLACEMENT

When something moves it is within the space the earth rotates. When anything moves above and beyond the earth's movement it excels as it exceeds the movement the earth provide and with it, it takes the space it holds in excess of the position the earth relates too. Thus by moving it form a space-time unit within the earth confinement but yet also out of and beyond the space-time the earth provides.

Between Π/Π^0 and Π^2/Π^0, the body finds itself in its circular displacement, which is the earth's linear displacement value. When an object exceeds the earth's circular displacement value, it is refer to as Mach 1, or the speed of sound. Objects fall at a rate of $7(3\Pi^2) = 208$ km / h while the earth gravitational displacement is $7(\Pi\Pi^2) = 217$ km / h. because the earth moves faster than anything can fall it will always reduce space and have all surrounding space confined to the earth. The object falling only holds space that is confined to the earth.

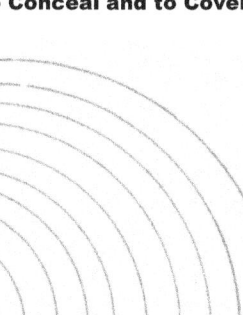

When the object or aircraft stands motionless according to the earth centre the object moves only by the movement of the earth and therefore in cosmic terms the object and the earth is one whereby the object has weight and holds a value in mass. The object holds a relative position to the earth by the value of Π. This is because gravity then links to Π^0 and this holds regard to singularity. When the object moves while being connected to the earth by singularity the earth holds space at Π^3 while the object moves in terms of $\Pi\Pi^2$. The earth represents Π and the object moves by Π^2. When the object gets airborne the relation becomes $7(3\Pi^2)$ but I will not go into that. As the speed increases the relevancy of linking Π^2 breaks when the Roche limit is exceeded. The Roche limit is Π^2 or movement divided by the four quadrants of the circle making it $\Pi^2/4$. However while the aircraft shares the atmosphere one half is still within the earth making the Roche limit $\Pi^2/2$

The Roche limit is:

The region surrounding each star in a binary system, within which any material is gravitationally bound to that particular star. The boundary of the Roche lobes is an equipotential surface, and the lobes touch at the inner Lagrangian point, L_1, through which mass transfer may occur if one of the components expands to fill its lobe. It names after the French mathematician Edouard Albert Roche (1820-83). The Roche limit is a limit stars hold that is closer or equal to 2.4674 times the diameter of the star. The value of $\Pi^2 \div 4 = 2.4674$ which is why it is gravity Π^2 that is divided $\div 4$ by the four quadrants of a circle. The "sound barrier" is just a division of this law, the Roche limit.

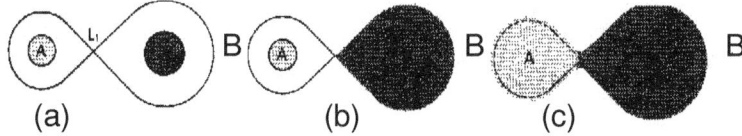

THE ROCHE LOBE: In a binary system, the Roche lobes of components A and B meet at the L_1 Lagrangian point. (a) In a detached system, neither star fills its Roche lobe. (b) In a semidetached system, one massive component, B, fills its Roche lobe. (c) In a contact binary, both components overfill their Roche lobes and share a common envelope. When the aircraft exceeds the Roche limit by half, the aircraft enters the territory of liquid where we have the Roche lobe and where two moving objects share space in motion or space-time within shared space.

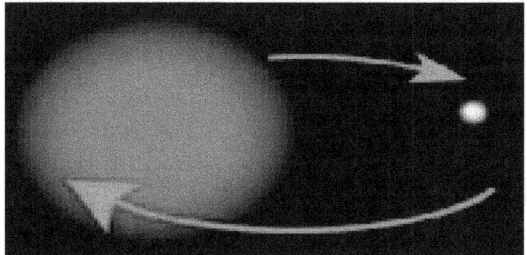

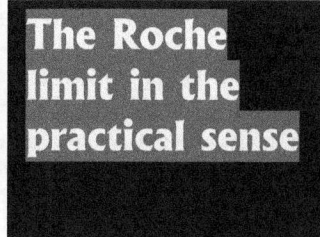

The Roche limit in the practical sense

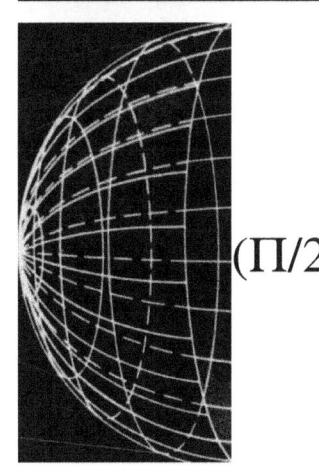

$(\Pi/2)^2$

The link of $\Pi\Pi^0$ connecting as Π^2 disconnects the link it has in Π. After the point where the Roche limit ends the earth hold a limit at gravity or Π^2 while the moving object also finds gravitational identity at Π^2. This then divides the gravity by dividing the connection in movement. In cosmic terms the aircraft then has separate identity dividing movement or Π^2. Then the stage of Mach 1 is entered where the movement in space $7(3\Pi^2)$ holds the limit $\Pi^2/2$ and the culmination is space-time displacement or another name for it is movement or another name for it is gravity. The Roche limit is sacrificed at $7(3\Pi^2)$ $\Pi^2/2$ which is anything from $7(3\Pi^2) = \pm 203$ to 207 and the Roche limit then is $\Pi^2/2$ making the sonic boom somewhere between 203 x $\Pi^2/2 = 1001$ km/h and 207 x $\Pi^2/2 = 1022$ km/h. The sonic boom could be as high as the movement of the earth 218 x $\Pi^2/2 = 1075$ km/h.

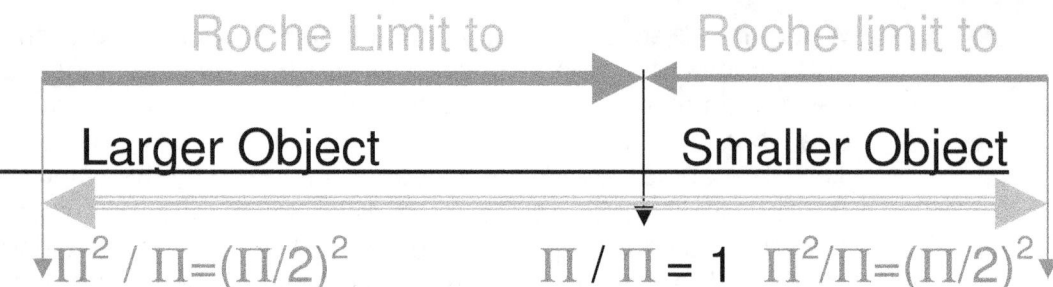

$$\Pi^2 / \Pi = (\Pi/2)^2 \qquad \Pi / \Pi = 1 \quad \Pi^2/\Pi = (\Pi/2)^2$$

The distance of $7(3\Pi^2)$ is not a measured unit such as a specific distance but heat increasing space might increase or decrease this value in terms of km / h because km/h holds 1000 meters of earth surface. This does not apply to space as space is not fixed but varies. The argument is a lot more complicated than this because of more relativity and other cosmic principles forming part of the argument.

The linear displacement value surpasses the circular value, therefore it is $\Pi^2/2 \times 203 = 1001$ kilometers per hour, and that is Mach 1. Doppler's effect has nothing to do with the process, and neither has sound anything to do with the affair. Sound is merely an innocent bystander conforming to gravity in relevancy.

Above the Roche limit of $(\Pi/2)^2$.

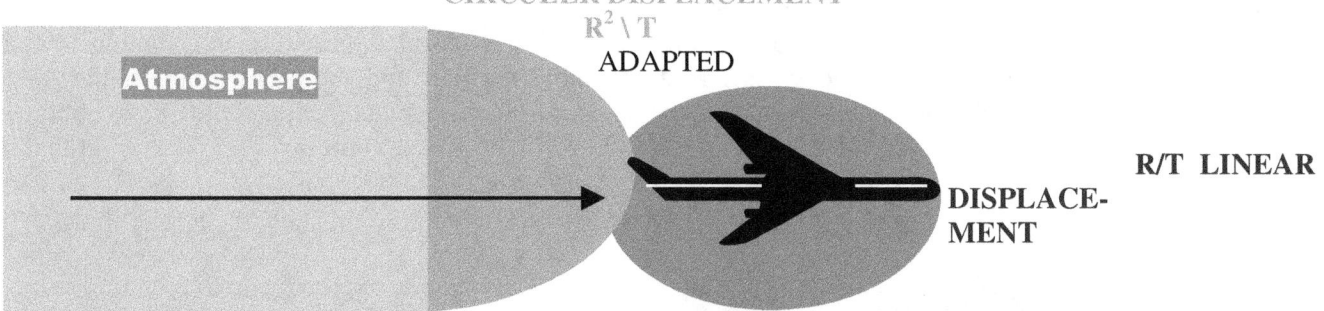

This will come into affect when the projectile reaches Mach 1. I choose to use the accepted term Mach 1 at this point as to limit any confusion that may arise from the new arguments. I have to stress the fact that the Doppler effect is, once again, merely a co-incidental but duly related by product. However, the Doppler effect, as such, plays no part in these phenomena, or in the outcome of the application. It must be seen in terms of cosmology and not from a human perspective. This is the principle applying when a newly formed star escapes the heat envelope within the centre of a galactica as a star is born. There now are two singularity factors within one space-time unit and the two goes into a battle of existing.

There is the same principle applying between stars sharing space-time. When two stars are at the Roche limit, the linear displacement reaches a value of one, and

$\Pi^3 /\Pi^2\Pi = \Pi^0 = 1$ is equal to singularity and in this we find space-time forming a value. This formula does not impress the most learned Physicists such as Professor Doctor Friedrich W. Hehl, which you will learn from a little further on in this book since Professor Doctor Friedrich W. Hehl thought this was to use his words "With a lot of words and some simple algebraic relations, there is no way to "explain" the world of physics." However notwithstanding Professor Doctor Friedrich W. Hehl not being impressed, on this rides the entire cosmos formed by gravity.

The circular displacement reaches its full complement of half Π^2 which is the Roche limit.

To this reason, stars would form massive binaries, where they share a common combined circular displacement, separated only by each stars Roche limit with no linear value. As the Titus – Bode Principle comes into effect the linear displacement would once again grow, or the common spin value will be to grate for either one, or both, and their structural composition will collapse, forming smaller structures with less space to occupy the time in which they are.

The solid part of what forms gravity will be Π^3 while the liquid part that moves holds a value of $\Pi^2\Pi$ and this relates to singularity $\Pi^0 = 1$. Whenever there is a conflict of relations between objects within the Roche limit of $(\Pi/2)^2$ the one would play the part as liquid and the other would play the part as solid. Where both have equal properties the one would be a liquid to the other while the other would be the liuid to the first and there would be a binary star developing.

Double Star System
Orbiting around the Barycenter (X)

This is not where it ends. Layers within a star also form these same relations and when the layers go array and the borders of the Roche limit is violated, a Super Nova erupts. When the relevancies are bridged the control of the star would not be valid any longer and the star would expand in Π that will lead to the compromise of $(\Pi/2)^2$.

The two stars forming the binary will be in a cosmic duel until both of the binary stars group together to unite as a black hole that is if they did not destroy one another. This would depend on the Titus – Bode law sets in where the linear space-time starts developing through the Hubble constant and they would spin around at greater distances. The circular distance, however remain valet as one con see from the "gravitational pull" that has nothing to do with jerking each other around.

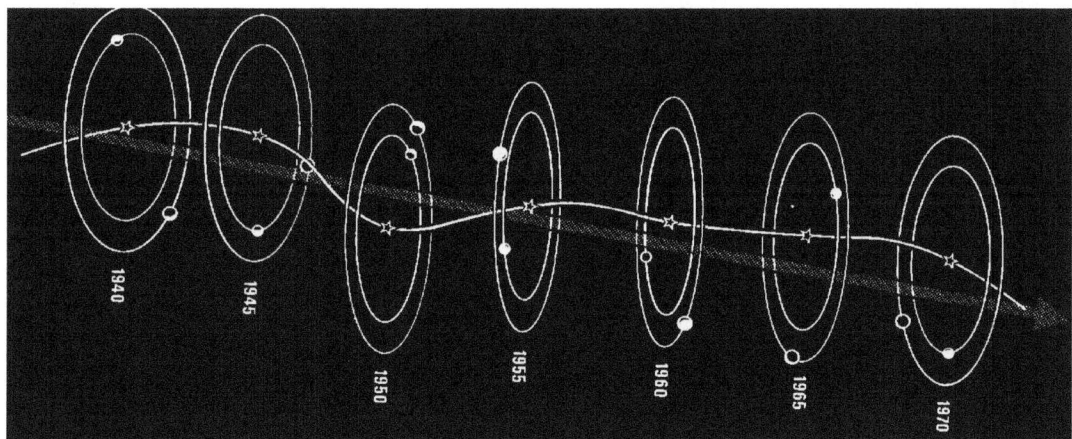

If one can illustrate the Universe and its relation with space-time, the following illustration would fit like a glove. The so-called mass pull is the effect that the Titius Bode law plays on the allocated rotation position of the planets.

$$\Pi^3 \qquad\qquad \Pi^2$$

$$\Pi^3 \qquad\qquad \Pi^2$$

This same space-time value ratio is also applicable where matter is moving through space-time as Einstein stated. However space-time is not the rhetoric that Newtonians apply to confuse their thinking even more than confuse the thinking of others. There are two distinct values in the cosmos. One is tome within singularity $(\Pi^2/\Pi) / (\Pi/\Pi^0) = \Pi^0$ and the next is space that time leaves behind as an afterthought, which mathematically is $(\Pi^3 / \Pi^2)/ (\Pi/\Pi^0) = \Pi^0$. The one says the movement Π^2, which is gravity, puts the relevance Π in place where the relevancy Π confirms singularity Π^0. That is how gravity applies singularity to form time and the result that time leaves behind is space. In this we have space-time.

There are only two energy forms in the Universe. The first is heat and the second is life. No other form of energy exists in the four dimensions of that the Universe exists. No force except for life is to be found in the universe, only balancing values.

The matter of the aircraft lays claim to the unoccupied space-time, which then replaces the occupied space-time (air particles) to another position. This is obvious in the cloud that surrounds the aircraft as it enters another stage in the sound barrier a stage long before the sonic boom comes into practise.

The Doppler effect does not cause the sonic boom, as the aircraft overcomes the sound wave. The sound wave is not broken, but merely relocated to another position relating to the aircraft. If the sound wave was broken, there can be no sound travelling through air being on a wave of matter.

The sonic boom is the relocation of occupied space-time (air particles) as the body frame of the aircraft lays claim to the unoccupied space-time. The sudden jolt of the relocation of air particles, leaves a densified strip of air that are then in a process that relocates the air and for that matter the centre point of the sound wave. The second sonic boom comes into effect the minute the matter relocates in another sound wave.

If one looks at the transmission of sound, it too depends on the relocation of matter, but to a very small degree, and in this process lies the transmitting of sound. To make the error of judgment in confusing the process with the breaking of the Doppler rings are quite understandable.

One more point of interest is that light is the only particle that can remain a part of space-time and not be in the wave. $(\Pi^2/\Pi) / (\Pi/\Pi^0) = \Pi^0$ which allow us to use it in observation.

THESE TWO COMPONENTS FORM **SPACE-TIME** (Π^3/ Π^2), THE DIMENSION WE FIND OURSELVES IN, AND WHICH IS NOT FLAT.

This is the reason why a spacecraft when launching does not "break the sound barrier". The circular space-time is growing faster than the craft can apply linear displacement.
The higher up in space, the less the prevailing circular space-time would be therefore the more linear space-time will be required to penetrate the circular value.
One of the two most important values in the Universe is the Roche limit.

On this and the Titus –Bode principle rests the growth of the universe, as space relate to time. The Roche limit comes into effect when the linear displacement factor reaches a value of one and part of the circular displacement value. In this is the value $R^3/ T^2 =1$

The photon relates to space being behind time while the wave as such connects to time Π^2 with the three connecting the wave to space in accordance with the Titus-Bode law (10-7=-3). Therefore, light moving in the wave, (coming with the wave) will all ways be RED, not blue.

What I cannot understand for the love or money is how this confusion extends to the speed of light? Nothing can come close to the speed of light, never-mind scramble the wavelengths of the speed of light.

Light reflecting from the rear of the wave will therefore be BLUE, not red as presumed.

Sunlight made of all colors

The Scattering of Blue Light by Gas Molecules in the Atmosphere

The Atmosphere

This HOGWASH is what intelligent teachers tell students. Because the Earths atmosphere "scatters" light, the sky is blue! If such were the case, the sky would be a rainbow of colours, because a rainbow comes about as the water droplets scatter the wave of the light. The earth's sky is blue, because the earth's atmosphere re-concentrates the wave's photon distribution in the opposite direction, as to the light wave.

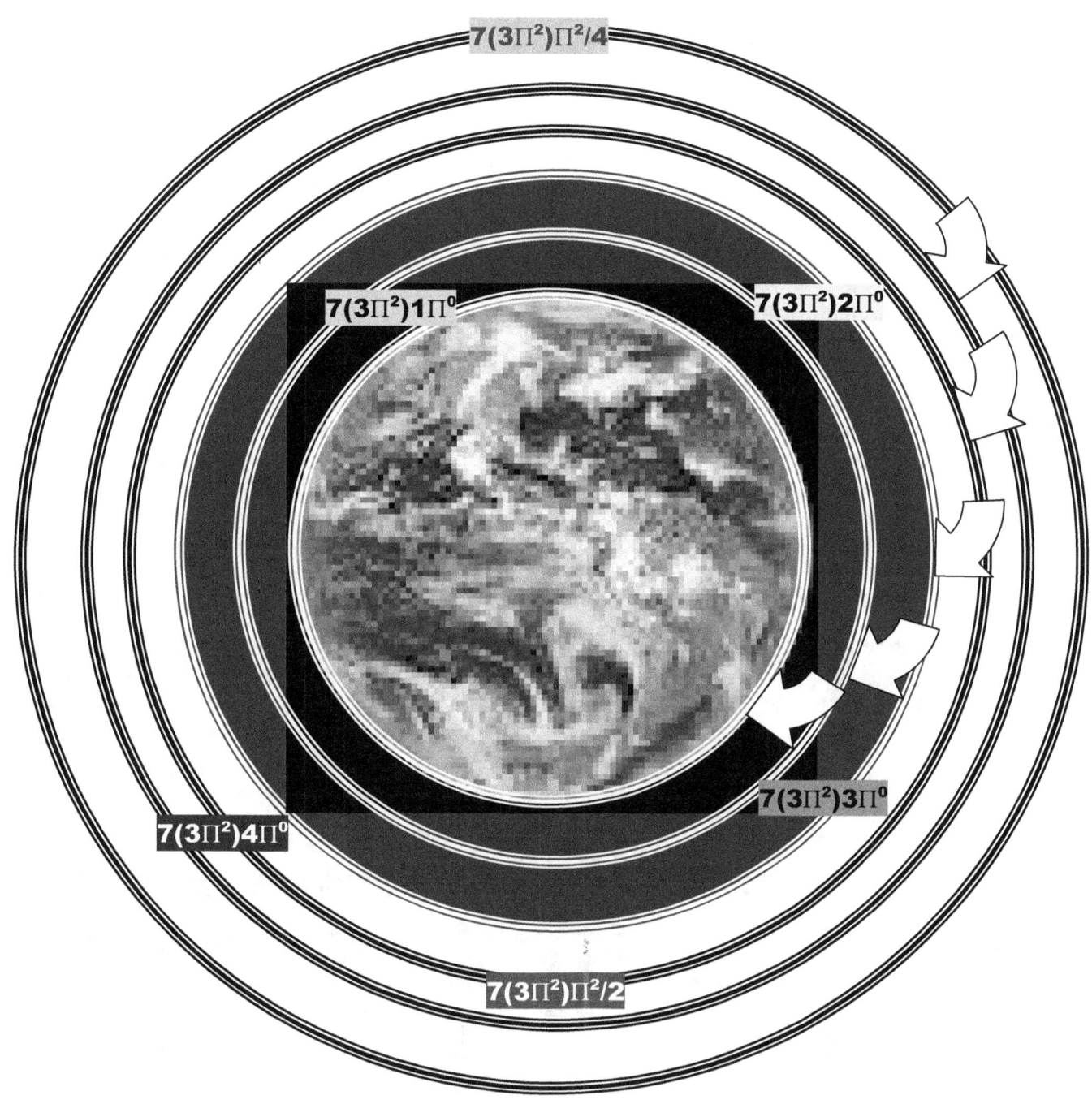

$7(3\Pi^2)\Pi^2/4$

$7(3\Pi^2)1\Pi^0$

$7(3\Pi^2)2\Pi^0$

$7(3\Pi^2)3\Pi^0$

$7(3\Pi^2)4\Pi^0$

$7(3\Pi^2)\Pi^2/2$

Welcome to
www.sirnewtonsfraud.com

Have you ever had doubt sitting in class while listening to your lector telling Newton's ideas? Have you ever got an eerie feeling something doesn't add up? While you are listening to your lector or teacher rambling on about Newton you are becoming brainwashed and with mind control and in applying this method for centuries they are drawing you into the biggest scam ever devised by any man ever. You are becoming part of the biggest hoax ever created and it is called Isaac Newton's physics. Have you ever thought you will hear any person with an apparent sane mind tell you Isaac Newton is fraud…well, now you have… and if you keep on reading you will learn what a lot of deceiving is part of physics and it is all about covering Newton's Fraud. All I ask is someone to show me that mass pull and to prove how mass pulls. Prove to me that the Universe uses mass or applies a pulling power in accordance to the value of mass. I accept that there is weight but how that weight arrives at a value I explain somewhat differently.

This book may lean more towards explaining than expressing evidence, but due to the nature of the theme it demands intelligence and concentration nonetheless. I attack the heart and sole of the most intellectual group of persons that forms the upper part of any civilised society albeit of whatever ethnic group or culture or race notwithstanding, those with the seemingly highest intellect have duped the entire world for three hundred years and that includes you and everyone that came before you and walked the road you are walking. No one can go and bash on their scheming in science and have the ethos be as easy to understand as a nursery rime. It calls for a lot of information to be sorted and digested and only a small part of readers might understand or be able to perform what is one of the most challenging tasks you ever called on. As unbelievable as it now sounds you are going to read about the most shocking conspiracy.

This book delivers thoughts and gives questions no one ever asked. This book demands you to think in terms of what you thought was more correct than the religion you practise. Every person on earth thinks of science as that something that is correct beyond question and could never come to doubt. It is what your entire understanding of everything is built on since the day you entered school. You now will be forced to think about what not only you never thought about, but also what no one thought about since the days of Newton's gravitational principle introduction was first thought about and accepted. You will have to loosen the shackles that tied you down since you were a child because you and everyone else was brainwashed in a system by overbearingly important persons that became sublimely self-important in what they thought they knew, and what they presented as the absolute truth and therefore they do not even tolerate a thought that there might be some lowlife that would ever dare to question them. If you feel you have the mental capacity to question what became religiosity and moreover a culture we call science, go ahead and be as surprised as you have never been before. However, while you contemplate the many facts you must study the meaning of what is said for it is no storybook or fake conspiracy theory: this is the real thing.

Everyone knows there is a problem about gravity in physics and no one can put a finger on the problem. This thought is lingering on… from generation to generation… without anyone ever finding a solution. Everyone is scared of admitting a problem so it is left for the next generation. There is a suspicion lingering in the back of everyone's mind that something is not quite correct about the approach physics take on the matter of gravity and only those academics seasoned with years and salted with time seem to miss this feeling.

Something about the way gravity is presented just doesn't add up as it should and does not quite reach the answers it should conclude. There is this vague unspoken question hanging in the air without ever finding words to express the question…and yet the question remains however unspoken it seems.

If you are a student studying in physics this web page is detrimental to your future, as you can remain part of the problem physics have or you may join the solution that came to physics.

The problem is that the faculty of physics are covering up a culture driven by the mentality to brainwash and control the minds of students to accept what should never be accepted.

Hidden under a cover of "understanding Newton" or "not being able to understand Newton" they force certain incompatible arguments to join that never can join and make sense at the same time.

Every generation find an itchy feeling but never is there a place that no one can secure the very point where it is apparent enough to scratch and rid physics of the undetectable itch. Believe it or not but this itch is in place because of centuries of brainwashing going on and is employed in physics from generation to generation for centuries on end.

This book is dedicated to bringing honesty into the faculty of Astrophysics and Astronomy as well as show the Physics student on what corruption and deceit does physics base their facts which they proclaim as being such well proven, and godly accurate facts.

Students listen to this:

Read the next pages and you are about to learn how students are brainwashed into accepting the baseless and ridiculous as truths. The Custodians of Physics have nothing better to offer than presenting you with unfounded corrupt and distorted facts. Doing what they resort to is simply mind controlling students by introducing baseless concepts and therefore manipulating student's thoughts. If you are a student then read in my website www.sirnewtonsfraud.com you will learn what they do to you. They are defrauding you by exchanging your institution fees for corruption so confront them about the dishonesty. Force them to be honest and to stop corrupting students with intentional malice. They offer no truth. All they have are misconceptions and incoherent facts. Custodians of Mainstream Physics stop your practice of mind abuse, thought control and criminal behaviour towards students immediately!

If you are one of the Academics and Custodians of Mainstream Physics I challenge you to show one piece of evidence where I am incorrect, exaggerate facts, produce incoherent and distorted views about your physics being based on corruption and misconduct.
If you Academics think you are innocent, read on.

Fortunately for the future and unfortunately for Academics in mainstream physics the truth is out and it is published and your dishonesty as Academics will come to be no more. Your entire generation and all the generations that came before you will be washed away as dirty mud on soiled linen. You will be remembered by the coming future as those in the past that was not worth remembering.

Students do not have to suffer your abuse and cheating. Students no longer have to sit and wait for thought control as you wittingly force feed the students all the distortions about the fundamentals on physics and expect then to allow you to get away with academic murder just because your professional position allows you to.
Now students have the choice to insist on you telling them the truth because the truth is written in

"*Newton's Fraud*" *The truth is finally out and your culture of deceit has eventually been detected* You will stop your mind abuse on students because now they can know more about gravity than did all those that came before you and all those that came with you.

Again I challenge you to come forward and tell your students the truth about what I uncover in the articles that follows and as the articles progress in introducing information...then you explain to them how you deceived their blind trust in you.

To Find Out More About **The Fraud science hides in A Conspiracy to Commit Fraud on a Cosmic Scale Click on www.sirnewtonsfraud.com To Learn About**

The Brainwashing And Mind Control in science

The facts you are about to learn will astonish you and it will seem unbelievable but notwithstanding it is true.
Unbelievable as it is nevertheless, I challenge any one to show me that the least of any or all facts I uncover is not true.

Sometimes you will wait a time for some of the documents or a page to open because they are extensive but you will lose less time in waiting than you lose by unwittingly learning the deception and then afterwards finding out you were deceived. It will still be quicker waiting on the response than losing the time wasted when you are being tricked by the Academics in science holding positions in Astro Physics and Physics. Sometimes the reading is slightly extensive but to find the truth you must dig deep and overcome corruption there is a lot of material that came in place over many centuries and it is the lie that detours from the truth that then is extensive.

A Conspiracy to Commit Fraud on a Cosmic Scale

Part of

www.sirnewtonsfraud.com

WRITTEN BY PETRUS S. J. SCHUTTE

©KOSMOLOGIESE EN ASTRONOMIESE TEGNIKA

Should you download you then will learn of the mind games…and…

The Practice of Brainwashing by thought manipulation in practising Mind Control on students studying in Physics

Let me show you what is The Practice of Brainwashing and Mind Control in Physics

The TITLE AND THE INDEX

Index of Chapters In Relation to Page Numbers

Science in the present and for the last three centuries placed all their focus on material. In placing the focus on mass while mass does not exist as a cosmic entity, science has been running around like a chicken without a head and the truth eludes them even after three centuries of lying and corrupting science. You think that this is harsh words, read and you will see those in science deserves much more that just that. Those in science concentrates on the material filling the space but gravity is about the movement of the space and not of the object in the space, but of the space the object takes with as the object moves faster than the other space. That is why objects all fall equal because the space the objects fill is the same, unconditional of what forms the material. All objects fall at an equal pace without size or mass becoming a factor. That is why I introduce gravity by a very new set of principles that no one ever heard of. When an aircraft goes through the sound barrier, the object fills more space per time unit by going faster through space in time. The object then in accordance with the space it holds stretches because it holds more space than when it went slower or when it stood still. Since the aircraft is solid and does shrink a little but not much, it has to concentrate the space around the aircraft and by concentrating it reduces the space. As the space concentrates through the movement of the aircraft a cloud appears around the aircraft because the water vapour in that space concentrates into a cloud that forms. All movement is part of the sound barrier because the sound barrier is the movement of space. However the sonic boom is confused with the sound barrier but the sonic boom is a small part of the entire sound barrier and only fills the centre spot or the middle in the sound barrier. If mass is anything then show me the role that mass plays in the sound barrier and how does mass conform into what we think of in terms of the sound barrier. Gravity forms as not by the mass of what any object presumes to have, which I prove in this book and in all my other books, is clearly not present as a cosmic factor. There is no such a thing as mass in the entire Universe and the conspirers that conspire to keep Newton's fraud disclosed, cover up this reality with all they have. I challenge anyone to prove where do they find proof of a factor such as mass. I found the place! It is in the imagination of physicists while they try to conceal that mass is only a product of Newton's imagination. They conspire to confuse everyone with the implication of weight. Weight there is but things pulling other things by the measure of their mass is a daydream and the thought could be funny if it wasn't that crooked. Gravity forms by redirecting the movement of space in circular flow by the spin of any cosmic structure that has the ability to do so. The aircraft goes straight while it also circles the earth and that forms movement within the earth's confinement by the earth's movement, which is what forms the sound barrier. The sound barrier is gravity and sciences present way of going on about the "Doppler effect" and "Mach's principles" shows how incredibly little those incompetent physicists know about physics. I challenge the lot to prove Newton by showing that the Universe contracts or where mass plays a part in cosmic physics. Show how the planets hold their positions according to the mass they have.

All objects move straight while at the same time circle around some other object. The object moves in a straight line while the spin the object hold diverts the space in which the object is into a circle. This circle brings the value of gravity to Π. By diverting the space in which the object is as well as the space surrounding the object, gravity concentrates space into becoming denser by circling and also hotter.

Gravity forms as the earth or any other cosmic body rotates by 7°. By diverting the straight-line movement by 7° a contraction forms in a circle. In my books I prove how this then brings about the value of Π by implementing the law of Pythagoras and gravity is the law of Pythagoras.

The reclining of space by redirecting the direction of travel from straight ahead to 7° reclines the space in a steady and sturdy flow. It is the space reclining or contracting and the space contracts albeit filled by solid material or empty of solid material. This is the reason why all things fall equally. It is the space moving down with or without holding material and the space has the same density in relation to the solid cosmic structure rotating.

In the centre of the rotating body singularity forms to the value of Π^0 and this extends to the curve forming Π in terms of the curve of the rotating object. This then forms part of gravity moving forming

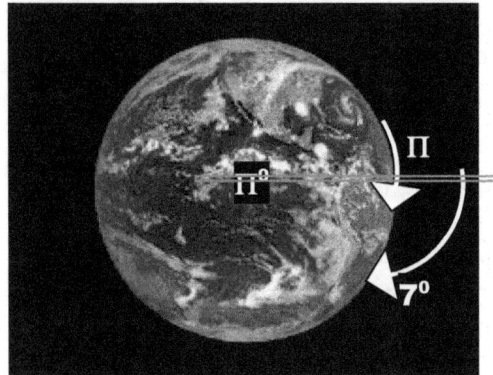

The extending of singularity goes from Π^0 to $5\Pi^0$.

singularity going square Π^2 and this form the relevancy Coming in from space while having no mass the coordinates change as the value of Π forms a relation to the 7°

When an object comes in from the atmosphere or flies the sky the object associate with the turn giving it a value of the rotation in line with the axis which puts a value of 21.991 / 7 on Π or on gravity. This is gravity and this is how gravity functions. It has nothing to do with mass in any way, shape or form and there is no factor such as mass in the entirety of the cosmos apart from being in the imagination of the Newtonian conspirators call

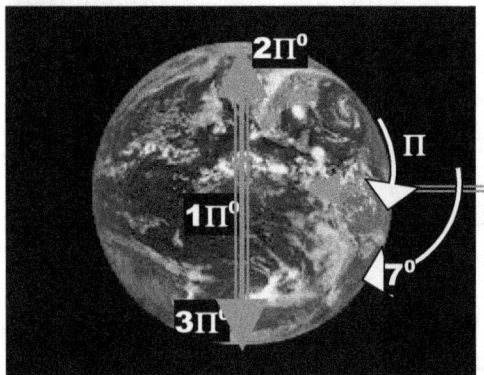

The entering by singularity that goes from 7° by $3\Pi^2$.

From these conclusions I prove that gravity is the result of four cosmic phenomena interacting to form the value of Π which by movement becomes the value of gravity Π^2 and gravity is equal to cosmic time applying. In order to understand the development of the cosmos and moreover the start of the cosmos and the progress in the cosmos as the cosmos formed, one has to understand the measure of Π. One has to microscopically dissect the measure of Π to find the cosmos in measure. One has to understand where 7 fit in Π. The fact that Π is 7 at the bottom and that 7 relates to a double value of 10 is a key issue. It is behind Π that we will find the four phenomena, which I named the four pillars performing as gravity as they form gravity. It is by the actions of Π that the Universe develops. The Hubble expanding goes by implementing gravity as Π in the square through the four pillars on which gravity and time rests. It is behind Π we discover the

meaning of singularity and how singularity forms the absolute and only building block as a form that forms the Universe. It is in Π we find the Cosmic Code unlocking the meaning of the Universe. Time is centralised in Π^0 that forms Π as space's limit that becomes space by gravity being Π^2.

What is in the Universe, is spinning and therefore what I am referring to, applies to everything holding a place in the Universe and therefore this which I mention directly links everything holding any space whatsoever in the entire Universe to one single point around which all spin, notwithstanding the allocation. In the **precise middle** of all **objects in rotation** disregarding size is a precise centre dividing the object into opposing sectors that will **start the spinning initiation** from that centre point. The spinning object will have a very specific **centre point that does not spin** and only holds Π as a specific value because no radius can apply at the point being one space away from Π^0 holding Πr^0. But also the one value such a line **cannot have is zero** because the line **is there and being unbroken, it holds**

contact with the rest of the material bringing about that **zero does not start any** line and therefore the **value of the line must be infinite**, just as described in **accordance** and by **the definition of singularity.** As I am introducing a very new idea, I wish to explain in better detail what I try to convey. While anything spins, singularity forms a line and when reducing the rotating line or radius progressively to the middle at one point all further reducing must end. As the rotating direction moves inwards, the rings forming Π will become smaller and smaller. Then we reach a point everyone thinks of as being the axis around which everything rotates. The line only forms when everything around the line spins by establishing a circle to the value of Π.

go to http://www.singularityrelavancy.com/ and also go to www.questionablescience.net Flying object is under this gravity control of movement and it is this that has crafts fly and cars requiring down force by the aid of aerodynamic devices.

Gravity is defined as a force that is present in mass pulling mass and it is that entire idea that there is not evidence of. When I refer to gravity everyone grabs on a cultural notion of a concept they formed and in that concept they link the smallest part of the concept to the become and represent the overall gigantic principle and by knowing one line everyone has the opinion that anyone then is the absolute master on the idea of gravity. When I freeze any substance the substance contract to a liquid and with more cooling it contracts to a frozen state of ice. The gas expanded more than what the solid did because the gas is hotter than the solid is. When we form the opinion that the outer space expanded to the limits the idea springs to mind that outer space is freezing cold. When I say the Sun freezes hydrogen to a liquid because my eyes see the liquid squirting from the Sun I am dangerously mentally impaired since the Sun is blistering hot. Then through this culture my effort to say gravity is motion and motion is the cooling of an overheating and thus expanding Universe goes wasted. Every one has the opinion that where gravity is the strongest such as the case is on the Sun or the centre of the Earth, such a place is extremely hot and where gravity is least that place is unbearably cold.

No person ever could understand the sound barrier because no person ever could **enter singularity** and see **what applies when the Universe goes singular** You will read what exactly happens when the "sound barrier" is exceeded but also to explain the "sound barrier" I will mathematically have prove what gravity truly is by proving the measured value of gravity. This I explain in simple understandable mathematical terminology. **It is so simple even I can understand it.**

I now explain gravity again and relate this explanation to the sound barrier functioning.

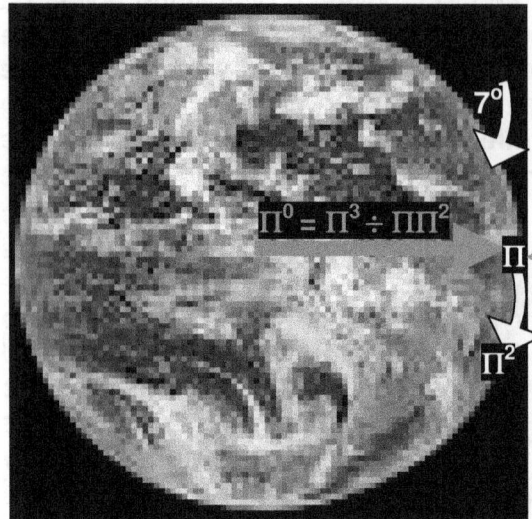

Gravity forms by association or relevancies that holds a value of coming towards the centre from the sky which then is 7(which is the curve of the earth) (Π associating with the curve of the earth) and Π^2 (associating with the moving of the earth) This is part of the Titius Bode law forming the measure of gravity which is not by mass but by forming Π.

$$\Pi^0 = \Pi^3 \div \Pi\Pi^2$$

$$7\Pi\Pi^2 \times \Pi^0 \text{ to } 4\Pi^0$$

Then when a line is drawn from the centre of the earth another value comes in place referring to the centre of the earth Π^0 and the curve of the earth Π. The value of Π is then shared by both disciplines (association from the sky to the centre ($7\Pi\Pi^2$) and association from the centre to the sky ($\Pi\Pi^0$). In reality it is ($7\Pi\Pi^2$) but I am not explaining this.

If the body does not connect to the earth by way of forming a unit or forming "mass" the axel line of the earth comes into affect giving the ratio a value of 3. Then when not connecting to the earth surface the ratio becomes $7(3\Pi^2) \times \Pi^0$ going up to $4\Pi^0$ all depending on the sped or the height. This is the way gravity forms by forming a connecting association with $\Pi^0\Pi$.

This forming of gravity explains the Coanda effect because the Coanda effect puts air or liquid in relative movement with solids and this movement (but not the mass part) is how gravity forms.

The condensing of the vapour surrounding the jet is a prelude to the sonic boom and is just one link in the sound barrier where the sonic boom is just another link in the sound barrier process. The sonic boom IS NOT the sound barrier but is as much part of the entire concept as the beginning of movement is part of the sound barrier. Waves in the sea are as much forming part of the sound barrier as winds howling or blowing is part of the sound barrier.

To whom it may concern and all others reading this document:
I am P.S.J Schutte, nicknamed Peet. Being a white South African my mother tongue is Afrikaans and my second language is English. I have per suiting this theory that I partly present in this book, of which the investigating research was done since 1977. First I located what was wrong in physics. Then I compiled my presentation thereof for the past going on to twelve years on full time basis whereby I was trying to introduce my findings to many academics without finding much joy from my efforts. This past eleven years plus saw me go without any income as I tried to get my theorem recognised. Going without a steady income left me almost destitute and in order to find a manner to get my theory across to the attention of influential readers, I decided to publish a theses of six books electronically as to try and get around the stranglehold of Newtonian bias controlling science at present worldwide. I decided to publish The these called The Absolute Relevancy of Singularity and then six separate thesis parts forming the theses published through LULU.com which I saw as way the only manner whereby I could generate funding by which I would be able to have the twenty seven books I already wrote linguistically edited and then to have the books published on a Print-On-Demand basis. With my first language not being English and the books not linguistically checked by an expert there are bound to be language errors that readers will notice. In the past I tried to check my work myself but after checking say one hundred and fifty pages for language corrections, instead of having corrected work I ended instead having four hundred pages of new written information which is still not language corrected but holds a lot more information. This is because my priorities lie elsewhere. I aim to spend money on correcting the work as far as language goes, as I receive money in the selling of my theses and in the hope that I will receive money. I will have all my work including the one you are reading edited professionally and corrected as I find money to do so...But first I have to get the public aware of the problem to get the academics to appreciate the problem.

Students, your professors are fooling you and you deserve to be their mindless monkeys just the way they think of you because you don't think about what they say! Cut the bullshit, force their ignorance about physics into the open and make monkeys of them, they deserve it even more! Let the professors that are so wise explain why the Titius Bode law is in place and is used by the cosmos instead of the Newton's mass idea that holds no legitimacy as far as cosmic evidence goes. It is what is used by the cosmos that has credence and not some surmising of Newton's fantasy-cosmic-principles. You and I should fight for the truth and not to uphold Newton's fabrication of the truth.

PROVE ME TO BE INCORRECT IN ANYTHING I SAY!

To whom it may concern and all others reading this document:
The above is my concluding about what forms gravity. I wrote to (at this stage) near as it might be two thousand Professors in Academic standings, Heads of Physics departments at Universities, professors teaching physics, Editors of Publishing Science books, Editors of Science magazines, Editors of publishing science articles and this is a summery of what I received as the general reply from our outstanding physics paternity world wide.

To christianseni@ukzn.ac.za <christianseni@ukzn.ac.za>
Dear Professor, I am P.S.J.Schutte called Peet. I send you the article attached in hope of finding publication as to open an avenue whereby I could launch a new cosmic theory. Should you require more background information or should you wish to have more information available before publishing, I would gladly send you a CD with all six books I offer holding the entirety of the work going under the title of The Absolute Relevancy of Singularity should you or any staff member that wish to do so you study the entire assembly of books so that your journal can rest assure that the claims I make is duly forthcoming by proof. I include a 15 page article in which I summarize the framework of thoughts that compile into what my theory entails.
Then I received a reply on which I answered.

>>> "Peet Schutte" <gravity@bosveld.co.za> 2009/03/07 11:23 PM >>>
Dear Prof Iben Maj Christiansen, I have received your e-mail reply and I will follow your advice to the letter. The book containing four volumes that I sent to you accompanying the proposed article, is the collection I named The Absolute Relevancy of Singularity and as such the collection forms a small introduction to the thirty-two or so books I wrote on various matters concerning physics, but as such does not officially even start to introduce the spectrum of every aspect of my work. I have been in contact with numerous Academics and about one in one hundred reply. When the one in a hundred reply, the academic always uses a most aggressive tone which I came to accept as what I receive from academics, and because of that I was most delighted to finds some kind remarks from you as a practicing academic, and might I add, the first such kind remark in ten years of my trying to contact any person in physics that would take note of what I have to say about a new line of thought.

The New Cosmic Theory I try to convey in total is much information and every time when publishers reject the publishing of any entire book I propose, the rejection was on the grounds that the discourse is not falling within the mainstream science discourse and therefore I was subsequently advised to write articles on the subject as to find recognition. I was told that only then could I achieve publication of any entire book.

Now I find that trying to publish articles has my work rejected on grounds as follows and the following is directly coming from the reply in which one of my articles was rejected recently: You submitted an article of 15 pages to the Journals Name Left Out. (I do not wish to include the name of the journal for obvious reasons but quote the reply as I received it). The content of this paper doesn't constitute a theory in physics. With a lot of words and some simple algebraic relations, there is no way to "explain" the world of physics. You seem to be out of touch with modern developments. This is also shown by the fact that you don't quote any relevant literature." It is not possible to introduce the totality of my work in 15 pages (or whatever a journal would allow) and remain coherent during such an introduction about anything.

I am trying to introduce a study I have done during twenty-seven years of research and there is there is not one word that I can quote from any other source since every word comes from conclusions that I make and I prove with the use of logic. All I try to do is to find a medium wherein I can tell some interested parties where to go to read my work and then for them to judge me on their merit and not be sidelined by rules set by academics in charge of publishing.

Let everyone read my work and then after that let all readers be opinionated by personal opinions applying.

Everyone goes on about the unfairness Galileo endured at the hands of the Catholic Church, but at least the Church allowed Galileo to publish his work so that the entire world could take note.

Every one in science as well as the Church thought Galileo was out of touch when he declared the science wisdom prevailing at the time was incorrect, and five hundred years later we know who was out of touch. I do not compare my work with that of Galileo but I find the same restrictions brought on me by the Powers of the day controlling science.

In this light I am most delighted by your attitude.

I have tried to acquire the addresses of the academics on the list I include, which I highlight, but was unsuccessful in doing so. Could you please show further kindness and supply me with contract addresses so that I could contact any one and try to arrange such a meeting as you suggested.
Peet Schutte.

Dear Peet,

Those Profesors on your list I know are in science education. You need someone in pure physics.

I am afraid that you will continue to get rejections if you do not relate your work to existing theories and previous work. While it is possible that a lay person hits on an insight that has been overlooked by academic trained in the field over many years, it is unlikely. We assume that work offering something new would be related to existing theories, either by building on top of them or by showing how and where they fall short. If you do not relate to existing work, it is repeatedly going to be dismissed as mind spin too easy to shoot down.
I am sure you understand.
Iben

I am going to show very briefly what I wish to introduce since the aim of this book is not to reveal my work but to reveal why there is a desperate need for my work. Whilst no one is aware that there is corruption present in physics on a cosmic scale, every sits back and feel very pleased with the situation going on at present. While I know that science at present is a farce, a fabricated hoax and manufacture conspiracy, Everybody conspiring to protect and uphold Newton's corrupted principles brushes me of the table and a life time of my efforts goes wasted because some criminals sit in Academic posts propagating the biggest load of bullshit that was invented by Newton and in three hundred years was never proven because it can never be proven. Unless members of the public realises the conspiracy there is to uphold the hoax they call physics I will lose out to criminals that should be in jail but now they are in command of students brainwashing the student into zombies they (the professors) became when they were students.

By applying my principles I take the reader back to the point where the Universe started and show why the Universe started because we can see it from how the Universe started. The big key unlocking the secrets is in the Titius Bode principal, the Lagrangian system, the Roche limit forming the Coanda effect These four principles is what forms Π where Π forms gravity. These phenomena are in place used by the cosmos in every picture, but since the phenomena corrects Newton's failings, the importance of their role is very well hidden. I show Newton's ideas are not what the cosmos uses. I show what the cosmos uses are there for a reason. Those physicist ignore what is in place in the Universe used by the Universe, discards what the Universe uses in favour of Newton where in the Universe there is not one point of evidence that the Universe applies what Newton said forms gravitational principles. With what is in place and by applying what the cosmos uses to apply gravity, I am able to show the place, the very point place we find why, how and where the Universe started.

Those professing they know everything about the Universe hide the importance of these principles because they don't know the working of these principles. They fail to disclose that the cosmos is built by applying the Titius Bode principal, the Lagrangian system, the Roche limit forming the Coanda effect By they're admission that they do not know what gravity is they then also admit being possibly incorrect about gravity but unfortunately mainstream physics do not see it that way (yet). I was able to use Kepler's finding of cosmic relevant values and as Kepler found it without Newton's meddling to decipher gravity by using the four cosmic phenomena.

That it does by giving gravity a very specific value. From such explaining what Kepler said without Newton changing formulas on Kepler's behalf I prove the Titius Bode principal also known just as the Bode principle conjoining the Lagrangian system of object formation. I explain how singularity forms the Roche limit and how singularity brings about the Coanda effect but most important of all Kepler showed me where to search for singularity.

My achievements came from my effort where I separated Kepler's work from the opinion that Newton formed about what he saw Kepler's work should contain and gave to the world his Newtonian concept about Kepler's work. For instance from Kepler's work I can explain the operation of the Black Hole, which not even Prof. Stephen Hawking understands. From my view a force is just motion applying and that is what Kepler said gravity is. Kepler said $a^3 = T^2 k$. The Coanda effect is the establishing of individual independent space a^3 by applying motion $T^2 k$ in relation to a centre point the motion of the liquid establishes. Einstein came to this conclusion but failed to refer his view back to Kepler and by not referring to Kepler missed the point he wanted to make. Just by my studying of Kepler this became possible.

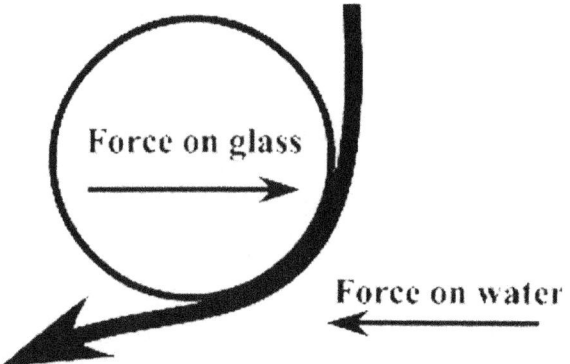

That shows that all movement is gravity or time related. This shows that when objects move then such movement of whatever is moving is moving extraordinarily because everyone always forgets about the fact that it is the earth that normally applies all of the movement while all else stands still in relevancy. It is not my manner to speak ill of the brain dead or the dead by other means, but in the case of the Newtonian academics I am left with no option. Their forces haunt me to death and it is their forces and ghosts and witchcraft I have to fight. The lifting of a body comes quite natural when a certain speed is exceeded. By exceeding $7(3\Pi^2)$ the body will start to lift no matter what the mass is. A 747 Boeing of multi tonnage lifts off spontaneously at excess of that speed. Newtonians are forever concerned with middle ages and with forces they can't explain but such forces and witches there are not, therefore they do not have to fear and can sleep well at night. In the sketch the circle portrays a glass and the arrow portrays running water. The Coanda effect is the water that does not drop straight down but follows the curvature of the glass.

The picture clearly shows the $7°$ inclination of gravity to the value of contracting $\Pi^0\Pi$. This is gravity!

This is the most vivid example of the Coanda effect and it is what gravity is! It is a whirl allowing the flow of liquid space around a solid centre in relation to the centre holding singularity or $\Pi^0\Pi$ as it contracts space into a denser liquid.

The Coanda effect is gravity and my explaining this statement is part of many other books in which I explain what gravity is. The Coanda effect shows how liquid attach to the solid by $7(3\Pi^2)$ and the solid attach to the liquid by a relevance value of $7(\Pi\Pi^2)$. That is gravity.

Should anyone require more or better explaining I would advise that person to purchase any of my books holding the title as an to go to http://www.singularityrelavancy.com/ and also go to www.questionablescience.net Flying object is under this gravity control of movement and it is this that has crafts fly and cars requiring down force by the aid of aerodynamic devices.

Gravity is defined as a force that is present in mass pulling mass and it is that entire idea that there is not evidence of. When I refer to gravity everyone grabs on a cultural notion of a concept they formed and in that concept they link the smallest part of the concept to the become and represent the overall gigantic principle and by knowing one line everyone has the opinion that anyone then is the absolute master on the idea of gravity. When I freeze any substance the substance contract to a liquid and with more cooling it contracts to a frozen state of ice. The gas expanded more than what the solid did because the gas is hotter than the solid is. When we form the opinion that the outer space expanded to the limits the idea springs to mind that outer space is freezing cold. When I say the Sun freezes hydrogen to a liquid because my eyes see the liquid squirting from the Sun I am dangerously mentally impaired since the Sun is blistering hot. Then through this culture my effort to say gravity is motion and motion is the cooling of an overheating and thus expanding Universe goes wasted. Every one has the opinion that where gravity is the strongest such as the case is on the Sun or the centre of the Earth, such a place is extremely hot and where gravity is least that place is unbearably cold.

No person ever could understand the sound barrier because no person ever could **enter singularity** and see **what applies when the Universe goes singular** You will read what exactly happens when the "sound barrier" is exceeded but also to explain the "sound barrier" I will mathematically have prove what gravity truly is by proving the measured value of gravity. This I explain in simple understandable mathematical terminology. **It is so simple even I can understand it.**

This remark I make is in response to physics academics not understanding something as elementary as the "sound barrier" which I prove is the most basic principle of gravity but they are forever telling me I am too stupid to "understand Newton" and therefore because of my lack in intellect I question "**the validity of Newton**".

To them and to those that says this I say: **Download what I give free of charge and then start to get wise...** What it is that these two phenomena has in common except sharing a Universe because he should know, after all he is the Newton-physics expert ... but I will bet you although being a physics expert, your professor will have no idea...mainly because he is the Newton-physics expert. This prevails because of singularity, which clashes with Newton head on.

This is my introduction and this is my prologue:

Gravity is the sound barrier because the sound barrier forms by four cosmic principles that form gravity and therefore gravity and the sound barrier are the same principles in conflict because of movement differences.

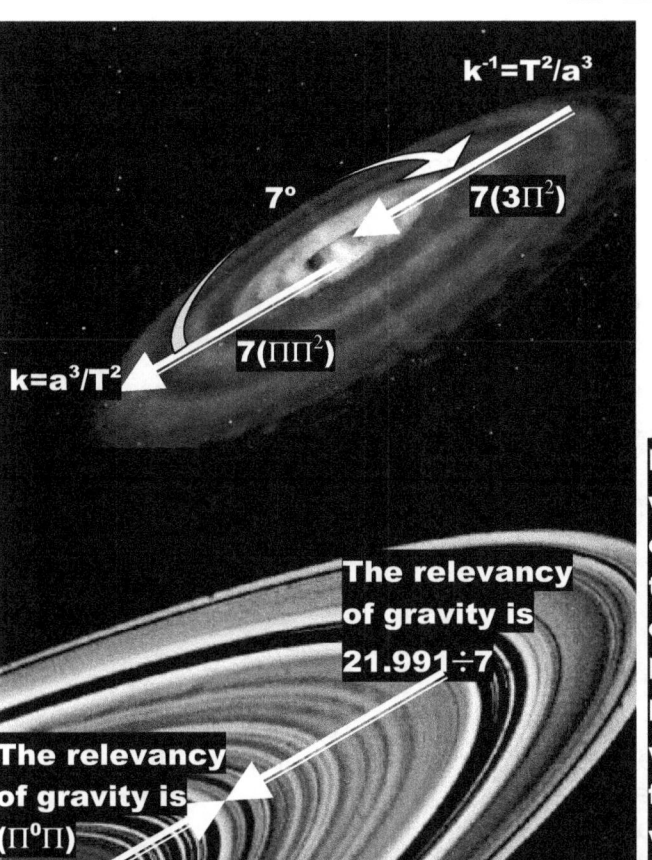

$k^{-1}=T^2/a^3$

$7°$

$7(3\Pi^2)$

$7(\Pi\Pi^2)$

$k=a^3/T^2$

The relevancy of gravity is $21.991 \div 7$

The relevancy of gravity is $(\Pi^0\Pi)$

The Universe is contracting by $7(3\Pi^2)$
The Universe is expanding by $7(\Pi\Pi^2)$
Gravity is the inclining of material spinning in relation to 7° that is part of Π
This is because Π is 3.1416 and spinning in space is 3 which makes time expanding 0.1416 x 7 = 0.991 larger than 3. The concept might appear simple when told in this manner but the entire philosophy is so much more complex when studied overall.

By observing the rings around the planets we can identify all four pillars forming the composition thought of as gravity. We have the Lagrangian points that is five and I explain why it is 5, we have the Titius Bode law and I explain in detail the Titius Bode law, we find the Roche limit and I explain why it has the value of $\Pi^2/4$ and we can see from my explanation I have provided above why the Coanda effect accumulates all the principles forming gravity.

Gravity has the value of Π. The value of Π has two measure where one is when the line gravity forms extends from singularity as $(\Pi^0\Pi)$ and when the line that gravity forms extends to singularity it is $21.991 \div 7$ because of the curve gravity associates with forming the value of Π To explain all these factors I have to take the reader back to the point where the Universe started as spot that formed a dot. I have to take the reader back to where the point where the Universe was numbers holding no space, which is way before the Big Bang era where the atom broke the Universe into space that formed.

Gravity forms as the earth or any other cosmic body rotates by 7 °. By diverting the straight-line movement by 7° a contraction forms in a circle. In my books I prove how this then brings about the value of Π by implementing the law of Pythagoras and gravity is the law of Pythagoras. The reclining of space by redirecting the direction of travel from straight ahead to 7° reclines the space in a steady and sturdy flow. It is the space reclining or contracting and the space contracts albeit filled by solid material or empty of solid material. This is the reason why all things fall equally. It is the space moving down with or without holding material and the space has the same density in relation to the solid cosmic structure rotating.

In the centre of the rotating body singularity forms to the value of Π^0 and this extends to the curve forming Π in terms of the curve of the rotating object. This then forms part of gravity moving forming singularity going square Π^2 and this form the relevancy

Coming in from space while having no mass the coordinates change as the value of Π forms a relation to the 7°

When an object comes in from the atmosphere or flies the sky the object associate with the turn giving it a value of the rotation in line with the axis which puts a value of 21.991 / 7 on Π or on gravity. This is gravity and this is how gravity functions. It has nothing to do with mass in any way, shape or form and there is no factor such as mass in the entirety of the cosmos apart from being in the imagination of the Newtonian

The atmosphere is liquid. Don't as a Cosmologist for they will most probably tell you it is extending nothing. Ask any Engineer and the engineer would tell you when you work with the atmosphere you work with a liquid that has a density 600 times less dense that water but it holds all the characteristics that a liquid has.

This picture shows the Coanda effect and while the Coanda effect is the most important principle in physics I know that less that one in a thousand that reads this knows about the Coanda effect.

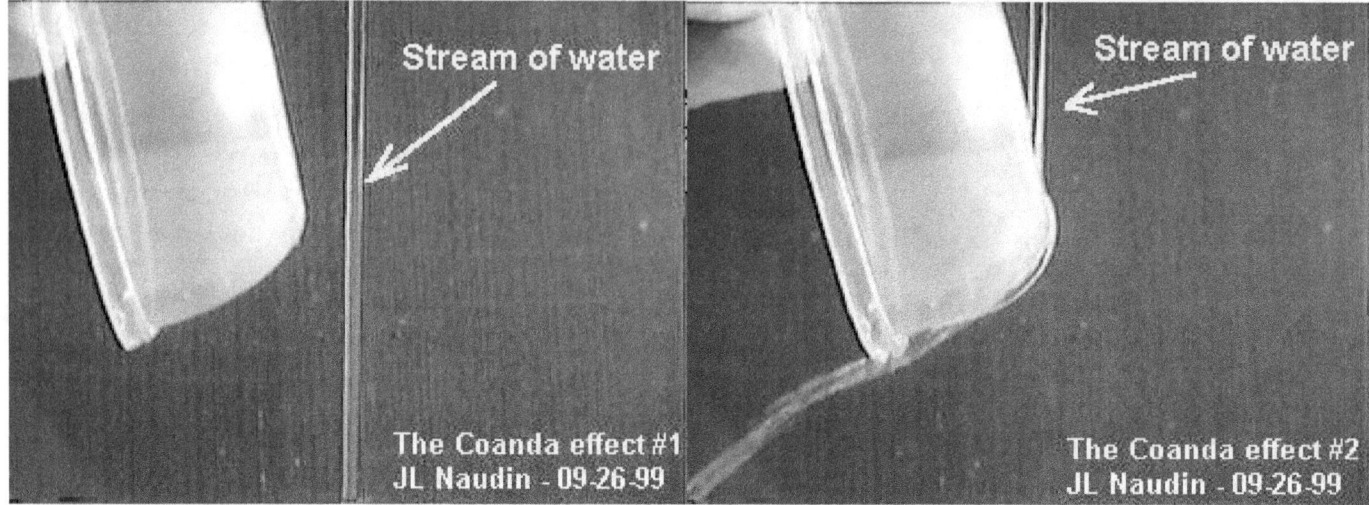

The water follows the surface of the curved shape, this is the Coanda Effect and the Coanda Effect works with any of our usual fluids, such as air at usual temperature, pressures and speeds.

The picture above is about a phenomenon called the Coanda effect but this is never mentioned in any physics handbook because Newtonian physics-religiosity is unable to explain or to understand this principle.

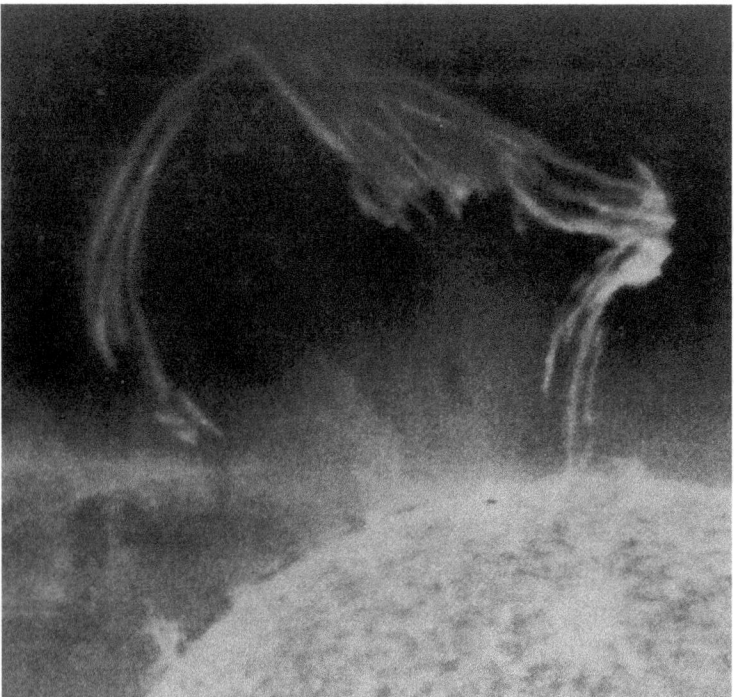

When liquids flow past a cylindrical object the liquid clings to the surface of the object rather than follow the "path of mass" and fall straight down to earth. Gravity is about the atmosphere that forms a liquid that is the same as the liquid running around the solid circle called the Coanda effect.

Gravity is the movement of the air in relation to the movement of the earth. If an object moves within the parameters the earth set the liquid to move around the earth and the object has the ability to maintain the speed, it will be just more liquid floating around the earth at a specific height fore filing a specific circle or rotation requirement. If the speed drops the object will fall notwithstanding mass or the lack thereof. The entire idea vested in gravity focuses on speed-differences. So what the hell has mass got to do with the entire affair of gravity?

If the circle is as large as the sun is, then the compressing of space turns the air into squirting liquid precisely as the picture of the sun shows. Again I ask what has mass got to do with the entire idea.

But remove mass from the equation and the Newtonian brilliance becomes idiotic stupidity because their entire mindset of playing God with mathematics becomes total invalid and their stupidity rings as load as a Cathedral tower bell. If you take away that which they use to cheat everyone, which is why they use mass as a factor, there is nothing they can show as an accomplishment this past three hundred years.

In front of your eyes you witness the sun freezing an atmosphere which the entire solar system shares and which is made up of condensed space that it freezes into squirting liquid that is the flow of heat, as pure as heat can be. This has noting to do with mass but those conspirers are so deliberately clinging onto their fictional Newtonian Universe that the truth passes them by and in total arrogant stupidity they refuse to use their eyes and see what the Universe shows it is.

The movement of space filled with material cools by movement of that which moves and this puts thermo differentiation between space that moves and space that does not move. As the space differences by thermo differentiation grows it would seem that one part grows in relation to another part shrinking. This is why large gravitational stars always seem to lose space while the Universe seems to expand.

The website http://www.singularityrelevancy.com-the-website will tell you why these phenomena happens and what cosmic philosophy prevails in physics allowing this to take place. This website will entertain you with arguments about physics you have never encountered and it will teach you what no other person knows... it is that everyone following Newton's gravity principles are in the dark. Do you challenge the remark I make with conviction about what you believe is physics? Do you know what the top picture has in common with this bottom picture wherein stars go supernova? The answer is gravity but gravity not as it is portrayed by Newton's vision of gravity.

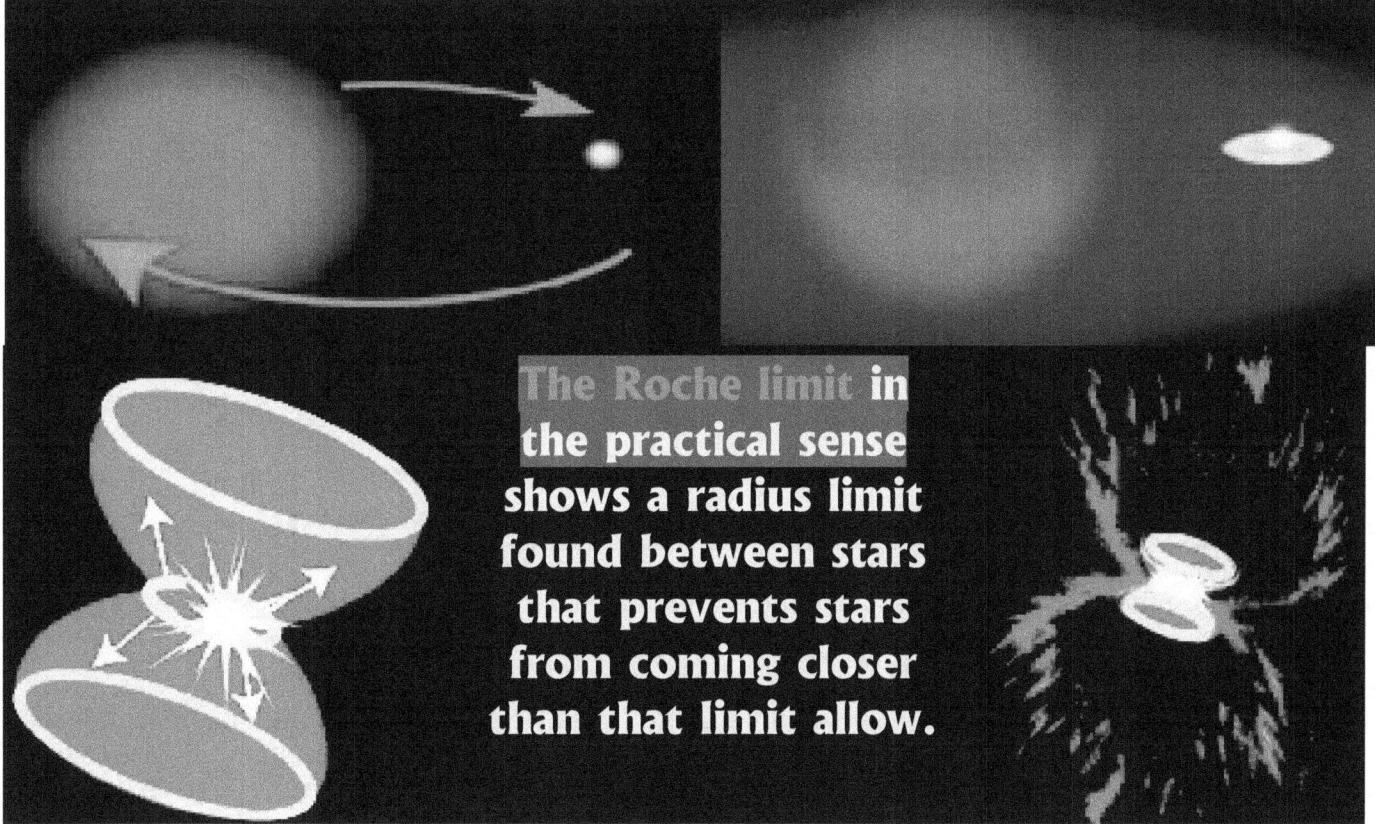

The Roche limit in the practical sense shows a radius limit found between stars that prevents stars from coming closer than that limit allow.

Do you know what this is? Those being experts in Newton's science or cosmic science have no idea why this happens. What happens in these pictures prove that stars could never collide because what prevents it is a cosmic law that science at present can't fathom because their understanding is too limited. This website informs you about the facts behind the lovely pictures. This website will tell you why the cosmic physics principles establish these phenomena...and because the information does not salute Newton, academics in physics despise what is said. Any star coming closer than 2.4674 times the radius of the larger star would liquidise the smaller star and this law and the Coanda effect are a precise duplication of true gravity truly applying. If mass doesn't pull stars to a point of collision then Newton's vision is wrong!
I FOUND A WAY TO USE THE FOLLOWING PHENOMINA AND BY THAT I COULD PRE DATE THE Universe past the Big Bang era. These principles apply as the building block used to form the cosmos ands without any knowledge about the phenomena there can be no success discovering the origins of the Universe. These principles are
1 The Roche limit
2 The Lagrangian system
3 The Titius or Bode law
4 The Coanda effect

That is only achieved if the following phenomena are used:

1 The Location of singularity
2 The Roche limit
3 The Lagrangian system
4 The Titius or Bode law
5 The Coanda effect
6 The sound barrier.

The location position and value of singularity as a factor forming space-time
Finding space-time, proving space-time and aligning space-time with gravity
The working principals behind and manifesting of gravity as a cosmic occurrence.
The Roche limit, and explaining the resulting as a law coming about from singularity.
The Lagrangian system and how and why that becomes the building form of the Universe.
The Titius Bode rule and how gravity comes about from that
The Coanda effect and the producing of gravity through applying space-time.
8)The sound barrier that is coming about by duplicating relations in space and time

Using the phenomena I found that I could enter the pre-Big Bang era, which was before the atom formed. The problem I encounter is when using the four cosmic phenomena is that I have to denounce Newton to do so and in that I find the rejection I encounter, not in my mathematics I employ, but in discarding and rejecting Newton's claims on gravity! Working with the four cosmic phenomena calls for applying true cosmic physics and that disallows the corrupted Newtonian Idea. With this being very new to everyone, I found that the Academics in physics denounce this idea because they become the equals to first year students and that thrashes their powerful egos.

1) An open letter Announcing Gravity's Recipe

Where the author links gravity directly to the Coanda effect by applying the four yet unexplained phenomena going by name as
1) The Roche limit
2) The Lagrangian system
3) The Titius Bode law and how these all combine to form gravity as implemented by the
4) Coanda effect.

2) An open letter Addressing Gravity's Formula

Where the author explains how the Universe came about at the first instant the Universe came about in using evidence on the four cosmic pillars and it matches the Biblical explanation in explicit detail

3) An open letter About Gravity's Prescription

Where the author explains how the Universe came about at the Solar system, as we know it took place. It explains why there are four solid planets, four gas planets and one cold structure. It also explains mathematically why all the debris is encircling the planets and where they come from. This is one part of another book entitled the Seven Days Of Creation

4) An open letter explaining Gravity's Rules

As the Author goes into detail about a new cosmos theory where the four cosmic pillars produce a cosmos everyone can understand. It puts time in relation to space and discovers what space is in relation to time. Never yet before was either time or space understood because everyone drooled on the misconception about mass and incorrectly interoperating that mass produces gravity.

In my work I call Creation by name and prove with science that we are in Creation. I employ science to prove that that controls Creation, which is not in the Universe but is noticeably because it is not in the Universe. In the light of all proof and when facing evidence I bring I dare an atheist to prove me wrong about Creation.

In mentioning this word in a science book I break a ground rule enforced by the atheistic dominated world of science. I challenge any person to bring proof about any part where any of my theory might be incorrect and furthermore I challenge any Academic in physics to prove that Newton's mass pulling mass is anything other than fraud. I charge any one to bring proof that the cosmos is contracting by the force of mass and that mass produce gravity as Newton advocated when he committed the biggest fraud of all times.

This series forms as a unit with four individual titles forms a prologue to a Thesis that introduces a whole new concept about Creation.

Matter's Time In Space: The Thesis in seven parts
ISBN 0984410-8-1 Written By Peet Schutte

For orders on the above books Got to e-mail
mailto:info@questionablescience.net

The above four books are only variable in print format because these books are too large to be e-book published

But before I can commence with that task I have another duty administer: I AM ABOUT TO WARN EVERY PERSON IN SIGHT OF MY WORK ABOUT MY SLENDER ABILITIES.

Therefore in the light of what the most respected academic group on Earth accuses me of, I therefore have to issue a most serious warning to any person with the intention of making some kind of inquiry to the content this book holds, then the most concerning matter involving any content within the pages of this book you hold are that you must please seriously consider that where the stating declares the possibility that the content in this book has been (written by...) then don't take the announcing Written By Peet Schutte (Petrus S. J. Schutte) very seriously for there are grievous doubts leaving considerable dispute about the possibility, which underwrites the authenticity of Peet Schutte achieving the (written by...Peet Schutte) status. Please take note of the following dehortation. In the light of the reference to me serving in the capacity as being responsible for authoring, (written by...) in line of keeping fairness and justice to members of society, where all civil beings should carry reputed honesty, then: Please be warned before any reader starts reading about the following extremely serious admonition: I am bound by my conscience to warn all intended readers that I am placed under caution by the Academics in Physics. Those most esteemed members responsible for the guardianship and maintaining the ethos in physics are of the opinion that I, Peet Schutte, am unable to write any book on the science of Physics as well as Astrophysics. Therefore I, Peet Schutte, must declare that I should be considered as not very able to write anything, because I am incapable thereof. I suppose, I merely generate new information, which I establish as thoughts and then gather as concepts. I further collect the result as words, which I put on paper using alphabetic symbols. I then compile that in a format that others may confuse with a book, but a book it cannot be, since the Masters in science found me unable to write a book. But before you go further and follow my arguments, I first have to level with you about how academics view me in the position I hold. Please do not allow me to fool you, for this then cannot be, or represent a book. Now I have done my duty in warning everyone and in that, I denounce further participating with any purposive intention to wilfully bring down the crux of civilization by acting unacceptable and irresponsible.

I didn't write any books since I am not schooled to do so. It is my guess that I merely generated uninformed thoughts, which I collected as alphabetic symbols and plotted that in ink on paper. This effort I achieved from harbouring my delusional ideas spawned by a dehumanised brain. It only proves my weak and under developed mentality, due to my lack of an informed insight that is a typical symptom that all those have that is suffering from a disadvantaged past that one can only have when the person obviously lacks formal education. While you are reading the letter deciding to regard or dismiss my work, then also please keep in mind when reading my language used and also please give credit where it belongs…if you do find linguistically improper use of words or misspelling, then remember that I am a feeble minded motor mechanic and not a literal giant. I do find much pride in my status as being Afrikaner and would like to have my names used by pronouncing it in the manner Afrikaans dictates…therefore I would sincerely appreciate the courtesy when readers will take note that my name and last name are pronounced in Afrikaans, which is originally from Dutch and must be pronounced that way. Peet one would pronounce "here" which is the closest English to the pronouncing of the "ee". The "Sch" in Schutte is pronounced exactly as school is where both actually are pronounced Skutte or "skool". By pronouncing my name in Afrikaans you do me the utmost courtesy any one can. Being an Afrikaner is what I am most proud of.

I submit article to well known physics magazines but my articles are rejected on the most unappeasable grounds and for the most outrageously ridiculous reasons the Newtonians can think of.

One such an article I may use because I said I was going to use the material as an open letter I gladly show. I submitted an article in which I show what the manner is in which gravity conducts movement by means of singularity.

I sent my new concept just as I showed it in this book to publishers but with much more explaining detail. It would be the biggest understatement to say that I was not very well received as one of the publishers e-mail shows. Notwithstanding the detail supplied and still I am ignored like Michael Moore was when he tried to enlist the Senate in the Iraq war. This is the biggest cover-up mankind has ever devised and let one academic prove it is not a cover-up as much as any academic prove that Newton is absolutely beyond suspicion and as faultless as the academics in physics are blameless.

The following is the response I received when I submitter a paper on how gravity forms by singularity forming.

Dear Dr. Schutte,

You submitted an article of 15 pages to the Annalen. The content of this paper doesn't constitute a theory in physics. With a lot of words and some simple algebraic relations, there is no way to "explain" the world of physics. You seem to be out of touch with modern developments. This is also shown by the fact that you don't quote any relevant literature.
I am sorry to say, but the Annalen is not able to publish your work.
I am sorry for having no better news for your.
Best regards,
Friedrich Hehl

Co-Editor Annalen der Physik (Berlin)
--
Friedrich W. Hehl, Inst. Theor. Physics
* University of Cologne, 50923 Koeln _____/_____ Germany
fon +49-221-470-4200 or -4306, fax -5159
hehl@thp.uni-koeln.de, http://www.thp.uni-koeln.de/gravitation
* Univ. of Missouri, Dept. Phys. & Astr., Columbia, MO, USA

Should anyone wish to read the entire dialog that went on between Annalen Der Physics and me it is with the article and the explaining of the article it is THE ABSOLUTE RELEVANCY of SINGULARITY: THE WEBSITE http://www.lulu.com/content/e-book/the-absolute-relevancy-of-singularity-the-website/7517996]

Dear Prof Friedrich W. Hehl, I have received your e-mail reply and I wish to respond on your letter. The article of 15 pages to the Annalen had in mind to introduce a very wide-ranging concept contained in many books. I wish to promote books in which I introduce a much larger and much more detailed cosmic picture. It is four books that actually form four volumes of one theme supporting The New Cosmic Theory. I wish to unveil a totally new approach to the thinking in cosmology. The concept is proposed in the article I sent to you which is "revealing" The New Cosmic Theory In the article as much as the theme I wish to go where no one ever attempted to go before. I introduce the Universe of singularity, a state in which the Universe still is because it is a state from which the Universe grows. It is where material in a dimensional dynamic does not apply because it is where Einstein said "the Universe goes "flat"". I show you how and where the Universe goes "flat" I will guide you to the point where I go…so that you may see where my books and the article lead you. It is in the domain of singularity.

When you read work about the Big Bang you have to go right down the development (in reverse order) to the point where the Theory of the Big Bang points at a spot named singularity. It shows the very start from where all material developed. At that point one will find The Absolute Relevancy of Singularity and there has never been any attempt by any person ever to venture beyond the dimensional birth of the cosmos, which is called the Big Bang by going into the era where singularity prevailed. I take you there in my books as well as the unpublished article. However, going there requires a very high degree of concentration and calls for understanding that a very little number of persons are capable to show. I try to show how the Universe goes "flat" as Einstein said the Universe goes "flat". Even by completing this unimpressive letter you will also know how the Universe goes "flat". Even where you failed to read the article I sent you, then by just reading this letter you will be able to find where singularity takes the Universe "flat". But it requires a mental capacity to understand because where I venture no one ever in the history of mankind reached into before. I do not speculate but even in the unpublished article I show with pictures and sketches as well as "some simple algebraic relations" where to go to where the Universe starts, but you failed to read that because you are opinionated as to what conditions should the Universe have before the Universe will allow any one into physics. That is a pity. One should learn from the cosmos and not tell the cosmos what it must be to qualify as the cosmos. Then in the article I show you by almost taking your finger to the spot, the very point where the Universe ends and that too I qualify. You might dispute my arguments and show me about what you disagree, but it shows very little understanding of reason on your part about qualities man should have before understanding the Universe. I go into a Universe that was in place before light was in place in the Universe and only darkness prevailed because light calls for space and in that era of singularity space was not even a thought yet. I show why the Universe goes "flat" and in a "flat" Universe only the value of 1 holds value since singularity is 1. If you can understand 1 or $5^0 \times 7^0 \times 3^0 = 1^1$ you have all the mathematical skills required to understand the applying concepts. To reach a value of 1 does not require big mathematical equations but to reach singularity requires 1.

The collection I named The Absolute Relevancy of Singularity: **The Theses** and the collection as such forms a small introduction to the thirty-two or so books I wrote on various matters concerning physics with gravity in mind, but **The Theses** as such in the entirety of the four books does not officially even start to introduce the spectrum of every aspect of my work. I have been in contact with numerous Academics and about one in one hundred reply. When the one in a hundred reply, the academic always uses a most aggressive tone which I came to accept as what I receive from academics, and because of that I was most delighted to find some kind remarks from you as a practicing academic, and might I add, the first such kind remark in ten years of my trying to contact any person in physics that would take note of what I have to say about a new line of thought, because the few others that replied were extremely aggressive about me confronting Newton. I only began to submit books to publishers after twenty-seven years of studying Newton and the role Newton play in cosmology and thereafter which was ten years ago I began promoting these ideas. The New Cosmic Theory is a process wherein I try to introduce a study that is ongoing for about thirty-seven years, give or take a few and I did not jump into the frying pan having my first thought about the matter published as an article when I sent the article to the address of Annalen der physics.

This is the ridiculous concepts that Professor Doctor Friedrich W. Hehl, Inst. Theor. Physics supports. Because I don't conduct tainted mathematics and highly suspicious Newtonian concepts covered by completely ridiculous malfunctioning concepts, I am despised as the ridiculous small-minded novice that is openly very feeble in thought. I do not support complete rubbish as this formula indicates the Newtonian vision of physics represents, but for my failing to go along with total trash my work gets rejected time and again by every science publisher I approach. **But before I can commence with that task I have another duty to administer:**
When you read this and you get to some mathematical equations that you don't understand, just ignore the mathematics because in the end it is the message in the arguments that is important and the mathematics just form part of the fraud science has committed for centuries. I use the mathematics THEY use to show THEIR fraud and if you just skip it, that more the better for you. Then you won't pollute your mind by taxing it with fraudulent garbage.

You are going to find mathematics used in arguments. The mathematics is not important. Understanding the mathematics only proves that THEY, the Newtonian academic mathematical genius don't understand the mathematics in principle and that is all. I show they say they are smart enough to formulate a Black Hole, but they haven't got a clue about the most basic principles in mathematics, which is the relevancy

used in $F = \dfrac{r^2}{M_1 M_2}$ and then to transform to $F = G \dfrac{M_1 M_2}{r^2}$.

To all those readers not very knowledgeable with mathematical principle: doing that which the equations show in principle that transpires to mathematical fraud and let any of the super-genius-in-mathematics correct me on this accusation or prove how it is done! So if you don't follow the mathematics then keep to the argument and skip the mathematics as you brush over it because you can do so much better when you don't understand the mathematics. Newtonians also don't understand the mathematics because if he did he would not cheat in this manner. Then for three centuries Newtonian Mathematical genius took it in principle to echo Newton's mathematical incompetence but by pretending they do "understand Newton in full" so that they would look smart, they fooled all of man for three centuries. Then they cheat on behalf of Newton that also clearly understood nothing about mathematical principles and went about also cheating the first time while Newton pretended to understand mathematics and in the analysis at the end he, Newton understood less about Mathematics than you do that admit you don't understand mathematical principles. At least you don't pretend to understand and then go and cheat as much as you can just to pretend you are smarter than the rest that don't understand mathematical equating principles.

When I said the arguments is what is important and the mathematics is the footprints of idiots that walked through physics for centuries, then me making that statement too their faces was for some physics Professors in South Africa (and I can name them) over the top and they blew their tops. They could ignore me afterwards but they never could challenge me in a debate to prove me wrong!

Physics students, it is your duty to pull the plug on the powers of the ALL-POWERFULL Academics in Physics and stop their dishonesty. It is your task as the as the next physics generation to stop the criminals that are filling the corridors and the lecture halls of physics departments throughout the world. Stop their teachings by forcing them to stop their criminal fraud. Force them to explain the deception called THE CRITICAL DENSITY, which is a conspiracy to commit fraud. Let them explain how an expanding Universe can suddenly and abruptly turn in direction of developing and start to contract as Newton stated it is doing at present, and when facing all other concluding evidence showings it was expanding since time began. Tell them to prove that the cosmos will begin to contract doing its turn about by using other proof than merely Newton's say-so. Tell them to bring proof with evidence that the cosmos is contracting as Newton said. Then force them to admit to the fraud they are precipitating in, which is THE CRITICAL DENSITY conspiracy. In THE CRITICAL DENSITY conspiracy all they say is that they are waiting to see when the cosmos would stop its criminally insane behaviour and start to listen to the laws of Sir Isaac Newton.

A Conspiracy to Commit Fraud on a Cosmic Scale

PLEASE ALLOW ME TO INTRODUCE THE THE MEMBERS OF YOUR CLAN THAT COMMITS COSMIC FRAUD ON A UNIVERSAL SCALE. THESE EXCEPTIONAL GIFTED AND UTMOST BLESSED ARE ALSO KNOWN AS THE SUPER-EDUCATED- WISE-AMONGST-THE-WISE-CON-ARTISTS THAT HAS NO EQUAL AMONGST ALL HUMAN IN ALL OF HUMAN HISTORY GOING BACK AS FAR AS YOU WISH. NEWTON SET LAWS UNTO WHICH THE UNIVERSE HAS TO COMPLY. THE UNIVERSE MUST CONTRACT AND REDUCE BY GRAVITY PULLING EVERYTHING AND THAT IS A VERY SPECIFIC STIPULATION THAT NEWTON LAID DOWN. THEN IT WAS FOUND THAT THE UNIVERSE DELIBERATELY DID NOT COMPLY AS THE UNIVERSE WAS EXPANDING ALL THE TIME IN STARK CONTRAST WITH NEWTON'S EXPLICIT ORDERS. HOWEVER THEY, THE ONE... AND THE ONLY, THE MAJESTIC MAGICIANS...THE NEWTONIAN WIZARDS, THE NEWTONIAN PHYSICISTS IN COSMOLOGY CAN FIND A REMEDY TO CORRECT WHERE THE UNIVERSE WENT WRONG AND DARED TO STRAY FROM NEWTON. THIS THEY CAN ACHIEVE BECAUSE THEY FOUND A WAY TO CONSPIRE BY WHICH THEY CONTROL THE MINDS AND THE THOUGHTS OF EVERYONE HAVING ANY KNOWLEDGE ABOUT MATTERS IN SCIENCE. THESE BRILLIANT SUPERHUMAN-MEN-AMONGST-MAN WILL FORM A COMPOSITION IN THE COSMOS COMPRISING OF AN UNSEEN, INVISIBLE NON-DETECTABLE-NON-EXISTING DARK MATTER THAT WILL PULL THE UNIVERSE TOGETHER TO GET THE UNIVERSE BACK ON TRACK WHERE THE REBELLIOUS UNIVERSE DARED TO STRAY FROM THE ORDERS OF NEWTON. THE PROCESS BY WHICH THEY WILL SUCCEED WAS DONE THE PAST THREE CENTURIES WITH MUCH SUCCESS WHERE IT IS CALLED BRAINWASHING BY THOUGHT CONTROL IN FORMING MIND CONTROL AND IS ALSO MUCH BETTER KNOWN AS TEACHING PHYSICS. THE PROCESS HAD COUNTLESS SUCCESS IN TEACHING NEWTON'S UNREALISTIC PRINCIPLES.

Physics students, it is your duty to pull the plug on the powers of the All-Powerful Academics in Physics and stop their dishonesty. It is your task as the as the next physics generation to stop the criminals that are filling the corridors and the lecture halls of physics departments throughout the world by acting as if they know and all they know is to fool the next generation of students. Stop their teachings by forcing them to stop their criminal fraud. Force them to explain the deception called THE CRITICAL DENSITY, which is a conspiracy to commit fraud. Let them explain how an expanding Universe can suddenly and abruptly turn in direction of developing and start to contract as Newton stated it is doing at present, and when facing all other concluding evidence showings it was expanding since time began. Tell them to prove that the cosmos will begin to contract doing its turn about by using other proof than merely Newton's say-so. Tell them to bring proof with evidence that the cosmos is contracting as Newton said. Then force them to admit to the fraud they are precipitating in, which is THE CRITICAL DENSITY conspiracy. In THE CRITICAL DENSITY conspiracy all they say is that they are waiting to see when the cosmos would stop its criminally insane behaviour and start to listen to the laws of Sir Isaac Newton.

With The Critical Density shambles the modern Newtonian set out to defraud the world in the same manner as their Master Sir Isaac Newton has done centuries ago. Newton said the cosmos is contracting. When Hubble proved the cosmos is not contracting, Newtonians looked where the cosmos went wrong by not following Newton guidelines he so clearly set the cosmos to follow. It has to contract and not expand.

These whom I named in Honour of Sir Isaac Newton as the are the guard of the Newtonian High Priests carrying the name as the Newtonians are Men amongst mankind, that charged the Universe with not applying to standards set by Sir Isaac Newton, and then went on proving how incorrect the behaviour of the Universe was in not adhering to the direction gravity has according to Sir Isaac Newton. Since Sir Isaac Newton can't possibly make a mistake, it then was presumed the cosmos made the mistake by not following the gravity settings laid down by Sir Isaac Newton, the one that cannot falter nor could his teachings fail, carrying the illustrious name of Sir Isaac Newton. It must be the cosmos being at fault by expanding without seeking the approval of Sir Isaac Newton to do so in contradicting Sir Isaac Newton

Up to this point in science and in despite of an array of evidence pointing to the cosmos growing by expanding in every sense and with all pieces of evidence gathered by science from all over the Universe (including the solar system), the theory of contraction is still hailed as the infallible Newtonian truth. Every one that is part of physics, shares the Newtonian vision of a contracting Universe where the lot would one day again come together and Creation will end where Creation started some time ago. The Universe has mass that is pulling mass towards one another and we are in the centre of an ever shrinking Universe. That is what the lot of us can see… we are forming the centre of the ever contracting firmament having the entirety of the cosmos where every Newtonian can vividly see with his or her eyes through any telescope that all Newtonians minded scientists are sharing the centre stage of the ever collapsing Universe. The Universe is about to end where all mass contracts into one huge lump of material, and this conclusion contradicts al evidence gathered by science. If you don't believe me marry Newton's contraction with the Big Bang and see a divorce in place before any Church consummation of such a union could begin…but then again just as unlikely union in principle marriage between Galileo and Newton is in place and the mindless masses never once frowned on that! The contraction idea was never questioned and was accepted as being truer and much more believable than the presence of a living God Almighty was.

Students in Physics, it will serve you well to read the following arguments very carefully and come to a conclusion about what gravity is and what mass is and how it is impossible for the concept carrying the idea of mass then become responsible to form what we think of as gravity. Mass can't ever and doesn't bring about gravity.

If you are one of those members of society that never thought you would hear the name of an accomplished person as Sir Isaac Newton being associated with fraud, corruption and brainwashing, then these books are specially written to inform you about the truth there is lacking in the correctness about science.

Everyone knows that planets orbit around the Sun. Planets circle the Sun which is the same as saying planets orbit the Sun. Just by calling the circle motion in terms of what applies, that statement nullifies

Newton's claim of mass that attract mass and put to question the reliability of Newton's dogma. Prove to the world that mass pulls mass to form gravity because that was never proven…accepted yes, on the say so of Sir Isaac Newton, but is you finish this book you will learn how awfully Sir Isaac Newton was mistaken about his entire cosmic principle.

If mass did attract mass, what kept the balance where the planets do find a balance in orbit and rather than moving towards the Sun. Planets orbit in ratio so precise we can set that on gears and yet the planets all so very different values in mass although the planets are randomly allocated and not according to the mass factor each holds. This is evidence of the fraud and a cover up. If mass did attract mass, then what is pushing the planets to remain in orbit?

Planets do maintain positions not according to mass that pulls but a balance foresees orbit. The idea of proof is automatically placed at the door of Newton. We talk of planets orbiting and that is what planets do. Planets don't creep up to the sun by the value of mass. If normal speech contradicts Newton, then it is this task Newton have to prove his supposition is correct and the claims about attraction Newton made can be substantiated, although Newtonians will deny this fact as if they deny he is correct the honour of their Master Newton and that is what they have to do. Think of what planets do… and you think that planets orbit. It is connected to the brain. No one thinks of planets spinning or planets basking in the summer Sun. When hearing about planets the first thing that comes to mind is the rotating of planets while circling around the Sun.

However, just using the term orbiting is in total defiance with Newton! Newton said gravity draws or pulls or moves in the direction…, which would have one understand that the two objects in example the Sun and any of the various planets will be moving directly towards each other.

The term pulling does not suggest any circling because no one can be pulling towards and does that while circling around the object. When pulling anything it must take place while using the shortest line possible. That serves the term pulling. Then the saying goes that planets orbit indicating they follow a circle. That is not what Newton said. However, wrong that may seem but circling is precisely what planets are doing.

In conversation we speak of the planets orbiting. If Newton was correct we should be speaking of the planets pulling, but talking about pulling would be blatantly wrong according to the normal spoken word. Never do we refer to the planets pulling the Sun or the Sun pulling the planets, but we speak of seasons coming from orbital positions. Being in orbit has to neutralise the pulling and then cancel the pulling concept that also became culture.

If there was a pulling, and the word orbit cancels such an idea, then there has to be some sort of prevention taking place that disallows the pulling to commit the direction of travel.

I know it is said that the orbiting object falls as fast as it circles and by falling while moving to the following side on position it never reaches the Sun, and yes, it makes sense, but there has to be some form of resistance replacing the planet in the next side position and preventing the falling or the pulling from taking place.

Using the formula $F = G \dfrac{M_1 M_2}{r^2}$ as Newton provided, disallows any other concept other than moving towards. The person Newton got his ideas from and the work he raped completely, that of Johannes Kepler explained this very well, but Johannes Kepler makes no room for any pulling of any sort. In the work of Johannes Kepler he said that the space being the orbiting route a^3 remains at a specific distance k while the orbit T^2 takes place…and in all my other books that addresses more information I take Newton to task on his dismembering of Kepler's formula by corrupting Kepler's work and with what amounts to fraud, Newton takes science on a goose chase that holds no truth. There is no pulling by mass of mass in any way.

We have either one of two that has to be incorrect. If Newton is correct, then the normal way of functioning of the is incorrect. Then we must start saying planets are pulled to the Sun. If the normal form of speech is correct and the planets are merely orbiting the Sun, then Newton is wrong. The planets can't orbit the Sun while at the same time we have Newton's accepted scientific presumptions being correct.

The fraud part is in the accepting of the Universe expanding while still insisting that Newton is correct in his dogma of contraction with mass. This web page is an effort to show how Mainstream Physics brainwash students into accepting Newton's hypotheses of mass attracting by force while the entire Universe is expanding at the rate the Big Bang indicates.

NEWTON'S MYTHOLOGY Written by Peet Schutte

Students, read the following message about my book I named **Newton's Mythology** and learn how you are brainwashed and how your mind is pre- conditioned into believing in Newton's myth of pure deception which Academics call physics. If you are a student in physics who don't believe that you are subjected to unlawful brainwashing, then read on.

Let's start surveying civilized principles by evaluating what lawfulness means and what would constitute as morality. Let's determine what makes the crook in the book?

If any person, notwithstanding what reason is given in justifying such depravity, tells a lie or conveys untruths to further whatever humble cause, it is seen as fraud. To convey information that is not substantiated as a verified fact then the mere conveying of such information becomes fraud.

When any person, notwithstanding what reasons given, repeats such a lie unabated while being well aware that the information passed on by such a person is incorrect, then the person commits deceit. When anyone is repeating the information that is passed on as being unblemished factual substantiated and verified truth while such a person knows very well that such information is void of proof or lacks proof, then committing such an act is a criminal enterprise. Academics in physics commit every one of the above indignities and yet see their actions as being lawful and even much praiseworthy and hold their role in society in the highest esteem imaginable.

They fail to see the crime that they commit while tutoring physics. Whatever motivation they may claim to have which they offer to serve them as forming their driving force, the fact that they perpetually perpetrate in unlawful behaviour, by spreading untruths, such actions on their part put those academics holding such highly regarded positions in the league of ordinary cheats, gangsters and common criminals. By willfully and constantly falsifying facts to further whatever humble cause and produce illegal claims repeatedly, remains derogative behaviour and is unlawful by nature, notwithstanding what morality it should serve.

A Preacher or Pastor lying on behalf of God is not lying on behalf of God and to think the Preacher or Pastor improves or underlines the Greatness of God by lying on behalf of God is very mistaken, because in reality such a Preacher is falsifying the truth for his or her personal benefit and trying to impress the congress about his importance and not the importance of God. Lying is wrong and doing so even in the name of God remains despicable.

The same applies to academics in physics. There is no argument that can change this truth about falsifying the truth and when doing so there is no hiding behind any excuses of ennobling to benefit mankind that will change such truth into righteous conducting.

Newton said centuries ago that gravity is the force of attraction there is between objects that hold mass and it is the mass factor that brings about this attraction, which Newton claimed there is. The Universe does not contract and all the proof we require to disprove such a statement we find in the Hubble constant as a guarantee.

Moreover, it is true that the Universe never contracted even for a brief instant and proving that is the Big Bang concept with all the proof that this concept brings in backing the principle of expansion in the Universe. Planets never moved closer, are not moving closer and will never move closer to each other and this is backed by all information collected this past century.

The Moon is not coming closer but the distance between the Moon and the earth is widening. Studies about the Universe reveals every time that space in the cosmos increases constantly. Studies find all things are moving apart and away from one another.

Any and all the proof about this is beyond what any doubt may present to counter this knowledge. Notwithstanding this irrefutable findings, science still regards Newton as the only person that ever lived whom no one ever could prove wrong…and this is upheld by Mainstream Physics in spite of the cosmos proving Newton wrong every instant of time.

The basis of what science holds as its foundation we find to be the Newtonian principle of $F = G \dfrac{M_1 M_2}{r^2}$. The foundation used by science promotes this argument and backs up this argument well knowing that in the cosmos there is no evidence backing up this proposal Newton suggested.

The Newton formula $F = G \dfrac{M_1 M_2}{r^2}$ used as basis for science sees gravity as being a force of attraction and the force of gravity is being in place between all objects in accordance with the mass factor that the objects have as presented by Newton in the formula $F = G \dfrac{M_1 M_2}{r^2}$

What we find as we gauge all evidence found while studying the Universe, is that reality shows there is no attraction between objects in space going on anywhere in the Universe, that the entirety of such a concept is a myth and the outward moving of the Universe has been coming from and since the time of the Big Bang and maintaining this flow of material is substantiated in a concept named as the Hubble constant, which proves Newton's perceptions to be a myth.

The Hubble constant proves that space everywhere is growing ever since time began and the growth never stopped ever since. Knowing this irrefutable fact does not deter science from under scribing Newton as the sole basis that underwrites all the correctness of all of science known as physics. However, Hubble and the Big bang and all other investigations contradict this attraction Idea Newtonian dogma holds. Therefore, any further believing that there is attraction going on as Newton claimed has to be viewed for what it is and that it is a fairy tail.

The Big Bang Theory proves Newton's idea as not only being wrong but Newton's idea of attraction is a joke. If the Big Bang is expanding the Universe, then how can the Universe contract at the same time? Any contraction by nature would have the Universe collapse back into infinity the moment the Big bang moved out of infinity.

Ask your professor to show how an expanding Universe can also contract and your professor will tell you about Einstein's Critical Density theory. This theory I prove is the biggest fraud ever devised by any group of persons in the history of civilization! This is perpetrating fraud and conducting in upholding deceptions instituted by Newton that then formed the institution of lies they call physics. The Universe does not contract in any way, means or form and even such a suggestion is incorrect! The Moon and Earth are not moving closer but are moving apart. The entire Universe is growing in space and nowhere is space depleting by any norm used.

Academics are very aware of this misconception Newton had and still academics in physics are promoting the ideas of Newton as the unwavering truth. Academics teaching these misconceptions are committing fraud, notwithstanding the portraying of their role in society being unblemished, spotless while they are covered in a lily white blanket making them being whiter than snow and having such a holier than thou attitude.

Teaching Newton is participating in deception and promoting Newton is criminally deceiving the public and while doing so, is committing an act with criminal intentions.

Then, in the face of all this evidence contradicting Sir Isaac Newton, they remain upholding the correctness of Sir Isaac Newton and keep on teaching students about the unwavering correctness of Sir Isaac Newton. They put down conditions of learning to this effect and are expecting students to repeat these untruths and unproven facts by forcing answers to that effect in examinations.

Forcing the acceptance of this untruth about physics is equal to preposterous subjecting students to physiological torture and heinous mind conditioning, scandalous thought control and brainwashing. This applies to everyone serving as a tutor in physics notwithstanding whatever status the torturers might have in society or the morality they attach as a reason to commit such atrocities.

If you are a student, then you are conditioned by academics in controlling your thinking by enforcing pre-mind setting and in which they methodically force you into believing in Newton and this is an on going process conducted for centuries in the past, while it is the truth that Newton is completely void of any tests that may secure any form of confirmation and in securing proof then also by that establishing proof.

Read this book **Newton's Mythology** and then use the information I supply in the book to insist that Academics who are teaching physics, prove to students that Newton's statements of attraction are correct. Let those academics explain the method mass uses.

Let them with precise detail show when mass is applying, that gravity is produced by mass and such producing of gravity that then would establish attraction! I show precisely how gravity produces mass but mass can never produce gravity. I show with explicit detail when, how and where gravity forms mass but mass can never form gravity. What I prove annihilates every Newtonian claim.

They never prove Newton's philosophy on gravity but those persons conducting teaching in the subject of physics force all physics students to learn Newton's gravitational concepts and accept the facts as if it has been proven beyond all other facts. Students have to believe that Newton is correct or academics will see to it that they fail their examination. The condition of being accepted in physics is to accept Newton without questioning the proof that is never supplied.

Let those academics now prove precisely how mass brings about gravity and then afterwards test you on how Newton is proven correct and not on you repeating facts about what they say is true about what Newton said, which they say is true. The manner they present Newton is completely hearsay and that method may not be used in any court of law.

Let your professors now prove how it is that Newton's teachings are correct and then examine you on the process they use to prove Newton's concepts. At present they say Newton is correct and then they test you on your ability in repeating that Newton is correct without ever proving to you that Newton is correct. Let those physics professors now prove Newton and then test you on the manner they use to prove Newton to be correct.

The truth beyond all other truth is that Newton's gravity has never been proven (because try as you may it is not possible to prove Newton's formula forming gravity mathematically) and because academics know that, academics require the blind acceptance of Newton by students. This unconditional acceptance of Newton's correctness relies only on the pre-conditioning of students' mind set and academics depend only on the student trusting the academic "say so" about the institutionalised correctness of Newton. That Newton is correct nevertheless and notwithstanding that there is no founding proof about this matter, is what students should be accepting blindly.

Pre-conditioning students into blind acceptance depends on the academics' insistence that students approve Newton's concepts without pre judgment or students insisting on scrutiny of any sorts. In examination students have to outright and blindly follow academics' say so only because academics say so. Academics depend on students never questioning their say so or demand proof about what academics teach. Those academics in teaching positions insist that all students accept Newton's accuracy.

This is methodical mind control as much as it is the brainwashing I show that they enforce. If you are one of those believing that Newton was ever proven, then what you believe to be true is a lie because Newton can't be proven and that is the truth! The time has come to face your teachers and force them to stop the ongoing old culture of bullying students and conditioning their thoughts by enforcing on them dogmas which is mind control! In order to get students to accept Newton's hypothesis, academics resort to brainwashing pupils and students.

They teach you that the Universe contracts and to state their case they force students to learn that gravity is proved by Newton introducing the following formula $F = G\ \dfrac{M_1 M_2}{r^2}$ They say that M_1 is the mass of the Earth and M_2 is the mass of the individual in questions mass and the multiplying of these factors with the gravitational constant produces the force of gravity when this gets divided by the square of the radius. Please let you lecturer put in all the values of the formula and prove Newton is correct. If he can't and I know for sure he never can fill in the symbols and calculate the force of gravity, then read the rest of the web page that follows to see how far academics in physics go to brainwash students into believing in Newton's fraud.

This is a fair test to see if Newton's contraction theory underwritten by Newton's attraction formula $F = G\ \dfrac{M_1 M_2}{r^2}$ is valid, then force your professor to use this formula as it reads and show WHEN the Moon and the Earth is going to collide. If he fails to do it by using Newton's formula as $F = G\ \dfrac{M_1 M_2}{r^2}$ then you will know who is conning you, him or I and who is truthful again him or I. I charge all academics to prove what I say is being wrong in any way or even that I exaggerate in the least. I challenge Newtonian academics to prove that mass does indeed form any force of any sorts and in particular gravity!

To those professors claiming Newtonian ideas are substantiated by proof, I say that notwithstanding your personal academic qualifications and while at the same time disregarding your status and previous achievements as well as ignoring your many admirable abilities you may have and however superior they might be, I shall teach you about gravity. I say it is time Students learn the truth about physics notwithstanding the status academics will loose.

Students read **Newton's Fraud** and challenge those academics depending on their ability to brainwash you into submission.

Welcome to

www.sirnewtonsfraud.com

http://www.lulu.com/content/e-book/wwwsirnewtonsfraudcom-part-2/8132511

Everyone knows there is a problem about gravity in physics and this far it seems as if no one can put a finger on the problem. This suspicion that there is this feeling about a certain concern and doubtfulness that is lingering on in the minds of many… and also is lingering on from generation to generation… without anyone ever finding a solution…

There is a suspicion lingering in the back of everyone's mind that something is not quite correct about the approach physics take on the matter of gravity and only those academics seasoned with years of studies and salted with time seems to miss this haunting feeling.

Something about the way gravity is presented just doesn't add up as it should and does not quite reach the answers it should conclude. There is this vague unspoken question hanging in the air without any one ever finding words to express the question…and yet the question remains however unspoken it seems.

If you are a student studying in physics then reading this web page is detrimental to your future, as you can remain part of the problem physics has had for centuries or you may join the solution that came to physics and start to heal the wound.

The problem is that the evidence behind the facts of physics places an awareness of the doubt and this everyone has and this all the generations of academics had all the while they were students but are now covering up a culture driven by the mentality to brainwash and control the minds of students to accept what should never be accepted.

Hidden under a cover of "understanding Newton" or "not being able to understand Newton" they force certain incompatible arguments to join that which never can join and while joining also make sense at the same time.

Every generation finds an itchy feeling but never is there a place that any one can secure the very point where it is apparent enough to scratch and by detecting the itch then rid physics of the undetectable irritation.

Believe it or not, but this irritation is in place because of centuries of brainwashing going on and is employed in physics from generation to generation for centuries on end.

This web page is dedicated to bringing honesty into the faculty of Astrophysics and Astronomy as well as to show the Physics student on what corruption and deceit does physics base their facts which they proclaim as being such well proven, and godly accurate facts and is unwavering depicting only the truth. *Students get wise or you are going to get screwed so listen to this:*

Read the next pages and you are about to learn how students are brainwashed into accepting the baseless and ridiculous Sir Isaac Newton puts forward as truths. The Custodians of Physics have nothing better to offer than presenting you with unfounded corrupt and distorted facts…and by doing that they resort to mind control on students and introduces baseless concepts by manipulating the student's thoughts.

If you are a student then read in this web site what they do to you. They are defrauding you by exchanging your institution fees for corruption, so confront them about their dishonesty. Force them to become honest and to stop corrupting students with intentional malice. That which they offer has no truth. All they have are misconceptions and incoherent facts.

There is always a thousand and one conspiracy theories going around and the one tries to be bolder and more sensational than the next theory flying around. However, the biggest conspiracy is going on in front of every person's eyes and is committed by the most respectable persons in any respectable upstanding society. Even if you had your personal favourite conspiracy theory, try and match it to the one that I have!

The figures who charge the highest form of respect in our communities and those in charge of the most dynamic part of society and those who stand beyond and above any form of suspicion are the very same persons that I accuse of betraying the ones trusting them. They misuse the positions of trust by participating in the conspiracy I am about to tell you about.

In other cases where such a conspiracy theory can be ascribed, and where there is doubt about blame or innocence in the mind of the beholder, doubt can always be raised on grounds of admissible accuracy of the scurrility in proving or disproving the case of the innocence of the accused and how far the blame can stretch before the blame becomes beyond any doubt.

In this case the blame is openly dedicated to those being blamed, named and shamed and they have to defend years of lying and contribution to cover misconduct that was committed by them as well as their predecessors in the name of science. They now have to explain why so many evidence were falsified to keep their noses clean while mud colours their lily white image in hogwash and the stink of their lies equals the pig pen it needs to cover such aroma. It is time that they reveal the truth.

Now students may tell the custodians of Mainstream Physics to immediately stop their practice of mind abuse, thought control and manipulation which is a method of practising unlawfulness by resorting to unleashing criminal behaviour towards students!

If you are one of the Academics and Custodians of Mainstream Physics I challenge you to show one piece of evidence where I am incorrect, exaggerate facts, produce incoherent and distorted views about your physics being based on corruption and misconduct.
If you Academics think you are innocent, read on.

Fortunately for the future and unfortunately for the pretenders posing as Academics in mainstream physics the truth is out and it is published and your dishonesty as Academics will come to be no more.

The lies and deception your entire generation produced including what all those generations that came before you produced will be washed away as dirty mud on soiled linen. Your method whereby you practise in corrupting the minds of students will become what you are remembered by as you will be remembered as failures by the coming future being those in the past that was not worth remembering.

Students do not have to suffer their abuse and cheating any longer for now they can get wise. Students no longer have to sit and wait for thought control as you wittingly force feed those distortions about the fundamentals on physics and expect them to allow you to get away with academic murder just because your professional position allows you to. Now students have the choice to insist on you telling them the truth because the truth is written in

"Newton's Fraud"
Or
"Sir Isaac Newton's: A Conspiracy to defraud science"
The truth is finally out and your culture of deceit has eventually been detected

You will stop your mind abuse on students because they are now able to know more about physics and the application of gravity than did all those that came before you and all those that came with you.

Again I challenge you to come forward and tell your students the truth about what I uncover in the articles that follows and as the articles progress in introducing information...then you explain to them how you deceived their blind trust in you.
To Find Out More About

Newton's Myth ISBN)))))))))))))))))

or Newton's Fraud ISBN))))))))))))))))

or the largely and the even more informative book that not only deals with the corruption but also explains what gravity really is and why gravity is what it is

SIR ISAAC NEWTON:
A Conspiracy To Defraud Science.

ISBN))))))))))))))))

All the above mentioned books clearly show what mental abuse is inflicted on physics students.

Students Learn About
Academics brutally subjecting you to ruthless brainwashing and Mind Control.

The facts you are about to learn will astonish you and it will seem unbelievable but notwithstanding it is true.

Notwithstanding as unbelievable as it may seem, I nevertheless challenge any one to show me that the least of any or all facts I uncover is not true.

How can a Universe contract as Newton said it does while that same Universe is expanding? In accordance with the Big Bang theory the Universe expands as the while according to Newton the Universe is contracting? If the Big Bang is correct then Newton is wrong and if Newton is correct then Hubble expanding ands the Big Bang is wrong.

How can there be a Universe that holds everything there ever can be and has whatever could be expand and therefore become more while the Universe already holds whatever there will ever be? To expand something has to increase, in order to expand that which already is. What there is must then become more of what there is in order to hold more of what already is available. The Universe already holds everything there will ever be, so what can become more in order to expand? If it is understood that it is space that increases and space is filled with nothing, then how can nothing become more when nothing is the defined by what is the absolute absence of everything.

To have "nothing: as a filling of tangible substance in the Universe becoming more in order to have the Universe expanding, then the Universe must reduce what it already has to increase having nothing whereby nothing then becomes more because more nothing applying means the removing parts of what has tangible substance of some of what already is. How is it possible to have "nothing" filling the Universe and still have distances between objects? It is said by those very smart and distinguished brilliant academics in astrophysics that there are 149×10^9 km of **_nothing_** (please note the word **_nothing_** that is specifically used) to hold any distance. The word **_nothing_** states a detail of what is total absence and the lack of anything present. How can that which can't be present because it is absent such as the term **_nothing_** does specify, how can that then fill space in terms of distance measured in kilometres or astronomical units. If there is 149×10^9 km of **_nothing_** filling the space all the way from the Sun to the Earth, then how long is one specific point holding **_nothing_** in terms of a measurable unit?

How is it possible that mass can be responsible for objects falling by creating gravity by which objects supposedly fall and then have all things fall equal as Galileo proved. All things do fall equal…

If you think this is nonsense, then you are reading official Mainstream astrophysics. You will read about things they find "on the edge of the Universe" while everyone knows if the Universe does have an edge the Universe must end there and any end must have a new start. How could the Universe that can never end have an edge where having an edge means it must end and having such an end is having a start? Whatever ends must start to bring an end and the Universe can never end.

However, please note that this following information mentioned is supplied as mostly being a part of the four books entitled as An Open Letter On Gravity Part 1 and 2 Volume 1 and 2

How and where and at what point can we see did the Universe begin… and if you think answering this is impossible you haven't followed the trail of thought that Kepler left. You were only misguided by Newton's misinformation. Why would you be most likely correct when saying the Universe is a sphere… as it always is depicted in pictures?

Have you thought about the following? Because these I can answer?

…where is the cosmos coming from…?

…where is the cosmos going to…?

…and most of all…

Why is the cosmos travelling through time …?

…what brings about the direction of expanding?

My studying Kepler helped me finding answers to all the questions, which was deemed impossible to answer. The following is only a few of the many questions that I do answer.

Where is the centre of the Universe?
How did time and space begin?

That I can answer…and I also can answer…

…why is the Universe still growing since the Big Bang…?
…why did the Universe start so very small…?

…why did the Universe fit into a neutron at one time…?
…how did everything expand from fitting into a neutron…?
…why does space grow from small to large?
…where is it going while it is growing …?
…why was the Universe any specific size…?
…what was everything before the Big bang…?

I grew tired of apologising for my (as they see it) having the audacity of being correct on matters of Newton's incompatible religiosity, which I bring to their attention. When being in contact I am expected to show the utmost humble attitude acknowledging their supreme posture with me being in their surreal presence. I have to feel honoured to be in their presence when I mention to them their mistake about Newton being mistaken about a Universe that is contracting according to Newton while it never ever contracted in the least.

I am quite fed up with the attitude of those academics looking down their noses at me or worse still are those ignoring me whenever I show that Newton's facts just don't add to a conclusive believable answer. I am at my limit with those academics ignoring me because while they can ignore me by using their all-powerful status they continue with the conspiracy with them never having to prove Newton and therefore disclaiming even my presence when I try to disprove Newton.

I have reached my peak with stomaching the corruption they hide behind a lily white cover of dishonesty while they sit in their mighty towers and live in a bubble where not even God can touch them less having me point a finger at their despicable ignorance about their mistaken Master they portray as a God. They can say the Universe is made up of nothing and go unchallenged for making a most senseless statement and when I bring this to their attention in a book, I am the person they condemn as being incoherent with my arguments about their nothing they have in place.

I say this again: any person telling a lie is committing fraud be it in the name of God or of science; such a person is a despicable liar. When they tell a lie to distort the truth and find financial compensation while falsifying facts, even if it is in conducting science then they are behaving criminally. That is distortion and is equal to the behaviour of the Mafia.

From my view and from my perspective, I honestly can't see any difference between the Mafia's racketeering and corruption and what academics commit in the name of being honourable scientists. Those in charge of Mainstream physics feed students lies in order to be compensated for their misrepresentation of the truth. They are being paid enormous salaries from student fees to ensure that students believe in the impossible and accept what can never be proven and force students with methodical examinations to repeat the unproven or be expelled from the institutions and branding those expelled students as failures. That is a rip off whether it is justified as science in the process of learning or if it is plain legal criminality; it remains the same because they fly the same banner.

Those academics in key positions of academic credibility keep certain facts and evidence away from students and give other facts that were never proven before, prominence as well as credence while applying their trade in brainwashing to give their Newtonian views undeserved credibility and from these proceedings they earn substantial incomes. That is the same as racketeering. When you deceive by conveying untruths and cheat to mislead, then your behaviour is criminal.

If you purchase any of the following books you will come in contact with the truth for the first time in centuries. My work is about uncovering the truth and blaming the shameful conduct of those persons no one expects to be criminals.

When you purchase my books, I don't sell ink on paper. I do not sell material with questionable information, holding facts that were repeated so many times that it is accepted as the truth because it became a culture to believe Newton.

The information my work caries, which you will read in the event of purchasing my books, you have never seen, it was never yet mentioned or the facts I divulge has never been printed by any person, ever before.

I put untruths about the work Newton claimed as correct in question and that might have been published before but never published as questionable evidence. The rest I bring is new.

The following books that I offer for sale in this web site are unique in every sense. Only the information I question has been published before. The books take on Mainstream Science, uncovering facts that were never touched by any person for more than three hundred and fifty years. I act on behalf of the students in protecting those students that at this time is studying, or those that studied physics in the past. I am giving students information that is hidden by Mainstream Science.

On the book entitled Newton's Mythology I am making no profit. I give the book at a price that is going covering basic cost to furnish students with facts with which they can challenge academics torturers. I sell this at cost to show I am truthful in the hope of selling my other books

The book entitled Newton's Fraud is showing profit. The profit is going towards my attempt to get the books published privately which is done at a considerable cost. As the Purchaser will see, there is a significant number of sketches in the books and private publishers charge money for every sketch printed in the book. It is said that a picture is worth a thousand words and with the explaining; I offer this saying is certainly true.

The book entitled Sir Isaac Newton: a Conspiracy to Defraud Science is showing much profit and that profit is going towards covering the expenses of what the publishing costs will be that this book holds. The information in these above mentioned books has never been published but moreover has the information appearing in the book Sir Isaac Newton: a Conspiracy to Defraud Science seen the light of day. In the first two books I deal with the problems academics in physics are hiding whereas in the third book I bring the solution to the problems. This information has never been in print or given in printed form to any member of the public and that information I do not intend to divulge free of charge. Should you wish to know what the problems are in the attraction by mass theory of Newton, and you wish to learn the truth about the working of physics plus you wish to find the answers concerning the truth about physics and in particular gravity, then you will have to pay for learning my decades of research.

All proceeding will eventually go towards having these books privately published and have the information freely available to members of the wider public. I have run into a brick wall called Mainstream physics and there are (I suppose) hundreds of reasons for preventing me from having these books published of which no less is the finically motivation to restrain the publishing of my work since this work will condemn many other books with profitable titles to the incinerator. If my books do get published, thousand of books that are already selling on the commercial market will have the same informative value as the fairy tale story of Cinderella. However, publishing costs are astronomical and this is the only manner in which I can circumvent the academic blockade placed on my work and have my work published.

These books I offer for sale on this web page are not yet linguistically edited or controlled and I mention this because in it there is a chance that some grammar errors might lurk in the content. I found the most difficult part of writing is to correct one's own work in grammar because it is done by the method one applies to speak. Please also keep in mind that Afrikaans is my first language and English is my second language. It is also partly in order to have my grammar controlled that I require funding. Part of the funding raised by this effort will go towards editing before I can have them published and it is also for that reason I turn to this effort of marketing the books privately in order to obtain funding to have these books published and marketed through normal channels. I do not intend to challenge William Shakespeare in creating a masterpiece of linguistically prudent magnificence and with Biblical grandeur, but my aim is to get students to see what academics hide from them and to show academics what Newton hides from them. The brainwashing that is going on and the mental control that is inflicted on students is criminal. I challenge any person and every person to prove that I exaggerate or that I show falsified facts.

When any person tells facts that prove to be untrue such a person is a liar notwithstanding the motive in doing it. When any person supplies facts, which prove to be untrue with the motive of distorting money in the process, that person is a con artist and an untrustworthy individual. When any person supplies facts, which prove to be untrue with the power to black mail students and distort students' mental abilities, such a person is a criminal that should be locked away behind bars.

Again I repeat that academics teach students untruths about Newton that can never be proven and those very same academics know they are teaching blatant lies while being paid to do so. They charge students institution fees only to have students pay for being brainwashed in accepting Newton. Whenever the students never dare to challenge Newton's statements because if they do challenge Newton Professors would simply fail their exam papers and banish them from further studying. I dare you to challenge your lecturer just on the facts I give in this web page and see how he or she will react. Academics in physics are paid to minimize your questions, limit student thinking, conceal other ways of reasoning that might be unfavourable to Newton, control all information they give students, have students accept the facts about Newton that they give unconditionally, have students never question or doubt Newton in any manner and accept what they are told to write in examinations. What do you as a student think would become of the

student that ask the professor to prove Newton's claims that $F = G\dfrac{M_1 M_2}{r^2}$ is believable when all

evidence points to an expanding Universe?

Module Four One
Mass and Anti Mass

It is commonly accepted that Physics demand respect because the general idea going around has the understanding that Physics only work with proven facts that cannot be in dispute or be disproved in any way. Well…I wish to bring to mind some of the facts that physics work with when academics as scientists only work with facts. Remember they are the ones boasting that if facts are not proven then it is fables and those very important academics don't waste time with fables because they only work with facts. Students, it is your liberty to ask them to explain what they say is such correctly proven facts. They maintain it is a fact that we have to have mass in order to produce gravity. Mass is responsible for gravity. If you don't have mass you're not going to have gravity. Mass is equal to gravity and gravity is only where mass is. If mass is anywhere it should show its presence otherwise mass is absent. If a body falls it is the mass that allows the body to fall because the body receives gravity by ratio of mass and mass that produces gravity in relation to the mass available. It is mass that drags you down because the mass is in charge of the gravity and the gravity finds the value from the mass available. So what happens to balloons? Have they got anti mass or anti gravity? They are moving up when the air is heated. Mass pushes you down by the gravity it forces onto you. If mass drags you down then what are lifting you up in the balloon? If mass gives the gravity to drag you onto the Earth then why would the hot air lift you up? Is the hot air causing anti gravity or anti mass because gravity and mass drags you down or so Newtonians say. The balloon is lifting the passenger and all that is in the bag plus the bag plus the balloon into the air. So what is then pushing the lot up if it is mass that drags you down. Has the air not got mass because then the air can't have gravity and then the air must escape into the blackness of outer space because by going up it shows a resilience of either mass or gravity. We have seen that it is mass that pulls everything onto the ground.

Why would the air defy mass and allow the balloon to go anti whatever. We find mass being the equivalent of that which brings the object to the ground. The object has mass to produce gravity. Why then would hot air allow the balloon plus everything in the balloon to lift into the air? The balloon lifts in relation to the hot air that blows into the sack. The more hot air and the hotter the air is the more lift and the swifter the lift will be that the balloon provides. The issue sticking out is that the balloon then must not have mass because with anti gravity it is pulling up. Remember mass drags you down and mass can't pull you up and drag you down at the same time. Then what is pushing while mass is pulling or is mass pushing while what is pulling? The object is not going in the normal direction where it is dragged down by gravity and in all my life I have never heard one Academic mention anything about gravity lifting and that makes the lot very confusing. What is lifting up when the lot should be pushing down and why did everything connected to the balloon lose the mass and if it has mass why is it not dragging down the balloon? If you think this is a little confusing try what is to follow.

They teach you that it is mass that produces gravity and gravity makes you fall because while gravity makes you fall mass drags you down. It is because those mind controllers are lying through their teeth with a menace in which they are the experts, as they know just how to pull cotton wool over you eyes. Take a truck of 15 tons into an airplane. Put next to the truck a petite little dancer weighing 45 kilograms. Put next to her a frog weighing 150 grams. Then get this lot into the air by airplane and let them jump. Take note that you are told by the wise amongst us that it is mass that produces the gravity that pulls you down. We have just had a lovely debate on how it works and how mass drags you down and wondered if it then is anti mass or anti gravity that lifts you up with the hot air balloon, well take note of this as your airplane reaches 11 thousand meters which is eleven kilometres straight up into the air.

Now we drop the truck and the girl and the frog at the very same time from the airplane. The frog then pretends he drives the truck and the next scene he is dancing with the girl while the truck is falling as fast as a truck can fall. Who do you think is lying? Remember only one group can tell the truth and the other must be lying. Have you thought why one party is lying while the other party has to tell the truth?

The academic Brainy Bunch are telling students all over the world that mass is in charge of gravity and it is mass that's pulling you down. Then the mass is pulling the truck of 15 tons down since the mass produce the gravity and the gravity produces the fall which is three hundred and thirty three times more in a down direction than the mass of 45 kg is pulling the dancer down. The mass providing the gravity that pulls the truck down is doing the pulling down of the truck one million times better than it is pulling down the frog. If the mass is doing the pulling by establishing the gravity the truck must fall 333 times faster than the girl and one million times faster than the frog. It is either that or the three has the same mass because they are falling at the same rate.

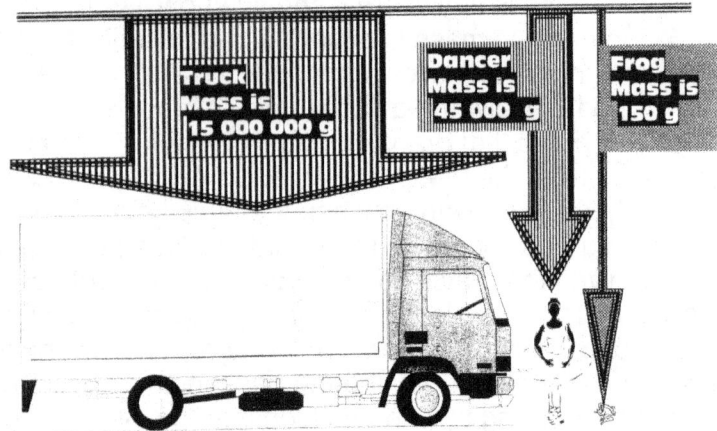

If the Brainy Bunch all too wise is correct the frog can fly to America and have a pizza in New York while the truck has a few micro seconds to get down if the girl is going to fall during the normal falling duration of a minute or so.

Everyone has seen skydivers jump out of airplanes next to cars and trucks and bags. Every one has seen they all fall at the same rate. The girl can do tap dancing around a jumping frog on top of the truck or below the truck and they can be inside the back of the truck galloping on fresh air inside the truck because the lot is falling at the exact same rate.

The academics wishes to brainwash you by mind control in accepting that it is the mass that the falling takes place and that mass is responsible for the gravity and by mass pulling you down it is gravity that makes you fall. Where is the proof of mass that according to them is that which is producing gravity. They tell you Galileo said all things fall equal and we can see from the TV monitors how all things fall equal. Where is the mass that makes the gravity to let you fall if all things fall equally? They tell you that the truck has a mass of 15 tons and that mass is making the gravity that is having the truck fall while the truck is falling at the same speed and distance than the frog does.

If you take that as proof then they got you. Then they brainwashed you into a zombie. Then if you don't repeat after them and echo every word test after test and exam after exam they will fail your papers and kick you from campus. That is mind control, better than what even the KGB is able to implement. You repeat after them and you live an academic life or you disagree and you go home to play with your toes. If mass is in the picture then mass must be represented by a factor of more than just one because if mass is not part of the overall picture then mass has a factor of one which proves that mass is not part of the equation since mass can't change the results. With all the objects falling equal mass has no role and if mass has no role then for my money academics in physics can't just go and put everything in as their hearts desire. If it is Galileo that is correct and if all things fall equal then mass has no part in gravity. If mass is the inspiration behind gravity the truck must fall a million times faster than the frog and in fact the frog should almost land in another country because that is how slow it falls.

The fact of the matter is that I don't wish to be near when any of this lot hits the ground because the truck will cause a quarry and the dancer will be a splash of red fluid while the frog might not be that worse for wear if the truck or the dancer doesn't land on the frog. But that is mass. The differentiation of having mass or having equality when falling and then not having mass and between individual differences in mass by each component that enters the equation when the objects touch the ground. Then every one gets the mass it has. Only when they touch the ground and land on the soil is mass as a factor awarded. While they fall they all fall equal and there is no distinction between the falling at all. What then is gravity? The gravity is the falling. The gravity is the motion. While the object is in a state of mass it is not moving. The tendency to move and apply gravity is the part that the mass restrains. The mass is preventing the falling from continuing. It is the role of mass to prevent further falling and independent motion to continue. Some of then might even still honestly believe it is mass that produces gravity because they were taught that it is mass that produces gravity and never thought about the matter again afterwards.

They were brainwashed by their tutors as their tutors were brainwashed before them. You don't need the brainwashing because you now can find out what the answer is to gravity. You are the first generation that can receive the light of knowledge about what gravity really is, or you can be the last generation that will live in the lie. You are in a position where you can teach your tutors the truth about gravity if you read what is in the books. The truth is there and the truth is out and the truth will be because the truth is written for all that wishes to read. The academics on the other hand have ignored my work and my being on Earth for the past six years while I was writing them letters about gravity. They ignore me as if I am a rattlesnake because to them I am a rattlesnake. With what I say I will have them tumble down from their pedestals because by accepting my work they suddenly find their position equal to yours as students, and then they will have to learn my work in the same manner as you learn my work because to them everything is as new as it is to you. The Academics of the day have too much to lose to recognise my work and therefore have to protect their interest with all they can muster. For that reason if no other they will rather go on lying to you and cover their corrupt fraud than face up to the truth and admit their work is lost.

The truth will be whether it is recognized by them and they can become the first to admit and repent or they will be the last of the laughing stock that those in the future will refer to as the bunch that couldn't see when things fall equal they cannot have mass and when things do not fall by mass then one can know mass has nothing to do with the falling and the gravity.

It is up to you as students to rattle their cages and make them admit they've been lied to as they are lying to you. Or you can be the last of the fools that couldn't see that when things fall equally they have no mass by which they fall. My book is written and those that read it first will know what gravity is. If you do not accept the role as being zombies that is brainwashed then confront these academics that treat you with disgust and betray your trust. They might tell you the mistake is not that serious and the damage is small but then how will they know how big or small the damage is if they don't even know what damage there is or what the damage is.

Science has stayed so far from the truth that they can't even see the truth any more. If you carry on you will learn about some of it and when you read my books I will entertain you with many more than you ever believed. My books will serve as the light switch that brings the light to you.

I charge your young minds to confront those fraudsters about the truth. I wrote to them in the last letter where I informed them that they protect the criminality of their corrupt teachings because when the corruption is removed then nothing remains because they have lived a lie for too long. If you reach the need you may down load it because it is a fair bit of information.

If I come to you with a proposal about something I wish to share with you on condition that you pay me an amount to share with you what I know then I am an academic wishing to teach you. Have you a name for such a person that will force another person to pay him to be brainwashed and be mind controlled because the tutor has absolute control over the life and death of the academic future of the brainwashed being and therefore is willingly forcing this unfortunate creature in accepting what will never amount to the truth? I think they are called Physics professors and rule Universities as draconian authoritarian dictators bent on sadism.

Let's investigate the falling as such and see what happens during the fall. The truck falls at the same pace in which the girl falls, which is the same pace as that which the frog falls. If the truck falls at the same pace as the girl and as the frog there has to be a common denominator in this process and since the common denominator eliminates size form and shape we can eliminate mass. Mass brings distinction and the falling eliminates any form of distinction.

When I fall down a waterfall with a boat I travel the same pace, as does the boat. That could be because I am fixed to the boat by sitting in the boat. But my sitting in the boat has certain condition and one is that I can remain sitting because I fall the same pace as the boat is falling.

I fall down with the boat and the boat and me forming a distinctive unit falls at the same pace as the water that forms the waterfall falls. Should I at the time of my falling hold an empty mug in my hand and I wish to fill the mug with water, and then I will have to move the mug against the flow of water streaming down the waterfall. I will have to thrust my mug upwards at a faster pace than my descending is casting the mug down and therefore I accompanying the mug down the waterfall. My mug will not automatically fill with water or if there was water in the mug my mug will not automatically empty with water just because the emptiness filling the mug will be at a different pace than the content that is otherwise the filling of the mug.

The mug being empty falls as fast as the boat and I. The empty space in the mug is falling as fast as the mug will fall when the mug is filled to the brim with what ever can fill a mug to the brim. Notwithstanding the content within the mug or the content within the boat or the content within the water being within the waterfall, the very lot is falling at a similar pace. By lifting the cup while falling the cup will fill with water.

I am not putting the water into the cup but I am exchanging the space that the water holds with space that the empty cup holds and my action in truth has no bearing on the water filling the space, which I then transfer into the cup. I am filling the cup with space that at that point holds water but the holding of water has nothing to do with the transferring of space.

If I leaped from the boat and fell I would fall alongside the boat. The boat will be empty but will fall at the same pace and as the same space as I fall notwithstanding being empty. The mug being empty will fall at the same pace as the boat being empty which will fall at the same pace as the water in the waterfall and I would fall. The space in the boat, which is empty if I do not fill the space, will fall at the same pace as the empty space, which fills the mug, and the mug will fall at the same pace whether the space in the mug contains or doesn't contain whatever can fill a mug. The space filling the mug is falling the same as the water that would fill the space in the mug should the mug be filled with water.

The space in the boat is falling at the same pace as I would fall whether I am filling the vacant space in the boat or otherwise filling the vacant space next to the boat. It is the space that falls and not the object filling the space that are falling. It is the space that is filled or not filled that is dropping down because the space being filled is in decline. If it was not the space that fell the space within the mug would fill first as the mug and the boat fell because the empty space would first fill before it could take anything down. But since the boat falls as fast as whether it is being filled or not we can assume that the space which the boat fills or does not fill is falling as fast as it would fall whether it is holding the boat or I or the boat and I. The space not filled by mass also moves just as fast as space filled by mass.

When the object such as the mug or the boat or I connect with the Earth the Earth disallow the object free motion by taking any more space the object claims through to the centre of the Earth. The object now has to give up the space it claims and take on new space that the object claims to flow by contraction to the centre of the Earth. In forming a blocking it resists the flow or the gravity or space lining up with the centre of the Earth. The flowing of space by contraction is gravity but the object being in the space that flows becomes and obstacle through which the oncoming space must drag in order to flow to the centre of the Earth. It forms resisting of allowing space claimed to release to the normal flow when the object will not relent form in favour of gravity. This resisting such relenting of form and consequently forming a frustrating barrier that blocks the free flow of space towards the centre is time displacement of space and this relenting of space-time flowing freely becomes the mass factor. The density and the resistance that the particles show forms the mass that implicate the degree of the frustrating or preventing or disabling of such free flow of space through time and the displacement of space during time is space-time notwithstanding what ever irrational connection Newtonians wish to add too space-time. Allowing space to displace through time to form time is space-time and that is gravity.

All this is not new and I am not the first and the big genius that thought this out for the very first time since Eve had a bite on the forbidden fruit. In around 450 BC a man going by the name of Empedocles killed the myth that it is nothing that fills the bowl when water runs from the bowl. It was named the clepsydra meaning water thief. He proved at the time that something other that water fills the container while the container is emptying of the water.

The water will start running only when a finger lifts from the pipe where the finger before the time blocked the intake of the pipe and therefore prevented something from entering and therefore releasing the water from the container.

 This he interpreted as being that the water was not running out from the container but was being pushed out from the container. The container was filling with something as it was being emptied of the water by something being anything other than nothing. This experiment was done some almost two thousand five hundred years ago and still Newtonians have to find a manner in which their grasp will accommodate these facts. It is the air filling the space that pushes the space filled with water from the bowl and the filling process of either space is named gravity. How difficult can it be to grasp an experiment that was understood two thousand five hundred years ago?

The space that holds the water is running down and away from the bowl and the water thief or clepsydra is stealing not the water but the space that holds the water. Once the water is out we can presume this displacement of space continues because there is no reason to think that the factor of nothing suddenly enters the scenario and the process stops because Newton saw

Empedocles' Clepsydra of 450 BC

Connected pipe allowing filling of bowl by water

Round Container Filled with water

Water running from outlet at the bottom

nothing where something had to be. The space keeps repeating the process of displacing the space it follows as it is displaced by the space falling. It is a continuous cycle never ending and the space flows notwithstanding it being filled with what Newtonians can understand or the nothing they do understand. They understand nothing so well they filled the entire Universe with the entire nothing they do understand.

If it was purely atoms replacing atoms in the clepsydra as Newtonians wish to think and nothing else as they say because they still think of space as nothing, then how can so little number of atoms that fills the air replace so many atoms such as the water has. If the atmosphere was vacuum in the sense as they see vacuum being the stuff a Newtonian would normally have between the ears, and that stuff is nothing in as much as something being present in whatever space filling capacity anything being nothing can fill and also can form nothing, then the vacuum had much less particles to fill the space that is filling the clepsydra than what the water has that is emptying the clepsydra.

Let them count the number of atoms leaving the clepsydra through the holes where the water sprouts from and compare that to the number of atoms entering the clepsydra at the top and see how those figures add up in an explainable argument. If the finger closes the hole in the pipe on top the space entering stops completely but so does the water escaping also stops. If the pipe on the top is restricting the flow of space entering the clepsydra then in relation there will be restriction in the water leaving the clepsydra. It is the space that fills the space emptying the clepsydra and whether the space is filled with particles or sparsely filled with particles or not filled with particles at all, it remains space replacing space filled or not filled. It is space entering and it is space departing and the filling of space with material has no bearing on the matter what so ever. Newton had every wire crossed in his head when hew put falling of any object down to the mass which the object supposedly should have.

It is so clear that it is space that is moving and the fact of being filled by material or not filled by material has no merit in the process. It is not the apple that Newton saw that fell but the space he saw as nothing that fell and the space that was falling took the apple with while the space that followed still fell whether it had an apple to accompany down or was filled with a Newtonian filling of nothing. The space holding the truck being next to the space not holding the truck is falling as fast as the space holding the girl and the space next to the girl which is not holding the space or the frog where this lot is falling just as fast as the space holding the frog. The space is falling. The space is falling whether it is filled or whether it is empty and that means mass has as little to do with the falling, as the colour of onions has to do with the depth of the sea or the temperature of the shining Sun.

It is not as if I wish to condemn and reject that which is in place without placing something of worth back into the process. All I ask is to read what I bring. Don't be a coward and stop reading as soon as you reach the point where I condemn what is in place! Just move past that to the point where I show what is wrong and how it can be corrected! Just judge me not for condemning what now is so apparently incorrect but for showing why I condemn what now is so apparently incorrect and what I bring to the table and offer as a remedy. See what I have to offer and not only what I am taking away. Don't set your sights on what there is to lose but take a view on what there is to gain!

Do not reject me on merits you do not wish to instate because you have the fear you are going to lose what is instated. Do not judge me by using your double standards that are useless in the face of the truth. Rather look at the double standards you employ and do not judge me by using your double standards on me. Rather use your mind to detect what is double about your standards and then investigate with me what needs to change. Don't hide the truth. Don't hide from the truth and don't hide behind what you wish to portray as the truth. Rather come out into the light for the first time in three hundred years and admit to the truth. Follow what I say and see for yourself what there is to gain by trying to detect what is wrong because we all know there is much wrong. The comet does not collide with the Sun and the Moon is not on its way to collide with the Earth in time to come.

It is not as if I wish to condemn and reject that which is in place without placing something of worth back into the process. All I ask is to read what I bring. Don't be a coward and stop reading as soon as you reach the point where I condemn what is in place! Just move past that to the point where I show what is wrong and how it can be corrected! Just judge me not for condemning what now is so apparently incorrect but for showing why I condemn what now is so apparently incorrect and what I bring to the table and offer as a remedy. See what I have to offer and not only what I am taking away. Don't set your sights on what there is to lose but take a view on what there is to gain!

Do not reject me on merits you do not wish to instate because you have the fear you are going to lose what is instated. Do not judge me by using your double standards that is useless in the face of the truth. Rather look at the double standards you employ and do not judge me by using your double standards on me. Rather use your mind to detect what is double about your standards and then investigate with me what needs to change. Don't hide the truth. Don't hide from the truth and don't hide behind what you wish to portrait as the truth. Rather come out into the light for the first time in three hundred years and admit to the truth. Follow what I say and see for yourself what there is to gain by trying to detect what is wrong because we all know there is much wrong. The comet does not collide with the Sun and the Moon is not on its way to collide with the Earth in time to come.

On TV we find that all things fall equal and that no size or mass differentiation plays any part in the falling process. Every time Newtonians are cornered with this idea they come up with a variety of answers where they would try to convey the idea that Galileo and Newton used the same mind in thinking. If things fall equal then while falling all things have equal mass. There can't be any mass description or mass variation and if it is mass that produces the falling the falling has to be different with mass varying. Don't let them confuse you on this and don't let them get away with more cheating…they have cheated enough for too long and got away.

Expand science and no the Universe for the Universe is the only aspect that has not the ability to expand.

I challenge all of you Newtonians to prove $F = G\ \dfrac{M_1 M_2}{r^2}$ and not just to declare it proven because it is

in use since the Dark ages. Expand your mind and double check the formula you all so vividly underwrite and support. Prove why you support the formula in a modern and a scientific way. Explore the

correctness that this formula $F = G\ \dfrac{M_1 M_2}{r^2}$ underwrites. Be a true exploring scientist and journey with

me through the following pages while we venture on the quest to find and vindicate my incorrectness by

proving the truth vested in the formula $F = G\ \dfrac{M_1 M_2}{r^2}$ that carries the entire physics everyone uses.

Let us start where the lot should start and get two Masters together on one point of argument. Galileo said all things fall equal. That says all things fall alike. The first thing anyone brings in is the vacuum bit with the feather and the hammer and since we do not live in vacuum there is no chance of finding a feather that will fall as fast as a hammer. Since the feather does not fall as fat as the hammer we immediately jump to the conclusion that there are falling disparities because of the falling discrepancy we find between the hammer falling and the feather falling.

Then what would give the feather the time to fall longer than the hammer does. Everyone concludes about mass coming into play and they are correct. But they are half correct while Newton still is completely incorrect by attaching mass to the entire idea of falling. Take away the resisting of the feather and replace it with something far less air resistant and one will come to a different conclusion.

We have to dissect what factor consists of gravity and what factor represents mass. Then we have to dissect which part does mass play and what part does gravity play. The falling object experienced no mass while falling therefore the falling or moving must be gravity's contribution. While objects are in motion those moving objects is experiencing gravity.

The object show mass when the object has a tendency to move but the motion towards the centre of the Earth no longer takes place. That means mass is the restraining of the motion or is that which prevents the motion or gravity taking place. On Earth, objects experiences mass by restricting gravity or motion with the Earth giving mass but taking away free motion. By giving mass the Earth forces the object to become one with the Earth and move with the Earth as a pat of the Earth.

Persons falling will experiences weightless ness while falling and they have a weightless state while falling. One cannot then go on to declare that the factor, which prevents motion, is the factor that causes motion because that is totally contradictory. The motion takes place without the presence of mass because the frog and the truck are falling equally fast. When landing the motion of the truck and the motion of the frog ends. Then the two have very different mass values but neither shows the ability to break from mass and move further towards the centre of the Earth. Kepler said the space a^3 is equal $=$ to the motion in a line k as well as a circle T^2.

While experiencing unrestricted gravitational motion a body a^3 is $=$ to the motion T^2k as Kepler said gravity is: ($a^3 = T^2k$). When motion stops then only does weight or mass form as a result. While falling we find that gravity applies as individual separate space is moving and putting time in relation to the distance that the falling object travelled. That makes the falling factor the part that is the motion that confirms gravity. In the motion or movement we find the gravity because that even remains as a permanent attempt to move. Even when mass comes in as that which results in the ending of the gravity and in that gravity as a term is also forming the motion factor, still remains as an attempt to move. The while moving Galileo proved mass is not present because all things fall equal. Mass comes in when movement is retained and although the mass is present as a factor that factor that mass represents is what produces restriction of such a movement and not resulting in such a movement. The factor that mass represents is the containing of further downward movement. Looking at the factors separately it is obvious that mass as a factor cannot produce gravity. Mass is the restraining motion that leaves gravity as intending motion. Mass occurs only when motion is prevented and when mass prevents further motion resting objects leans against each other. When objects rest against each other they restricts individual gravity motion. Mass is a substituting factor, compensating for motion loss. When mass restricts motion gravity becomes the tendency of motion. Mass counters motion when the Earth restrains further motion of falling objects. When motion seizes, falling objects remains individual while still tending to move. The Earth resists further movement of falling bodies' movement restricting motion individuality. Having mass does not bring about.

Physics is Brainwashing by Mind Control

If you are a student in physics then you should read the following information. It is about the subject of gravity and is most important. Do you realise that it is an accepted practise that all students that are studying physics on all levels are subjected to the most intense brainwashing and thought control found any where on Earth? This must be some sort of a joke you may think but thinking that way in disbelief is just what those practising the mind control wish you to think!

Should you think this page is some sort of a prank then answer the following simple question to yourself in utter honesty?

The questions concern that which you are studying and that touches every aspect you are academically concerned with. You are taught that gravity pulls objects to the centre and obviously gravity then has to ultimately pull everything to the centre of the Universe. That is what the Critical density research that Einstein initiated wishes to establish.

When visiting the classes you attend in physics, has any one confirmed a location where one might find the centre of the Universe? This they have to do if they say that all objects are submitted to gravity. Then they must know where gravity is taking the Universe. If you wish to apply a Gravitational constant as a calculated factor in using the basic formula $F = G \dfrac{M_1 M_2}{r^2}$ then it is apparent that every one must know to where such gravity is pulling. Gravity is pulling to the centre and therefore the gravitational constant also is pulling to the centre of the Universe. If there is such a force then where is the force taking the pulling…if it is a gravitational constant applying through out outer space then where is it having a centre base? Tell your tutor to calculate when the Earth will collide with the Sun by using Newton's also accurate formula $F = G \dfrac{M_1 M_2}{r^2}$.

To calculate the following data is necessary:
Mass of the Earth = 5.974 x 10^{24}
Mass of the Sun = 1.989 x10^{30}
Gravitational constant = 6.67 x10^{-11}
Diameter between the Sun and the Earth = 149.598 x 10^6 km (remember that this has to square)

If he can't give you a ready answer it is because it is the biggest hoax man has devised ever. Using this formula $F = G \dfrac{M_1 M_2}{r^2}$ to calculate is complete rubbish because it can't be done. They are brainwashing you into believing the use of this formula is viable while it is complete and utter rubbish. Gravity does not draw by mass at all.

I wrote a book in which I found a means to define gravity. This feat I accomplish and by my effort it was done this for the first time ever. For the first time ever my investigating physics runs further back than since the time Newton introduced the idea of gravity. Before I achieved that discovery, I firstly had to find the centre of the Universe because it is there that I could locate gravity. I can now show how gravity forms because I have detected the centre of the Universe. But by my effort in finding the location I disrupted everything Academics in physics hold holy and for that I am most unwanted in the presence of the Academics charged with guarding the ethics of physics. Every time I try to indicate what I discovered about gravity, academics throw Newton at me and detecting from the information I discovered when I investigated another even much wiser Master, it is clear that Newton is the last person that knows what gravity is...he (Newton) even admitted to not knowing what gravity is...and yet they (the Newtonians) keep throwing at me the ideas of the man that admitted he had no foggy idea about gravity. During my research I discovered abnormalities and inconsistencies about mistakes Newton made long ago and it is clear that the Arch fathers in physics must be aware of such misconduct performed by Newton but they (the Newtonians) are hiding such information from students using all their considerable influence. I will come to a few of the inconsistencies but part of the discovery I made when I investigate the other Master I mentioned I was also introduced to a much better vision about gravity as well as many new aspects never before realised in science. The road was never smooth and the resistance I came across from the Newtonians was almost unbearable. Academics guarding physics will never allow an outsider to enter their domain without the intruder paying a heavy price and in this matter I am seen as the intruder. Intruding allowed me to find much that I was not supposed to find which was only allotted to the most inner circle and much I share with you.

In achieving goal to locate the centre of the Universe I had to step on some very important toes, which made me very unpopular. With my unpopularity rating this high as it does, I never qualified for help and found intolerable rejection as I tagged along while trying to convince those Newtonians about mistakes in their field of expertise. Because of this insider rejection I received so blatantly, I had to resort to private publishing because from the nature of my work I take Mainstream science head on and am confrontational on most aspects of astronomy. This is the only road to go if one wishes to lay an axe to the root of the insider corruption that all they (the Newtonians) are guilty of. In that sense there does not seem to be any publisher that wants to go head bashing with the Physics Custodian establishment of science on official science principles, which I have to do to convey my message in no uncertain language.

I argue that if it is the correct practise to use $F = G\dfrac{M_1 M_2}{r^2}$ to calculate gravity then the radius holding the gravitational constant must lead one to the centre of the Universe. I found the route that gravity takes when gravity goes to the centre of the Universe but it definitely does not apply when using the Newtonians formula $F = G\dfrac{M_1 M_2}{r^2}$ With nobody willing to publish my work as I contest science all the way and even at the most basic level, I had to go the road alone and fight the battle by my private effort.

I know the explanation I give in this seems cramped, but when reading the book it is much clearer. This is the point that I wish to make on this one issue and similarly there are thousands other unexplainable issues. If the Sun for instance has mass that is apart from the Earth and the Earth also has mass and there is a gravitational constant in between the Sun's mass and the Earth's mass we have the radius in that location. It then must be the gravitational constant $F = G\dfrac{M_1 M_2}{r^2}$ represented by "G" that fills the space that the radius holds. It is rather obvious that while the radius is filling the vacant space between the Sun and the Earth it is the only place left where the gravitational constant can hide. The space Newtonians give a value of nothing or emptiness must hold "G". Where is this gravity pulling if it fills space that is empty and is not filled? If it is the centre of the Earth that does all the Earth's gravity pulling and it is pulling towards the Earth's centre then that other gravity being "G" has to have a point whereto that gravity is pulling if it is pulling at all. To find the centre of the Universe I had only to find the gravitational constant that holds the centre. Through my venture I discovered one person that knows what gravity is! From studying that person's work I found the centre of the Universe.

If you think scientists know what gravity is do not be duped that easily because no one in science remotely knows what gravity is...not even Newton knew what gravity is except Kepler... and because of what Kepler introduced now I know I can prove what gravity is. Gravity is precisely what Kepler said gravity is and only Kepler new where to find the centre of the Universe because only Kepler knew what gravity is all about. Did you know that Kepler showed what gravity is decades before Newton even had a thought about gravity? Does your Professor know that Kepler found and proved what gravity is long before Newton had admitted he had no idea what gravity is? Newton had no idea what gravity was because Newton changed Kepler's work without ever studying Kepler's work and therefore never understood Kepler's work.

Students read the following message about my book I named www.SirNewtonsFraud.com and by reading the book you can learn how you are brainwashed and how your mind is pre- conditioned into believing in Newton's myth which is pure deception which Academics call physics.

Let's start surveying civilized principles by evaluating what lawfulness means and what would constitute as morality. Let's determine what makes the crook in the book?

If any person, notwithstanding what reasons given, tells a lie or conveys untruths it is seen as fraud. To convey information that is not substantiated as verified fact then the mere conveying of such information becomes fraud.

When any person, notwithstanding what reasons given, repeat such a lie unabated while being well aware that the information passed on by such a person is incorrect, then the person commits deceit. When anyone is repeating the information that is passed on as being unblemished factual substantiated and verified truth while such a person knows very well that such information is void of proof or lacks proof, then committing such an act is a criminal enterprise.

Academics in physics commit every one of the above indignities and yet see their actions as being lawful and even much praiseworthy and hold their role in society in the heist esteem imaginable. They fail to see the crime that they commit while tutoring physics. Whatever motivation they may claim to have as their driving force, the fact that they perpetually perpetrate in unlawful behaviour by spreading untruths such actions on their part put those academics holding such highly regarded office in the league of ordinary cheats, gangsters and common criminals. By willfully and constantly falsifying facts of what order repeatedly remains derogative and unlawful in nature, notwithstanding what morality it should serve. A Preacher or Pastor lying on behalf of God is not lying on behalf of God and to think the Preacher or Pastor improves or underlines the Greatness of God by lying on behalf of God is very mistaken because in reality such a Preacher is falsifying the truth for personal his or her personal benefit. Lying is wrong and doing so even in the name of God remains despicable. The same applies to academics in physics. There is no argument that can change this truth about falsifying the truth and when doing so there is no hiding behind any excuses of ennobling to benefit mankind that will change such truth into righteous conducting.

Newton said centuries ago that gravity is the force of attraction there is between objects that holds mass and it is the mass factor that brings about this attraction, which Newton claimed there is. The Universe does not contract and all the proof we require to disprove such a statement that we find in the Hubble constant as a guarantee. Moreover it is true that the Universe never contracted even for a brief instant and proving that is the Big Bang concept with all the proof that this concept bring in backing the principle of expansion in the Universe. Planets never moved closer, are not moving closer and will never move closer to each other and this is backed by all information collected this past century. The Moon is not coming closer but the distance between the Moon and the earth is widening. Studies about the Universe reveals every time that space in the cosmos increases constantly studies find that all things are moving apart and away from one another. Any and all the proof about this is beyond what any doubt may present to counter this knowledge.

Notwithstanding this irrefutable findings, science still regard Newton as the only person that ever lived which no one ever could prove wrong…and this is upheld by Mainstream Physics in spite of the cosmos proving Newton wrong every instant of time. The basis of what science holds as its foundation we find to be the Newtonian principle of $F = G \dfrac{M_1 M_2}{r^2}$. The foundation used by science promotes this argument and backs up this argument well knowing that in the cosmos there is no evidence backing up this proposal Newton suggested. The Newton formula $F = G \dfrac{M_1 M_2}{r^2}$ used as basis for science see gravity as being a force of attraction and the force of gravity is being in place between all objects in accordance with the mass factor that the objects have as presented by Newton in the formula $F = G \dfrac{M_1 M_2}{r^2}$

What we find as we gauge all evidence found while studying the Universe is that reality shows there is no attraction between objects in space going on any where in the Universe, that the entirety of such a concept is a myth and the outward moving of the Universe has been coming from and since the time of the Big Bang and maintaining this flow of material is substantiated in a concept named as the Hubble constant which proves Newton's perceptions to be a myth. The Hubble constant proves that space every where is growing ever since time began and the growth never stopped ever since. Knowing this irrefutable fact does not deter science from under scribing Newton as the sole basis that underwrites all the correctness of all of science known as physics. However Hubble and the Big bang and all other investigation contradict this attraction Idea Newtonian dogma holds. Therefore any further believing that there is attraction going on as Newton claimed has to be viewed for what it is and that it is a fairy tail. The Big Bang Theory proves Newton's idea as not only being wrong but Newton's idea of attraction is a joke. If the Big Bang is expanding the Universe then how can the Universe contract at the same time? Any contraction by nature would have the Universe collapse back into infinity the moment the Big bang moved out of infinity. Ask your professor to show how an expanding Universe can also contract and your professor will tell you about Einstein's Critical Density theory. This theory I prove is the biggest fraud ever devised by any group of persons in the history of civilization! This is perpetrating fraud and conducting in upholding deceptions instituted by Newton that then formed the institution of lies they call physics.

The Universe does not contract in any way means or form and even such a suggestion is incorrect! The Moon and Earth are not moving closer but are moving apart. The entire Universe is growing in space and no where is space depleting by any norm used. Academics are very aware of this misconception Newton had and still academics in physics are promoting the ideas of Newton as unwavering truth. Academics teaching these misconceptions are committing fraud notwithstanding their portraying of their role in society being unblemished, spotless while they are covered in a lily white blanket making them being whiter than snow and having such a holier than though attitude. Teaching Newton is participating in deception and promoting Newton is criminally deceiving the public and while doing so is committing an act with criminal intentions.

Then, in the face of all this evidence contradicting Sir Isaac Newton they remain upholding the correctness of Sir Isaac Newton and keep on teaching students about the unwavering correctness of Sir Isaac Newton. They put down conditions of learning to this effect and are expecting students to repeat these untruths and unproven facts by forcing answers to that effect in examinations. Forcing the accepting of this untruth about physics are equal to preposterous subjecting students to physiological torture and heinous mind conditioning, scandalous thought control and brainwashing. This applies to everyone serving as a tutor in physics notwithstanding whatever status the torturers might have in society or the morality they attach as a reason to commit such atrocities.

If you are a student then you are conditioned by academics in controlling your thinking by enforcing pre mind setting and in which they methodically force you into believing in Newton and this is an on going process conducted for centuries in the past while it is the truth that Newton is completely void of any tests that may secure any form of confirmation and in securing proof then also by that establishing proof. Read this book www.SirNewtonsFraud.com and then use the information I supply in the book to insist that Academics that are teaching physics prove to students that Newton's statements of attraction are correct. Let those academics explain the method mass uses. Let them with precise detail show when mass is applying that gravity is produced by mass and such producing of gravity that then would establish attraction! I show precisely how gravity produces mass but mass can never produce gravity. I sue explicit detail in showing when how and where gravity forms mass but mass can never form gravity. What I prove annihilates every Newton claim.

They never prove Newton's philosophy on gravity but those persons conducting teaching in the subject of physics force all physics students to learn Newton's gravitational concepts and accept the facts as if it has been proven beyond all other facts. Students have to believe that Newton is correct or academics will see to it that they fail their examination. The condition of being accepted in physics is to accept Newton without questioning the proof that is never supplied. Let those academics now prove precisely how mass brings about gravity and then afterwards test you on how Newton is proven correct and not on you repeating facts about what they say is true about what Newton said that they say is true. The manner they present Newton is completely hearsay and that method may not be used in any court of law. Let you professors now prove how it is that Newton's teachings are correct and then examine you on the process they use to prove Newton's concepts. At present they say Newton is correct and then they test you on your ability in repeating that Newton is correct without ever proving to you that Newton is correct. Let those physics professors now prove Newton and then test you on the manner they use to prove Newton to be correct.

The truth beyond all other truth is that Newton's gravity has never been proven (because try as you may it is not possible to prove Newton's formula forming gravity mathematically) and because academics know that, academics require the blind accepting of Newton by students. This unconditional accepting of Newton's correctness relies only on the pre-conditioning of students' mind set and academics depend only on the student trusting the academic say so that about the institutionalised correctness of Newton. That Newton is correct nevertheless and notwithstanding that there is no founding proof about this matter is what students should be accepting blindly. Pre-conditioning students' into blind acceptance depends on the academics' insistence that students approve Newton's concepts without pre judgment or students insisting on scrutiny of any sorts. In examination students have to outright and blindly follow academics' say so only because academics say so. Academics depend on students never questioning their say so or demand proof about what academics teach. Those academics in teaching positions insist that all students accept Newton's accuracy.

This is methodical mind control as much as it is the brainwashing I show that they enforce. If you are one of those believing that Newton was ever proven, then what you believe to be true is a lie because Newton can't be proven and that is the truth! The time has come to face your teachers and force them to stop the centuries old culture of bullying students and conditioning their thoughts by enforcing on them dogmas which is mind control!

I charge all academics to prove what I say is being wrong in any way or even that I exaggerate in the least. I challenge Newtonian academics to prove that mass does indeed form any force of any sorts and in particular gravity! To those professors claiming Newtonian ideas are substantiated by proof I say that notwithstanding your personal academic qualifications and while at the same time disregarding your status and previous achievements as well as ignoring your many admirable abilities you may have and however superior they might be, I shall teach you about gravity. I say it is time Students learn the truth about physics notwithstanding the status academics will loose. Students read **www.SirNewtonsFraud.com** and challenge those academics depending on their ability to brainwash you into submission.

I don't think so because your professor is an expert on Newton's work and Newton admitted he didn't know what gravity is. That means whatever your professor has expertise on, it involves Newton and Newton admitted not to know what gravity is. Then notwithstanding what your professor says about gravity and what he professes to know about gravity, he knows what Newton knew about gravity and Newton by personal admission said he knew nothing about gravity. When you read my work you will learn that the last thing gravity can be is that gravity is some force pulling objects closer! Try to tell that to a professor that says he or she is an expert on the work of Newton. They are experts on the work of a man that knew nothing about gravity except giving the idea behind a notion as gravity some grave connection and a ridiculous religious name. Try to get an answer from academics physics about gravity in detail or where does gravity originate or even about where the centre of the Universe is, and they stone wall you. Achieving that is more like trying to touch the moon. Talking about the Moon let's stay with the Mon for a thought. Tell your professor to tell you much is the moon moving closer to the Earth since $F = G\dfrac{M_1M_2}{r^2}$ is in effect, because such calculations and measurements are easy to measure.

With that information they can determine how long it will be until the Moon is part of the Earth. The Big thing is that the moon and the Earth is moving apart at the same rate as what human nails and hair grows. The Moon and the Earth is departing and not arriving at a point. The fact that the Mon should be moving closer and not be moving away is all part of that big scam and cover up that I write about in **www.SirNewtonsFraud.com.**

The best your professor could profess to be gravity is that they know that gravity is a force that works by magical powers pulling on whatever nobody can find. By merely putting gravity in the Universe that is acting as a mysterious FORCE that is pulling towards a common point in an allocated general centre is rather avoiding the question with simplicity because the question about how and why remains unanswered. Not knowing the answer will leave you empty and unfulfilled because of being a student and not knowing is the same as suicide n a mental level. Ask yourself the following: If gravity pulls towards a centre and gravity holds the Universe attached the question arising from that simplistic answer is then... where is the centre of the universe?

Should you decide to go to **www.SirNewtonsFraud.com** it will bring along a new perception about Kepler? Science sees to it that Kepler stays the least appreciated Cosmologist because **NEWTON'S FRAUD** destroys the entire substance of the work of Kepler where as in truth Kepler proved gravity, proved singularity, proved space-time, proved the Big Bang, proved every dynamic most of the wise persons that came up with all the various ideas afterwards thought about. However, even Einstein's special relativity theory was devised by Kepler beforehand! I can trace all of Einstein's work straight back to Kepler. Yet no one gave Kepler any recognition up to now because science denies Kepler his limelight.

Through my effort in investigating Kepler I came upon a mistake concerning physics.

This mistake is about the cosmic phenomena called gravity. Detecting the mistake is simple because it is uncomplicated to understand. One only has to look at the Big Bang concept and from that one can question Newton's idea that it is mass that is attracting material. To circumvent students detecting **NEWTON'S FRAUD** those cheats in physics came up with the biggest fraud scam any one ever invented. They still call it the Critical Density Theory where at first Einstein was called upon to calculate all the mass in the entire Universe but then found there was insufficient mass to pull the Universe into attraction. That again didn't support **NEWTON'S FRAUD** as Einstein was set at task to do so the cheats had to come up with more scandalous conniving. Then with this Critical Density of Einstein going to the dogs they found another way of trying to prove **NEWTON'S FRAUD** as being believable and correct while knowing Hubble proved the Universe is expanding and what Hubble found proved that **NEWTON** was committing **FRAUD**. They then had to find another way to cover **NEWTON'S FRAUD** but to do that they really extended all earlier criminal schemes by bettering all previous avenues of betrayal through which **NEWTON** committed **FRAUD** before. They started looking for so called dark matter, which according to their idea is undetected dark material that supposedly fills the cosmos while we are unable to see the dark matter. The matter is supposedly unseen but should be there and with enough of that the matter in the Universe the material forming mass will eventually start to pull the Universe into contraction. The question that shatters their deception is that if the matter is there dark or not dark, and mass attracts like they say it does, then what prevents the mass from energising the force of gravity in the present moment and through that the force must start to pull the Universe into contraction instead of expanding? What prevents the mass from committing the force of gravity if the mass is there? Why would the mass not produce gravity at present and why would the mass start to produce enough gravity to start to pull the Universe at a later stage as Newton said it has to do. Either the mass (dark or not dark makes no difference) is there and produces gravity to pull the Universe together or Newton is wrong with his contracting idea and the Universe is expanding like Edwin Hubble said it does. If the mass is there and is irrelevant now it will remain irrelevant through out time to come. Why is a fact such that the mass is not producing light preventing gravity from contracting the Universe? Why is this visibility being connected to the fact that it is not producing enough gravity to contract the Universe and then why will the mass at a later stage start doing so and when will the mass eventually going to start to do the job of contracting, because in this expanding Universe the mass, notwithstanding being there or not and notwithstanding being invisible or not is not, whichever way it is, it is not doing the job of pulling at present. With an expanding Universe as we have at present the mass can't contract as **NEWTON'S FRAUD** claims it does and that proves that mass doesn't attract at all. The Critical Density Theory is the biggest fraud committed by any group of persons and is put in place to cover **NEWTON'S FRAUD.**

I have found what gravity is. I have found what drives gravity and what the factor mass is. Be assured that It is not what Newton saw is taking place, but for my not supporting Newton I am rejected by Mainstream Physics. Mass is not producing gravity as Newton thought but mass rather is a method of restraining gravity where gravity is being blocked by a phenomenon Newton called mass. Mass prevents an object from moving and that prevention of movement is mass while the inclination to continue to move is what remains of gravity the object retains. But while the academics brainwash students into submission by telling them untruths and stupefying their brainpower with **NEWTON'S FRAUD** and in this way controlling student's learning thoughts, it is through this method of improvising the truth with **NEWTON'S FRAUD** that they get away with mind control. I am unable to crack their deception and introduce the truth because by the power vested in them they can prevent me from doing so. It now is the time that students start asking the correct questions and begins a process of unmasking those criminals that is teaching physics. Ask them to tell you students exactly how much did the Moon come closer to the Earth since the time of Kepler or even since the time of the Moon landing in 1969. Insist on them proving **NEWTON'S FRAUD**

$$F = G \ \frac{M_1 M_2}{r^2}$$ is correct and it is working by showing that the Moon is moving towards the Earth at any

rate. If they fail to tell you how much the Moon has come closer or at what speed is the Moon is coming closer or in the event when that they have to confirm that the Moon is moving away from the Earth, then you will know that they are lying through their gritted teeth and I am telling the truth. There is no contracting Universe and Newton presumption of mass attracting is one big farce those cheats use to

brainwash students into stupidity. Physics is based on $$F = G \ \frac{M_1 M_2}{r^2}$$ which is based on **NEWTON'S**

FRAUD which is complete and utter well placed fraud that is set in place by student brainwashing. Here is more of such deception students are lied about.

Academics in the science of physics say that a feather will fall with the same speed as what a large rock would fall, but the condition they connect this process too is that the fall of the feather and the hammer has to occur in an atmosphere filled with vacuum such as we find in place on the Moon. This gives the impression that such falling of a light object versus the falling of a heavy object requires a vacuum atmosphere and in that this atmospheric condition is completely alien to what we have on Earth. That is part of their cheating! What they never add is that the largest rock will fall at the same speed as a tennis ball and to achieve that there is no vacuum filled atmosphere required such as the Moon has. This can and this happens on Earth every time and thus the process of all things falling at an equal tempo is far from alien to the Earth. Why don't they say a man and a car will fall to the Earth at the same rate and then prove how mass is part of such a falling equation! Mass has nothing to do with any object falling because all things fall equal just as Galileo said and this applies notwithstanding size (or mass) differentiation also as Galileo said it applies! You can see on TV how cars, bags people and clothing fall precisely equal and that is the case notwithstanding that not one of the objects are even closely resembling an equality in mass! If mass created a force called gravity, then it must be true that objects could never fall equal because mass will create different forces in measure and every force will have the object fall to the Earth by the mass it weighs. ...And don't let them come out with the nonsense that there is a difference between mass and weight because they use that lie frequently and that is another part of the fraud science created to cover **NEWTON'S FRAUD.** If things fall equal but their mass is unequal then things do not fall by mass and then mass can't be responsible for gravity. Those professors hide behind inconsistencies to get you students confused while they are brainwashing you into submission to accept their betrayal of the truth.

That all objects fall even and fall precisely equal is an accepted principle according to Galileo and that was accepted as a principle in physics long before Newton thought of becoming as wise as he later on though he was. Newtonians never can see any difference between what Newton claimed happens when he claimed all thing fall because of mass producing the force of gravity by which all things supposedly falls and what Galileo claimed when he said all things fall equal thus ruling out any consideration of mass inequality interfering with a fall. To them it is the same when Newton said things fall by mass and Galileo said all things fall equal thus not by mass. To them these total contradicting statements are still the same thing and they have the task to brainwash you into believing Galileo and Newton said the same thing! It is their job not to teach you the truth but to force you to believe that Galileo ands Newton said the same thing. Newton said all things fall by mass. Galileo said mass has nothing to do with falling. Their job is to force you to believe this is the same idea. For the first time ever since the time Newton introduced gravity I seem to be the first and the only person that questions this interpretation.

Has anyone ever explained how the idea of a feather falling as fast as a hammer fits into the idea that mass pulls mass and how the falling by the gravity forms power that is exerted by mass as a hammer has much more mass than even what a large feather has. Test you Professors explaining ability and his skill to hide the truth from you when you ask him how mass that is the principle bringing absolute differences between objects moving influences all things to fall equal. In other words how does he console Newton's theory on differentiation with Galileo saying all things fall equal! How on Earth do these two concepts of a feather falling equal to a hammer or a car fall equal to a human, which every time it happens proves that Galileo was absolutely correct and Newton was absolutely incorrect fit into this interpretation they use of mass causing objects to fall?

How can a large mass pull as equal as a small mass pulls to travel equal covering an equal distance it descends to Earth at the same speed over the same distance and still be driven by the power of mass creating gravity. If it doesn't differ then there is no proof that mass is any factor in the falling because what does mass then bring into the equation. Have you given this idea a good thought? By me scrutinising this concept I disagree and by me disagreeing I am silenced by those academic frauds in power. Don't settle for any more brainwashing and mind control and start to insist on answers.

When any person disagrees with any academic in any lecture hall about mass not forming the factor that is responsible for pulling gravity and you come to a conclusion that you doubt the mass part that they bring into the picture they claim is responsible for establishing gravity, then the academics wipe you from the table with a swipe. To cover their crime they contemplate that you are so stupid because you are unable to "*understand*" Newton. By your "inability to *understand* Newton" you fail to see facts and you are too stupid to *understan*d physics. They even in some cases go on to say that physics is not for stupid people and only "*clever*" and "*informed*" persons would be able to "*understand*" Newton!

I have been at odds with academics for years and only because of the superior positions they hold in office are they able to bully me into submission and silence at the time. They hold the power of the reigns in their hands, but what they do not count on is that I can go to the public and bring my case and their fraud into the open. Academically I am not from their league and neither am I from their ranks and with me not being part of their ranks they form have they bluntly dismissed me. Because I am not part of their group they are of the opinion that it disqualifies me to have any opinion. They hold the opinion that only they filling academic positions are allowed to form any opinion and the rest is too stupid to have an opinion. In that way they were able to dismiss me for years, but no more…I will not tolerate their behaviour any longer. Now I fight back. They have this opinion about their positions that while being what they are that status they have gives them the rite that they may regard or disregard all opinions when they do not fancy other opinions. It is up to them to decide what the truth is for they control all judgement as to what is the truth! They may silence whatever I may say notwithstanding my correctness and validity. To summarise their attitude one requires three words which is arrogance, autocracy and megalomania. Absolute power corrupts and they are the living example of that.

Due to the important positions Academics hold in the huge academic institutions such castles of power gives them free sanctuary from where they can hide their criminal ploy of deceit. If you think I stretch the truth by accusing them of fraud, then please read www.SirNewtonsFraud.com and find out why I am blaming them of utter corruption. They sit so high and are so mighty that they do not need to explain anything but to themselves amongst themselves and their deeds go totally unchecked. That makes them be the untouchable and unapproachable powerful from where they rule with absolute authority. This unquestionable authority gives them the locations erect a cover and give them the opportunity to hide behind that wall of absolute superiority and suppress little persons such as I into silence and submission notwithstanding...Whatever I have to say can never go past their scrutiny and can never pass their sanctions.

Now I am taking my case to the members of the public so that the truth must be brought into the open. I have had the tour they give and then more came my way. I never got around swallowing the fact they claim about mass creating a pulling force Newton named gravity. I distance myself from witchcraft and soothsaying and their claims about the force pulling by mass is fiction. Let one prove that part where science is of the opinion that mass pulls as gravity is...after reading www.SirNewtonsFraud.com. Academics condemned my work and therefore me and for eight years where I could not get a publisher to come around and had not one publisher that bothered to read my work, let alone seriously proposing a publishing contract. There is no publisher that is willing to go against Newton and face the world of physics. The lack of support in the publishing world forced me into a corner where now I had to finally go private with the publishing as all doors shut in my face as soon as the academics read the content of my work because from the nature of my work I take Mainstream science head on and am confrontational on most aspects of astronomy. There does not seem to be any publisher that wants to go head bashing with the establishment of science on official science principles, which I have to do to convey my message in a no uncertain language. Fact is I know what gravity is. Fact is Newton doesn't know what gravity is and also admitted to that. Fact is that I saw Newton is miles off the road with his presumption about what gravity is. Fact is Academics in physics carry on as if Newton and they are the sole experts on gravity while Newton and they have no foggy clue about what gravity is...and no publisher is willing to agree that neither Newton nor academics have a foggy clue about gravity. If you also have doubts about the academic's indisputable correctness please read on and confront either them or me on everything you read here.

After reading www.SirNewtonsFraud.com you will have to take sides because you will know the truth.

Then you either have to become a partner in their crime when joining the academics in physics with the purpose to cover the truth from getting known or you will be part of the truth and become an activist fighting by helping me confronting those perpetuating to perpetrate in crime until the academics stop their criminal conduct and acknowledge the truth.

By not confronting the establishment, you give the establishment grounds to allure you into being sheepish. They declare facts and you sheepishly follow as so many did for centuries without questioning those in academic power. Because they see you, as being just another stupid senseless student they have the opinion that they can brainwash you into accepting these fallacies that I am about to tell you in **NEWTON'S FRAUD**. They do literally brainwash and condition your mind to and control your thinking in believing what is correct until you accept what they never yet were able to prove.

They are of the opinion you will swallow any of **NEWTON'S FRAUD** they throw your way just because every generation before you were mind controlled in the way they are about to control you. You may think this is big words but read **NEWTON'S FRAUD** and see after you come to know all the facts if I exaggerate even in the least. They see you as a slow-witted and mindless nobody with a mind to form any opinion because they think they are the academics being superior which is just other words to describe what is making you the inferior. If you are not aware of the facts beforehand and before they start to brainwash you by mind control they know you will follow their teaching without asking questions.

They think that your naivety makes you mindlessness and being vulnerable leaves you so stupid it will incapacitate your thinking ability which would lead you into their control. They don't want you to ask nosy questions about contradictions existing that they have to refuse to answer because there never was answers available since the time of Newton. This process of brainwashing and mind controlling in physics has been in progress for hundreds of years. Just answer how a small object such as a hammer and a large object such as a car would fall equally while mass forms the driving force that establishes gravity as a force. The fact that they fall together is images we see on TV everyday and is in modern society beyond doubt. If you can't explain how it is mass driving gravity that allows a car and a hammer to fall at the same speed while a hammer is so much inferior in mass to what mass a car has...well they can't either! Their task is not to explain but to mislead since they think you can't think while they think they know how to control you.

The following web page does not aim to represent the full entirety of the original book called **NEWTON'S FRAUD** but is reduced to aid any possible potential reader in the examining what the purpose is of the information this web page wish to announce. Anybody and everybody are aware that all objects fall at an equal rate. If an object such as a car weighing one ton falls at the same pace as a person weighing fifty kg while the two objects escort a hammer weighing one kilogram all the way to the ground, then how does mass come into the picture by committing a force to do the pulling in accordance with mass? Mass has to pull because according to their teaching it is mass that establishes gravity. However, mass is a factor that produces differentiation whereas all objects show equality during their fall. If it is mass that is establishing the force gravity, all objects must fall at different speeds. That they do not do as they all fall equal. That means physics is wrong from the start because mass cannot have any input in objects falling.

This is but one of many I mention in **NEWTON'S FRAUD** where all that I mention in **NEWTON'S FRAUD** are not the only untruths but forms facts that the Paternity called Mainstream Science is keeping concealed as a cover up that is wrapped under an airtight blanket of deception. If you sit in class and listen while also experiencing the sinking feeling that the facts you hear about what Newton claims are not adding to a total that you personally are comfortable with you are most probably correct because Newton is incorrect. If your professor makes claims about physics while you disagree with what is said then you better read **NEWTON'S FRAUD** because **NEWTON'S FRAUD** has it at task to show all that will read **NEWTON'S FRAUD** how much discrepancies academics lay on unsuspecting students that trust Academics with their future and their life. Do you as students realize the inconsistencies that physic Academics present you with when portraying that what they teach you as being the solemn truth.

Students tell your Professors to stop deceiving and tell them to stop trying to control your minds with their fraud. Those Academics tutoring you are telling facts about physics and about gravity that has never been proven.
That is mind control and giving selected information to have you form an opinion they wish to manufacture.

They wish for you to accept facts on physics and on gravity that they hold as the truth but in all of three hundred years were never proven…not even once. They claim those truths are beyond questioning yet with the least examining those truths they stand by then proves to be totally void of substance because it was never corroborated by one single experiment.

Should you question that mass produce gravity and insist on showing that $F = G \dfrac{M_1 M_2}{r^2}$ is not utterly flawed in principle, they will expel you from University by letting you fail your examinations and this will happen while the formula that Newton introduced $F = G \dfrac{M_1 M_2}{r^2}$ was never proven. They will expel you because they will claim you are unable to "understand" Newton and therefore lack the knowledge to present Newton accurately. Let them first show how $F = G \dfrac{M_1 M_2}{r^2}$ works when all the values of the symbols are processed and an actual value of the force is calculated by using strictly the mass of the Earth as well as the mass of any object divided by the radius between the object standing directly on the Earth as it is used in $F = G \dfrac{M_1 M_2}{r^2}$. But they will rather terminate you schooling and have you fail tests should you question their authority on the matter of gravity formed with the implementing of $F = G \dfrac{M_1 M_2}{r^2}$ while at the same time they can't for one second bring evidence in support of what they wish you to accept as the unquestionable truth.

That's brainwashing by mind control because if you don't accept their baseless fact as God given truths they dismiss your academic career.

It is either put up and shut up or be gone. Academics do put mind control to work on unsuspecting students by forcing students never to question the legality of statements they offer as being sound and correct.

Module Five
The Comet's Gravitational Demise

In the book the author, which is I explains gravity. This achievement is possible because I saw a way to break away from invalid concepts Mainstream physics hold. I recognised the impossible double standards Mainstream physics apply to promote their much shady explaining. The inconsistencies brought them double vision and to compensate their incredible theories they simplify issues to a level where what they embark on to understand is becoming meaningless. What they say can't be supported and authenticated any investigation even in simple terms. It is as if they never read with interest that they explain and they never scrutinise that which they advocate. They give values that are senseless and make that which they say meaningless.

In this article I am going to investigate how much truth there are in mass pulling by the force of gravity. To most if not to all of the persons reading this and just the thought about me embarking on the investigating of the issue is totally senseless to investigate. It is senseless because the concept it carries became accepted as household practise and life science.

Do you think of astrophysics as the department that is run by the wise and the level minded the sober thinking and the absolute trustworthy? If you are a student there is no other choice you have. If you think those in charge of astrophysics are the pillars of trust, then get wise and read the following. What you are about to read is simply mystifyingly simple and yet to this day I have not challenged one academic any where that had the honesty to admit to the fact of Newton being wrong. After you have considered the following you might agree with me that even small Children can reach a higher level of clear-minded logic and find more sensibility than what those scientists promoting astrophysics have because science lives in a make believe fool's paradise. If you are a student, then ask your Educated Masters please to explain the following abnormalities and inconsistencies they promote as part of official physics, which I present in this web site and get wise instead of brainwashed. I say brainwash again because they force-feed you fabrications, as you will come to see. They can't explain the facts as the facts but hide the fact that the facts are in fact untruths. Tell them to prove that planets have mass. Tell them to prove that it is mass that generates gravity that pulls the planets. Ask them to explain gravity in detail.
The idea of proof comes automatically to the door of Newton although Newtonians will deny this fact as if they deny the honour of their Master Newton and that is what they have to do.

Think of what planets do… and you think that planets orbit. It is connected to the brain. No one thinks of planets spinning or planets basking in the summer Sun. When hearing about planets the first thing that comes to mind is the rotating of planets while circling around the Sun. However, just using the term orbiting id in total defiance with Newton! Newton said gravity draws or pulls or moves in the direction,…which would have one understand that the two objects in example the Sun and any of the various planets will be moving directly towards each other. The term pulling does not suggest any circling because no one can be pulling towards and do that while circling. When pulling anything it must take place while using the shortest line possible. That serves the term pulling. Then the saying goes that planets orbit indicating they follow a circle. That is not what Newton said. However, wrong that may seem but circling is precisely what planets are doing.

In conversation we speak of the planets orbiting. If Newton was correct we should be speaking of the planets pulling, but talking about pulling would be blatantly wrong according to the normal spoken word. Never do we refer to the planets pulling the Sun or the Sun pulling the planets, but we speak of seasons coming from orbital positions. Being in orbit has to neutralise the pulling and then cancel the pulling concept that also became culture.

If there was a pulling, and the word orbit cancels such an idea, then there has to be some sort of prevention taking place that disallows the pulling to commit the direction of travel. I know it is said that the orbiting object falls as fast as it circles and by falling while moving to the following side on position it never reaches the Sun, and yes, it makes sense, but there has to be some form of resistance replacing the planet in the next side position and preventing the falling or the pulling from taking place. Using the

formula $F = G \dfrac{M_1 M_2}{r^2}$ as Newton provided, disallows any other concept other than moving towards.

The person Newton got his ideas from and the work he raped completely, that of Johannes Kepler explained this very well, but Johannes Kepler makes no room for any pulling of any sort. In the work of Johannes Kepler he said that the space being the orbiting route a^3 remains at a specific distance k while the orbit T^2 takes place…and in all my other books that addresses more information I take Newton to task on his dismembering of Kepler's formula by corrupting Kepler's work and with what amounts to fraud, Newton takes science on a goose chase that holds no truth. There is no pulling by mass of mass in any way.

In the book named "*an Open Letter on Gravity*" I bring the solution to the mystery behind gravity. I tried in vane to introduce the principles I find valid to the academics in charge of astrophysics. Facts that Science present as being the uttermost explicit and unwavering truth, fails to bring any logic answers to so many questions that it should address. It fails to have substance in addressing the most basic and simple questions about gravity and physics. Yet to every question science can't answer my approach does bring many solutions. The presentation and the delivery of my answers that I reach are understandable and simple where it serves both logical science and the truth. Since my answers do not match Newton and his misconception about gravity and that mass generates gravity, those in charge of science don't even bother to read my work. With their affixation to the corruption they portrait I can do little to the giants where they are in the mighty positions they have and just because of that they can go about to sideline and ignore my work and this is notwithstanding the correctness that my work delivers compared to the utter failing that Newton's work shows. When confronted with my evidence and they have to match my work with the hypocrisy and misleading nature of Newtonian cosmology their defence in substantiating their claims is to ignore me. Since I do not applaud mainstream science and the clear fraud they embrace and fraud it is that they embrace, I am silenced. Why is it that my work is going unrecognised or even in the least goes never debated and never commented on…it is because it will then trash every article anyone has ever written about astrophysics and cosmology. They show little integrity when academics with high standing such as they should have will rather protect fraud and save their skins than seek the truth. It is that when they begin reading my work they then have to back my work. Doing that will trash all work in cosmology delivered thus far and condemn it to the waste paper basket and render all work invalid and void. It will put all the Newtonian's bias and fraud into the place where it belongs. Considering that such acting will lose them money, those academics in controlling positions then will rather misrepresent the truth in order to benefit from continuing to corrupt student's minds further. If they wish to justify their inconstancies they have to attack my work and disprove the accuracy of my work. That they can't do. They then ignore my work because they can't attack my work. In that sense they also place their work beyond my approach, as they can simply ignore me as if I represent the plague while they carry on with little consequence to bother them. I challenge them to prove Newton correct and not just declare Newton being beyond reproach after all has seen the evidence I bring. After reading this all students must challenge them to defend what they can't or get honest.

This is the basis that Mainstream science uses as the foundation of all physics anywhere. If this is wrong then everything they have got to work with goes out the window. They put mass and the distance that parts objects in a relevancy, in other words the one is a ratio to the other. The one factor brings a measure to the other factor's value. The one cannot be without the other. The increase in one becomes the reducing of the other and the other way round also applies. When the distance is large, the influence of mass will be small and when the distance is small, the influence of mass will be overwhelming. Then they state we are in a Big Bang expanding of the entirety. Why then, when considering that if it is mass that produces an inclining force of contraction as Newton says there is going on then…why didn't the expanding stop before it started when the Universe was small. Today using hindsight after the fact of the exploding Universe became apparent by the studies Hubble brought to light did the lot of everything that is not implode as Newton would have us believe whereas, instead it did expand just as Hubble proved. The radius at the time of the first instant back then was no factor, which makes the gravity at the time a totality of unrivalled force. The radius being that insignificant leaves the mass unchallenged in asserting power in relation to the non-existing radius it had.

When the Universe was at the point where the Big Bang started, the radius was incredibly small. According to their studies it shows that the Universe was the size of a neutron. Having the size of a neutron proves my statement and it is science telling us that the Universe was the size of a neutron at that stage. That would make the mass that was at the event of the Big Bang that was producing gravity and the gravity charged a force by contraction, which then had to prove to be inconceivably large. This is because the mass was completely overpowering all factors with the small radius. But as Hubble proved this did not result in an implosion that drew all that there was into something even smaller. That is what the overbearing mass contraction was supposed to unleash on such a small Universe in the beginning. This understanding and accepting of the most basic brings us to the next inconsistency as far as I can gauge the situation. Maybe Newtonians are correct. Maybe I am too stupid to understand Newton but see how you fare.

When the most basic and simple mathematical law prevails it is clear that a small radius will bring about an incredible powerful gravity since the power that the mass establishes will then be overbearingly strong. The more the radius develops in time, the lesser would the gravity, be that the mass factor generates in relation to the advancing radii developing and the larger would the reducing be of all contraction. The effectiveness of the force that the mass are able to produce will tarnish as the radius that separates the material from each other increases as time moves on to create space.

Although it is presumed that the Universe was small at the dawn of the Big Bang, such presumption will bring validity to another presumption. The presumption is that if it is that produce the gravity then mass at the time of the Big Bang just had to be enormous. The influence of the incredibly small distance in radii and the factor such distance produced increases the influence that the mass factor can assert in relevancy by an exponentially large number, which elevated the mass to an enormous large factor.

An electromagnetic charged magnet typically demonstrates this example. If one holds a magnet far from an iron you would hardly feel the drawing of the magnet and the iron. When the magnet is a very short distance from the iron it draws so furiously the contact can't be prevented as it clings onto the iron notwithstanding the human effort to keep the magnet apart from the iron. The magnet didn't get stronger and the iron did not pull harder. The close proximity favoured the magnetic fields exponentially when the magnet was almost touching. This was then also true and what applied to the magnet also had to apply to the mass that produced the gravity that caused the pulling of gravity.

This same issue becomes true if an object is a distance apart; the radius is exponentially less influential in value, which then is by dividing into the mass, a factor that increases the influence that mass must have on the gravity pulling force. As the radius increases such growing by the distance will reduces the influence that the mass can produce by one, a ratio equal to the growing distance. Then the same force drawing would in that event be much less than when the objects in question are only a short distance apart. That is the most basic realisation about mathematics. It puts ratio to order and define coherency. That is what gravity is to the Universe as it puts respect to factors about the Universe in the Universe. It is what derives order in the Universe of mathematics.

We will find in a Universe that was so small where it had a radius of less than one kilometre, then at such a time with the Universe being that small it then must also be accepted that the gravity the mass charged was massive. Einstein said that where space is zero the gravity is ultimate. The extremely small radius that was only the size of one neutron in radii distance and with the factor that such a distance produces, it then in that case must have promoted the mass factor, which will support the gravity in having an enormous large factor by relevancy to what the case must be at present. The mass factor that produces the gravity at any given point during the event of the Big Bang, had to be eternally larger at the dawn of the Big Bang while having at the time almost an infinite radius, which gave gravity all the power it could ever have and which it will ever have.

We can see when only looking around us that in the event at the time when the Big Bang began no material growth could come in place. The with not sufficient mass to destroy the radius and prevent the expanding from coming about, immediately at that given instant of moving apart the expanding won the dual because the mass then in future will never become sufficient to launch a Universe of contraction as Newton stated. The contraction afterwards could never match the expanding that took place. If the Universe is expanding as the Big Bang concept promotes then there can be no contracting Universe as Newton had us to believe. If the Universe started a journey of parting objects by allowing the distance between the objects to grow then no amount of dark matter that might lurk in the night sky and is at this moment hiding from detection will afterwards produce the gravity required to stop the expanding from continuing. The moment the mass is not able by quantity to support and sustain a contracting Universe the expanding will carry on because the mass is losing influence as the radius increases and that mass then by ratio produces less gravity in force or influence. The declining of the influence the force experiences is directly in ratio with an increasing radius. At the start when the expanding became evident and as the radii grows the inclination of producing a contracting Universe will suspend as the balance progressively shifts in favour to decrease the influence of gravity as a factor. By the third microsecond after the direction was decided there was no chance any longer that the mass would ever find a manner to contract the Universe by enforcing gravity as a pulling power. If there was insufficient mass at the start in order to tilt the balance in favour of the reducing factor, no amount of mass can ever accomplish to turn about the direction flow afterwards. There cannot be a growth in mass because what is in the Universe cannot escape and what is not in the Universe does no exist. The Universe is the container that contains whatever possible things there are to contain and no adding or no releasing is in place. Then Newton's surmising was corruption, which is making that which all physics are based on a fool's idea that can only find proof if evidence is corrupted.

If you might be of the opinion that my accusing the greatest intellectual department in the world as being in misconduct and to your view such accusing is outrageous and far-fetched, then please give me the honour in being my guest and judge the following evidence that I bring with a clear and unbiased mind because when scrutinised with a clear view then the facts cannot fool an idiot. However, that is just what the physics paternity thinks we are. Academics regard us being the rest and being those that fill the sector that are not academics as the small-minded. We are to their minds therefore those that are the idiots they can manipulate. We being the part that is forming the general public at large are too incompetent to see the deception in their theories. If they don't regard us in dismay then why feed us all the rubbish they do and which they do for such a long time? Reading from their actions we can clearly see that they have the opinion that they can feed us being the everyday tax payer, us in the public arena any senseless rotten garbage they dish up because they see us as being inferior by thought and mind. Why else would they promote such shambles? It is either a case where they think of us as incompetent idiots with pea sized brains and which are thoughtless or they are the mindless incompetent idiots that truly believe their incoherent nonsense because what they promote is senseless rotten muck and hogwash.

With all this in mind did any one ever come to wonder about the reality driving the all too famous Einstein's Critical density theory and the fact that this idea was conceived to conceal the corruption of Newton in physics? Allow me please to elaborate and then make up your minds. The facts in truth are that the Einstein's critical density theory was a scheme plotted by those in charge to cover up and conceal corruption in the heart of physics. Hubble saw an inflating Universe that contradicted Newton's deflating Universe and this perception of a deflating Universe being a myth had to be most ardently hidden as to yet again compromise the truth about Newton and his theory. One minute ago I showed how the relevancy applies between the strength of the mass and the ever-increasing radius that keeps on diminishing the influence of gravity as long as the radius is increasing and the distance between objects are growing. This is so basic that primary children learning the basics of mathematics will understand! Yet Einstein proceeded in searching when the mass would bring a turn about in the direction that the cosmos is involved in. Einstein was looking for the moment the mass will become strong enough while the most basic principle indicates that an increasing radius leads to a decreasing gravity since the mass becomes less prominent. If Einstein was unable to recognise this most basic of mathematical principles, then what type of genius did physics create in him and what slur did physics promote. This idea of the two factors being in opposing relevance is so simple that children will recognise the principle, and yet those fathers of physics wants me to believe that the greatest mathematician that ever lived did not realise this principle…the principle that the radius and the mass stands related and the growth in the one will promote the decline in the other as a dominant factor. Can any one with this information including the information given on the previous page have any other conclusion than smelling rotting fish somewhere? It is obviously clear that having such a total idea that there might be dark unseen mass floating in the Universe which at this time does not generate gravity but will some day kick in to generate gravity in order to cover up Newton's deception about a contracting Universe and just because Newton has to be correct at some point in the future. Science wishes me to believe that since there is a lack of seeing material there then will be dark and unseen material where they are so dark they are undetected by all humans. That leaves another question to address…where can such mass be found. How will such undetectable mass be found, which will bring about contraction after all this expanding ends and the Universe recognises the infallibility of Newton once again? Why would the mass at present then not activate gravity and why would the mass at some point spring to life and start activating gravity? How much can the Physics paternity still hide the fact that **Einstein's critical density** is being used as a **cover-up to distort the truth** to **conceal the fraud Newtonians wish to cover**? The **uncovering** of the **Newton's fraud** by the Hubble constant is so simple to see. Hubble found the Universe is expanding and Newton said otherwise. Who is lying about what? Hubble's declaration was on track to blow the cover that was concealing the Newton fraud wide open and uncover the century old deception. To see this we have only too look at the comet behaviour when any and all comets again come around on a cycle by repeated visiting the sun. The question is if it is mass pulling mass onto mass, then why do we have comets left in the solar system? The mass of the Sun should by now at least have destroyed every comet going around.

Every one believes in Newton except comets, because comets fail to collide with the Sun.
However I can explain in some way…

Every indication that we so far received in vivid portraying from astronomy photography studies from outer space disputes a shrinking Universe concept. From the Moon increasing the radius distance between the Earth and the Sun, to the Hubble Constant indicating there is a space growing all over and any where in space wherever man may conduct studies. Since the end of the middle ages a force called gravity was identified, but further than that science did not take it. What is gravity, besides being a force? What forces the force? I introduce a cosmic theory that turns the missing questions to answers.

Let us for one second return to the science we all know.

$$F = G\ \frac{M_1 M_2}{r^2}$$ There is an undefined phenomenon in the cosmos, never mentioned (in public) because it obscures reality but is proven in using this foundation of science, the basic formula used to prove all science.
Let's put the mathematical formula into a practical context.

By reducing r would bring about the same result as enlarging the mass factor of the cosmic objects i.e. the Sun and the planets. It is a very drastic implication that will cause much more than just seasons changing. It must bring about that gravity changes through out the year…yet the radius does constantly change therefore… it is evident that Kepler's name was used when science introduced a formula as follows

$$E - e \sin E = M$$

That proves everything but that mass is responsible for gravity

$$F = G \frac{M_1 M_2}{r^2}$$ In relation to the next few arguments it is very critical to understand the arguments I present because on the soundness that these arguments represent in the arguments that I make I am in dispute with the arguments that Newton makes.

The entire philosophy of the science of physics rests $F = G \frac{M_1 M_2}{r^2}$ on the arguments that Newton makes. It is physics that base everything on the fact that mass produces gravity and therefore by the force that mass provides as gravity the entirety of all physics are founded. I do not dispute mass as a factor in physics but what I do dispute is the way Newton presented the fact that mass has any value between cosmic bodies.

The formula as Newton presented it is just not baking the beans any more and must be re-examined because there is a lot of contention in this statement.

Please follow my line of thought and scrutinise all that I say. That which I touch has resting on it the entire philosophy of astrophysics as well as physics.

What I dispute is that it is mass that is in control of the cosmos and that mass provide the sticky substance called gravity.

Also do I dispute that gravity is a contracting force by which the entire Universe is in collapse. However I begin by disputing the idea that it is mass that is producing the gravity that supposedly produces contraction that is there to pull planets towards the Sun.

Newton insisted that mass is the influence under which gravity becomes a force. He promoted the idea that gravity is a force instituted by the measure of mass because the mass unleashes the force of gravity on other unsuspecting objects and then begin to pull the unsuspecting objects in, in the same manner as what anglers will do to fish. Once the mass gets hold of mass the mass reels in the mass from both ends that carries mass and according to Newton everything in the Universe carries mass. We know the strategy behind gravity is to first get hold of a body containing mass. Then it natural to presume that everything in sight is pulling on everything in sight and the bigger the mass is the stronger the pulling is. Jupiter is pulling much harder than what Mercury is pulling because mercury is much less massive than Jupiter, which is more massive.

Another thing we cannot miss about gravity pulling objects around by the measure of the mass is that gravity pulls mass towards the centre of the other mass conducting the pulling by gravity. We therefore know that all objects are heading towards any object's centre and the pulling of going directly to the centre of the other object pulling by the force produced by the mass the body has. I ask the reader politely to control my facts since so many academics told me with much sympathy in their attitude that I just don't have the insight to understand Newton. Somehow it is suggested that Newton is far too difficult to be understood by a person with my meagre intellectual capacity such as I apparently posses and because I was somehow denied by God to inherit a strong brainpower with a capacity to crack the limits of the Universe in a Newtonian fashion I ask you please to check that I do follow these extremely complicated facts correctly. Those academics I confronted with the issue in hand are of the opinion that it as a God given fact that I am too simple in mind and comprehension about my surroundings to understand Newton and when taken into account that they are capable of understanding Newton due to the capacity they have in using the brilliance of thinking with mind clarity they can hardly be wrong about anything.

$$F = G \frac{M_1 M_2}{r^2}$$ Every time I confronted an Academic in the physics department I was told in a very polite and sympathetic manner but also in a very unmistakably fatherly and firm way when they share their opinion with a lesser being such as I am in their opinion is what they can see about my intellectual status I am a lesser-blessed individual. There are some people that are intellectually advanced and such

persons have a born ability to understand Newton. Then there are those with much less potential and with much reduced thinking ability and in the case of those or should I say we there is not enough grey matter in our skull filling the vacuum between our ears to follow Newton. I therefore have to accept I was born with much less capabilities and then with that much reduced mental capacity I had to accept my fait which is that I shall never have the ability to understand Newton. With my meagre intellect I do try my best but where it comes to the obvious then it becomes hard to follow Newton when it is Newton that apparently lost track of his senses. Be as it may please inform me where I lose track of reality and where Newton loses track with reality. They (our Super Professors in Physics) say I have no ability to grasp how mass entices gravity by mass that is producing gravity as a force forcing gravity to pull over a distance to the measure of r.

They must be of the opinion that I do not understand mass. Well I am not sure that they understand mass because there are as many definitions explaining mass as there are opinions about what mass is. It also could be that they presume I do not understand the gravitational constant through which the mass must move to reduce the distance between the objects. In that case they don't know much about it either because not one of them could up to now show me where the centre of the Universe is and where it is that that gravitational constant is pulling all the mass. To have gravity pulling what there is to be pulled, there also has to be a centre to which all are pulling all there is to pull. It could be that they suspect I don't know what a force is…I do know that…it is that which they burned witches for because they said witches and ghost are or have forces. Then finally they might suspect I don't know what a distance is that separates objects and there too I think I know what a distance is that separates objects. Other than what I just mentioned I truly don't understand what they say I don't understand about Newton. That means they are correct and that my mind is so weak I don't even understand what I don't understand about the complex issues they seem to understand.

This is what I do understand but apparently it is not enough to understand what I think I do understand. There is a force that puts relevance between the amount of mass and the strength of the force produced over the distance the mass produce with such a force.

If the distance increases, the influence of the mass coming across the increased distance will decrease. When the distance decreases the influence coming across the decreased distance will increase as the distance decreases. In that the force of gravity that the mass of the comet and the Sun produces will increase by the square of the distance that is diminishing due to the shrinking of the distance parting the objects. I suppose there must be more because that which I understand and I just shared with you are not enough to prove that I understand Newton!

Mass produces gravity in accordance with the radius distance between the two bodies. This is where my mind gets too weak to understand Newton. Looking at all orbiting objects there are always a wide part and a narrower part in the orbit where the planet in orbit seems to be closer than on the other side (E- e sin E = M) or on the orbiting structure side further away from the centre which was the position that the Sun claims. This does not quite fit the picture of a constant and never changing mass that all structures supposedly have. If mass is responsible for gravity why there is this flexing wobble in the radius. The lot should be rather constant because the mass is constant. But that is my weak mind as that is what I don't understand about what Newton understands…and with my weak mind I can get no one that understands that.

We know that the Earth has no mass increasing and decreasing and neither does the Sun have mass adding and then removing of mass. In that case when considering the practical mathematical implication the radius has to be at a constant with no mass fluctuating on either side of the factors.

We know from personal experience living on the place all our lives the Earth doesn't change mass as the year progress. From that as well as the evidence we know about all other things in the Universe with mass that the mass of all things are a constant and doesn't fluctuate. So the gravity that the mass of the Sun and the mass of the planets would generate should allow for a pretty round r to be in place that will keep the planets circling evenly. That is not the case because we know there is a fluctuating especially in the case of the comet orbit. We then have the task to find what would encourage the deviation.

$$\frac{M_1 \, M}{r^2} \, G$$ Or on the other hand

$$F = G \, \frac{M_1 M_2}{r^2}$$ It is my understanding when considering the implications about the relevancy there is between mass and distance parting mass when used as it is in the Newton formula that if the mass is at a constant then the radius too must remain at a constant.

When the object having mass asserted the gravity the mass employs and the radius parting the objects are large, the objects in mass asserting gravity over such a large distance must be small. If the distance is large then in that case the distance will reduce the force that the mass can produce to a trickle. The force there is between Mercury and the Sun must therefore be ten times weaker than the force that there is between Pluto and the Sun. It is not only that Pluto and Mercury has about the same mass and therefore they should peddle around the Sun equally but since Pluto 39.44 x AU or (5900139992.8 km) is 101886 times further than what Mercury 0.387 x AU or (57909 km) is if taken that one astronomical unit is the distance from the Sun to the Earth and that is measured in AU = 1 or in km 149 597 870 km. That means the force of gravity must allow Mercury to go 101886 faster than Pluto's 101886 slower. That is not happening. I surely don't understand Newton but that is also surely not because I am too stupid to understand Newton.

The gravity one must find between Pluto and the Sun is then concededly less that the gravity there is between mercury and the Sun. Mercury is so close that one might form an opinion that the gravity Mercury generates must be in the vicinity of one of the outer gas planets. The gravity Mercury has with mercury being so close would have Mercury spin at a rate that only Jupiter can mange since Jupiter has the enormous mass in its favour. No other planet should come near to the speed that Mercury and Jupiter have when going into orbit around the Sun. Jupiter has the enormous mass in its favour giving it momentum around the sun a boost and Mercury is so close that the mass it has is many times bigger than the mass it should have just because it is so close to the Sun. Then Pluto must be at a snail pace and hardly moving in consideration of everything the gravity applying to Pluto has to endure because it is a hundred times further from the Sun than Mercury is while it has the same mass as that which mercury has. Sorry I forgot that all planets move at an equal pace. I wonder if it is because they share the same mass or are they possibly at the same distance while in orbit around the Sun because there is no mass indication of any of them speeding up or being slower than the other. They all orbit at the very same and equal pace.

Now we get to comets and their bad behaviour. When taking this distance effect further as it affects something as small as a comet, the gravity increase on the comet has to be devastating to the comet. The force that can hardly move the comet out where the comet is hiding from the Sun must be billions upon billions of times more when the comet is closely approaching the Sun.

With all that in mind we now draw our attention to the comet and the way gravity pulls the comet.

When the Sun gets hold of the comet the comet is miles away and the miles stretches for miles without end. The comet is where the Sun can't shine. Think of the considerable small mass that the comet has and the enormous distance there is at first when the Sun gets its gravity onto the comet and start to drag the comet closer. The distance being that large must seem to make the mass incredibly small.

Yet with the massive mass the Sun has in its favour it not only gets hold of the comet but it drags the comet all the way through all the space and ever closer to the Sun.

Gravity the **force that mass supposedly provides is a hoax, which keeps the world believing in science and with which science holds the world at ransom for the past three hundred years.**

You might decide to ignore these facts but then you would remain part of the last bastion of the dark Middle Ages, or you may read with concentration and become part of the future.

Gravity is based on **Newton's presumptions and as clearly can be seen not facts**, which are **all incorrect. My work** is **based on** the findings of **Galileo and Kepler,** and **their findings** are very **inconceivable** with the **findings of Newton. Newtonians forever tell the cosmos what Newton orders the cosmos to do and never takes into consideration what it is the cosmos does.**

When an object swings around another object, the distance between the two bodies divides the force that exist between the two bodies. As this force is mutual, coming from both bodies together, the distance is calculated as if it is two; therefore this distance is multiplied by its own value. The bigger this distance is, the less force will be excerpted between the two bodies and the shorter the distance is, and the stronger the force will be. With mass forming the force, the only fluctuation in the distance must be as a result of the mass on either side growing or diminishing at certain stages of the orbit. That can't be the case.

Even in the event, where this force applies an even distance, the two structures will be evenly apart through out the year. This ultimately will lead to the perfect day, the perfect seasons and the perfect year (for some people). We may start by determining the influence of gravity on planets as we find them in the solar system. **First, let us concern ourselves with a comet**.

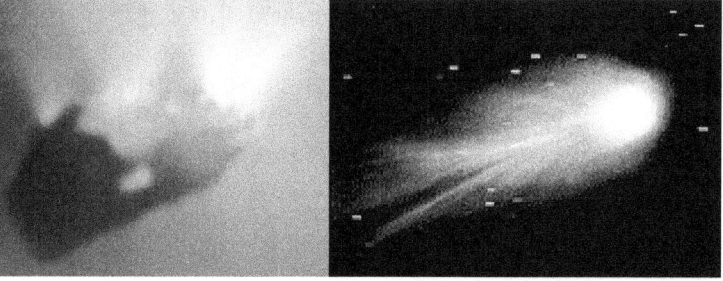

Gravity pulls of the sun on the comet It is common knowledge how the comet stands related to the sun's gravity. **Firstly, picture the comet at its farthest point, away from the sun.** Comet at its most furthest point from the sun

The **gravity** of the **sun pulls** the **comet straight towards the sun**, this we all know. Gravity always pulls an **object directly towards** the **centre of a cosmic body**: that too is common knowledge. Therefore, the comet is drawn directly towards the centre of the sun and throughout its journey the comet is picking up momentum directly related to the gravity that is centred in the middle of the sun, (**gravity is always**

centred in the middle of a cosmic body). As the comet is increasing its speed, the comet comes closer to the sun and therefore the sun's gravity pull is simultaneously increasing as the distance between the two cosmic bodies is reducing. Each instance the comet is drawn towards the sun, the gravity that the sun applies to the comet becomes larger progressively.

When the comet is at its **closest point to the sun**, **something odd happens which cannot be explained by Newton's gravity at all! Remember gravity should now be at its strongest point because of the proximity of the two objects.**

1.   The comet remains at an even distance encircling the sun.

2.   No longer does the gravity of the sun pull the comet towards the centre of the sun.

3.   At this very point the gravity that the sun applies on the comet does not pull the comet towards the centre of the sun any longer, in fact, it seems as if the effect of the gravity has been neutralized.

4.   The comet stays at an even space from the sun as it goes around to complete a half circle's orbit around the sun. It only completes a part of its rotation around the sun.

5.   After this, an even more peculiar event takes place. **The sun, at the point where gravity should be at its most dominant, suddenly loses its complete grip on the comet.**

6.   The comet brakes free from the sun's pull of gravity and speeds off towards its destiny into the vastness of the cosmic space, undeterred by the gravity of the sun.

As I said we now come to the difficult part where very few people understand Newton and apparently according to those in the know how I am one of those being incapable of making sense of Newton. Let's try again and see if I can manage to get it correct this time. The Sun has mass. The mass covers the distance between the comet and the Sun by the mass that produces gravity and the gravity is pulling the comet. Then the comet also has mass and that mass produces gravity that pulls the Sun over the distance between them. Because the distance is coming from both ways the distance is by the square where there it is including the gravitational constant and this lot is in a product to one another that gang up to destroy the distance parting this lot.

There is normally a diverting from the centre or the rotation axis, which the Sun forms by the orbiting planet but this, is much more so applying in the case of the comet ($E - e \sin E = M$). Considering the weakness of the force that initially reaches the comet as it drags the comet from the depth of total darkness where the comet hides in places the Sun does not shine it is remarkable that the comet does apply the gravity that will pull it on route to the Sun. It just comes to prove how correct a genius such as Newton is! He saw the mass that creates the gravity and the little comet stood no chance hiding from this domineering Sun with all the mass it can display

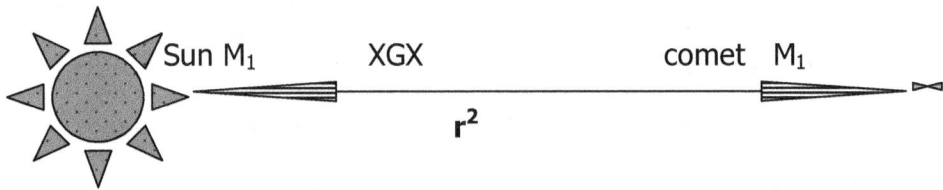

From the coldest and the deepest of space where not even the Newtonian imagination can reach, the Sun finds the gravity strength to draw the comet through the deep of space. One would imagine that if it was or is not for the shear size of the massive mass of the Sun, the Sun would not be able to generate a force so strong it could get hold of such a tiny piece but it does accomplish the gravity to bring the comet from where it is lurking in the depth of space and drag it towards the centre of the Sun. But with the resilience that only the Sun can muster as it is the only object that has the mass in ability to form the gravity that can extend all the way even to where the comet is lurking, such dragging would have been hopelessly inadequate. But the gravity that the mighty Sun manages with that enormous mass, the comet has little chance in defending its position. The comet is heading towards the centre of the Sun and towards its final doom since with such gravity coming from such a mass the comet has no chance to defend its position. If any one ever invented an unfair and totally bias fight then it is this fight between the small comet and the enormous Sun.

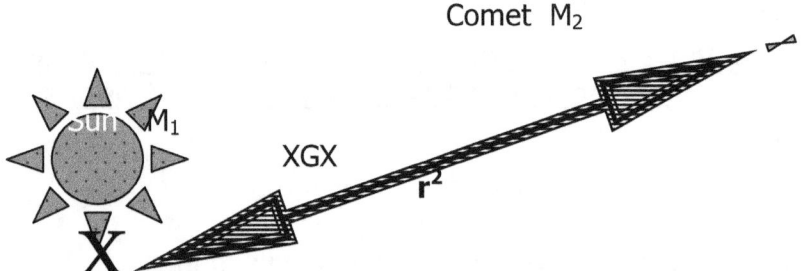

Alas it seems that the bookies once again had a say in the outcome of the fight because it seems they rigged the outcome or had some intervention about the outcome…well it is either that or the unthinkable has happened, Newton is mistaken…Newton is incorrect. The comet strayed from its pulling and the Sun lost its grip on the pulling. The mass of the Sun proved insufficient to grip a small little comet by force of gravity and pull it to the centre of the Sun.

Then after a pre-determinate and pre-calculated time the sun starts applying its gravity on the comet once more. At a point where the comet is at its farthest point, the gravity of the sun becomes strong enough to bring about a complete turn around to the comet's direction of travel. **However, the gravity between the sun and the comet is at this point, at its weakest point of influence.**

So, when the sun's gravity is at its strongest, the comet manages to brake loose and neutralize the sun's gravity pull in order to avoid its fatal collision with the sun and when the sun's gravity is at its weakest, the comet cannot escape the pull of gravity. There is definitely something very wrong, either with the comet's behaviour or the laws made up by Newton.

$$\frac{M_1 M\, G}{r^2}$$

The relevancy of the big radius by the square reduces the influence that the mass supposedly brings about. The further the objects are form each other the smaller will the gravity be that the mass can inflict A long distance has a small mass that produces weak gravity.

The comet aims at spot marked X and misses the Sun centre where all the gravity is concentrated. The mass of the Sun had the comet when the mass was small due to the enormous distance there was between the two but as the mass grew in potency with the aid of the shrinking distance and with it the force of gravity that the mass produced becoming ever increasing in the presence of the declining of the radius parting the two objects in mass, the force was unable to direct the path of the oncoming comet to the centre of gravity within the Sun. There is an unexplained diverting of direction, which Newton's brilliance never foresaw… that is if Newton's brilliance did foresee anything at all…and it goes back to where it came from and that oversight blows Newton into a barrel of shit.

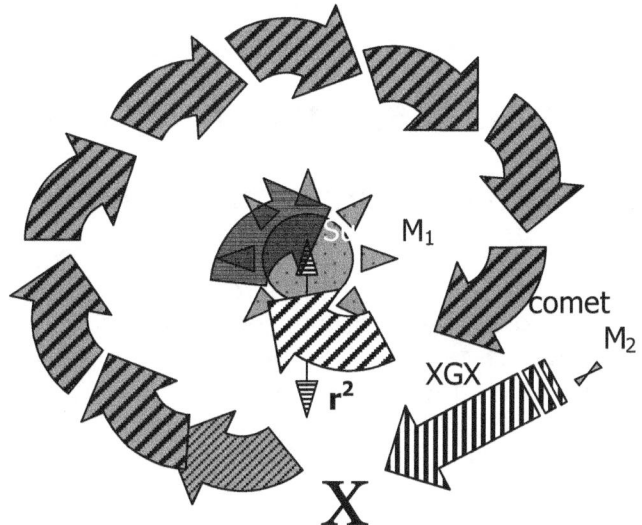

It started when the radius was big which made the mass be small that produced an insignificant force in gravity. That pulled the comet out of the shadows and into the Sunlight. The force was strong enough to grip the comet and pull the comet towards the centre of the Sun where the strongest gravity mounts an attack. When the force of gravity grew the influence the force made started to tarnish because it allowed the comet to aim for spot X, which is way outside the centre of the Sun.

$$\frac{M_1 M}{G}$$

Somehow with an ever-increasing gravity, the comet got brave and fighting fit. In the presence of overwhelming mathematical evidence to prove the contrary and in spite of mathematical laws and principles applying nevertheless the comet prevailed and got the better of the Sun. It missed the centre of an ever-increasing gravity force notwithstanding. This is not the end...it is the beginning of Newton's problems. Up to this point it seems as if Newton's was a profit sent by God to foretell us other far less important mortals about the future and about the future of the comet. The Sun had its mass providing a force called gravity and the force called gravity was going to take the comet to its eternal grave, as gravity will bring the comet its final demise. Alas the comet has other ideas because...

At this point Newton makes complete fools of his followers, the mindless sheep that follow his word through centuries without ever thinking about anything of his word that they come to follow. This is where the brainwashing kicks in. This is why only the mathematical minded excel in physics. It is those that cannot think that become the Brainy Bunch in physics. Those that are the Super-Educated-Masters-In-Physics are Master in the art of not thinking but through an excellent brainwashing that is the best tried and tested system for centuries those that underwent mind control before anyone knew about mind control became the world leaders in the art of physics that perform in matters not requiring the ability to think about concepts. That is where the mathematical mind is very powerful, in the process not to think but to be programmed to calculate.

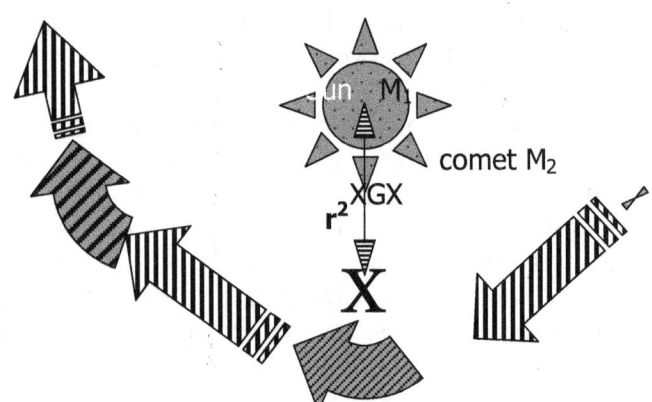

$F= \dfrac{M_1 M}{r^2} G$. **This is why Newton does not require or deserve proving, his followers can't prove or disprove, they can calculate...** $F= \dfrac{M_1 M}{r^2} G$ **Explains the first sketch but then?**

The mass of the Sun brought the comet in range and to the point where the strength of the force of gravity of the Sun is most and up to where the force the mass provides is at it's greatest and seems to be unconquerable. With the radius tarnished to being almost representing a factor of one the force of the gravity becomes eternal since the small radius will charge the mass into greatness which will unleash a force that will drag the Universe around if it could get hold of the Universe at such a close range. That does not happen because at the point where the force that the mass must produce is at its point of ultimate victory we find the comet avoiding defeat. The comet not only avoids the centre of the final destruction but it wins the fight completely. In the absolute defeat when the comet should have clinging to the ropes and living to survive the last moments, the fight swings unexpectedly in favour

$$F = G \, \frac{M_1 M_2}{r^2}$$

Sun and jaws of been desperate of the loser.

The comet rushes past the Sun as if the comet is Newton's predictions. The comet behaves in a Newton in doubt. That cannot be tolerated. Newton because Newton is always correct. The comet the distance where the dark is much and the light is

unaware of manner that leaves cannot be wrong cannot speed off into little. The comet cannot move around the Sun and then slip past the Sun because we know Newton said the mass of the Sun is pulling the comet as it is pulling everything there is.

The thought alone is more than what the mind can bear and the principle alone is outrageous. The man that has never been proven wrong proved himself wrong by predicting the demise of a small comet crashing into a big Sun and where the mass of the Sun produces a force of gravity that pulls the comet towards the Sun and then into the Sun, that is not happening. That which pulled the comet towards the Sun is now pushing the comet away from the Sun. If mass was pulling then what is now pushing because the silly little comet is in defiance of the great Newton and is rushing into the blackness again where the Sun hardly shines. If it is mass pulling then what is pushing?

Newton's vision gave us all the answers to the point where the fishing of the Sun starts but then as things really get serious Newton slips away like the comet and leaves us all without an explanation. Newton fails to provide answers and that leaves the academics looking silly. It is at this point that I don't understand Newton because I am too stupid and uninformed to understand Newton. Newton now gets beyond what I can comprehend and that is true.

One may even suspect that if the mass was insufficient to provide a force of gravity strong enough not to pull the comet into the Sun, the comet would at least begin to circle in a reducing fashion around the Sun until the comet falls into the Sun. One may argue that the comet may come at a speed where the Sun is too little to stop the velocity of the comet but that argument doesn't actually make sense. The speed will ultimately not change the direction from the centre to a point where it totally misses all apparent targets and coming at speed will benefit the Sun as well as the comet in aiding the effort of giving more of the gravity that the mass provides. I know such an argument is outrageous considering the mass of the Sun and the speed of the comet but hey, at this point Newton needs all the help he can get...and he needs help.

But without getting silly, all arguments we might think about in order to save Newton's reputation is bordering on madness. The comet is small. The comet is providing a force at best it can to enable the Sun to accomplish what the Sun set out to do from the start. The comet is pulling as hard as it can and in that the radius reduces by the square. It is hardly as if the comet is trying to preventing the seemingly inevitable destruction it is heading towards. The comet by mass is enforcing gravity as much as it can to self-destruct. This is in aid of the Sun's efforts to destroy the comet. While this is all going on it is at the same time not happening. That which is pulling is now pushing because the comet is escaping into the yonder.

If the comet's mass does assert gravity onto the Sun by measure of mass while the Sun does the same right back to the comet using the same grounds and mass principles in doing so, it must be conceivable and detectable that all the planets rotate around the Sun at different speeds seeing they all have different mass bringing about different forces of gravity. The rate of orbit must vary considerably as Jupiter must spin around the Sun billions of times faster that the little comet we just now observed does rotate around the Sun.

In a following article we are about to investigate such a scenario and then when it comes to light how much faster Jupiter spins in orbit than does Mercury on the very inside orbit and as Pluto on the very outside orbit we than can start to vindicate Newton and find the resolve about the comet behaving very awkwardly and out of step with the rest. With the massive mass Jupiter has in enforcing gravity there can be no more arguments about how the mass brings about gravity to have Jupiter move at 318 times faster than the Earth around the Sun because Jupiter is 318 times more massive than is The Earth. Then surely all scepticism must be nipped. The closer any two cosmic objects come the stronger the force should be, with eventually no force in the Universe being able to keep them apart. This is just not happening!!!

There is no indication of truth about a contracting solar system as Newton proposed and as Newton's followers promote a contracting Universe while when seen from the Hubble perspective where he introduced the Hubble Constant...and not from any other evidence seen through the Hubble Telescope Newton just is not matching reality notwithstanding fraud such as the critical density issue.

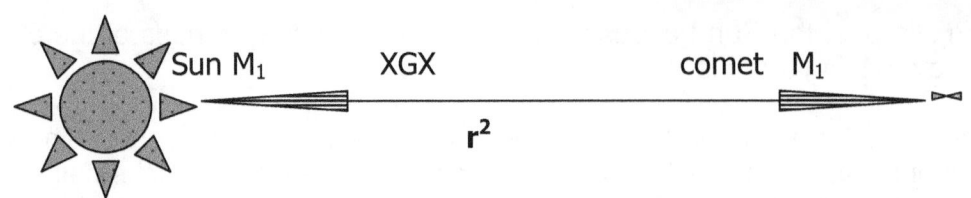

As explained, there is some discrepancies about calculating the force of gravity, because gravity would apply as nicely as it does if it was the perfect balance, a balance exists in space of equal measure bringing about equal seasonal time.

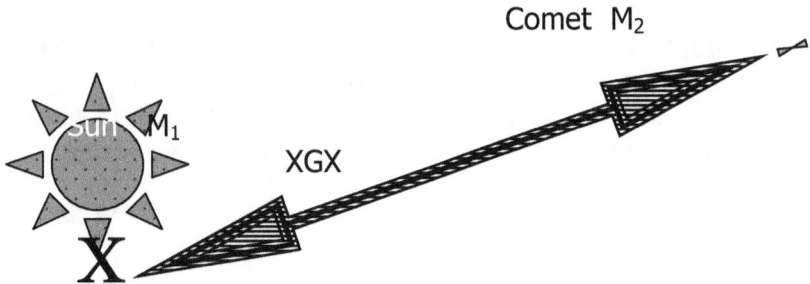

The biggest discrepancy and a practical denouncing of the official version of the comet's flight around the Sun,

The Sun gets a grip on the comet by mass inflicting gravity and as it gets hold of the comet it drags the comet through the solar system straight ahead to the Sun just as Newton predicted the Sun with all its gravity producing gravity will do.

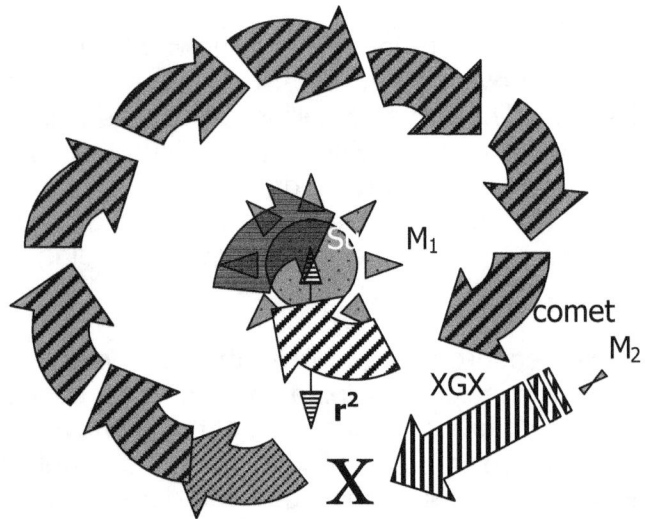

As Newton had said the gravity that the Sun and the comet mass induce will pull the comet to the Sun. this at first appeared to be true. As we all know the comet moves to the centre of the Sun just as Newton predicted but with a slight complication and a change in the venue, the comet no longer aims directly at the centre of the Sun but aims at a target outside the limits of the space that the Sun occupies. It follows a line aiming towards a point to the side of where the Sun is located.

Again this is the part my intellect is apparently insufficient to follow Newton because Newton is his formula never mentioned this apparent deviation from his prophecies. It seems the Academics are correct in saying that I don't understand Newton but they never try to explain this part that I don't understand about Newton.

The comet aims at some point gravity does not hold and misses the point that gravity secures the strongest force, which is right in the centre of the Sun.

Sun M_1

comet M_2

r^2 XGX

X

$$F = G \; \frac{M_1 M_2}{r^2}$$ **In over three hundred years not one Newtonian came forward and tested Newton's formula...and then the world must believe there is no conspiracy in science to brainwash students by enforcing mind control on the subjects called students they teach, or is it that they train to accept and believe!**

Should one give Newton the benefit of the doubt and disregard this missing of the centre of the Sun the position that the comet should aim at then one would reason that it could be that the gravity was not generated strong enough to locate the centre at such a great distance as the comet was at first. It could be that the reducing of the radius comes in instalments of a few circles. His formula does not indicate these presumptions that I make but hey, let's try and give the guy a chance. The comet might just take a cycle or two to wind down as the radius reduces and one should wait and see if it is not that which Newton meant when he said the mass by the mass is dismissing the radius between the objects.

Alas, it is not what the comets intends to do because the comet breaks the strangle hold of the Sun and what ever was pulling the comet at first is doing all the pushing at this point because the comet is surging into the darkness of the abyss. The comet speeds away from the Sun and also at the same pace it was heading towards the Sun and there is no altering to the speed in any way. The comet seems very much unaware of the comet behaving in opposition to what the great Newton predicted with his formula. Again I admit I don't understand Newton but I suspect the intelligence of those that do understand Newton because they might for the same measure believe the story of little Red Riding hood and other forces fairy tales offer. When we enter the world of forces what is to be expected?

Newtonians improved on Newton's force by creating so many more forces legally without the Church interfering in the process of creating forces that the witches went on strike. They don't work any more! The poor magicians and witches no longer has any authority when it comes to forces because from the one side the Church roasted then alive for having or allegedly having forces and from the other side physicists stole them blind by taking away all the legal forces flying around as forces. To some life is not fair!

Teaching students that it is mass pulling the planets around the Sun while the evidence that are supporting this matter is very skimpy and dodgy such teachings is a way of committing an act of folly by which then perpetrators are committing fraud. With so much evidence lacks and yet academics still insisting that students accept the fact that Newtonian presumption on the matter is correct is brainwashing. It is outrageous to force ideas onto students by disciplining thoughts and controlling the minds of the students into believing that the mass is in charge of the gravity, which is pulling matter onto matter while all obvious evidence so far

lacked any proof. A presumption remains a presumption until it is proven and Newton made a presumption in the case of his apple falling which until now was never substantiated by fact. In such an event the presumption remains a presumption until facts prove the presumption accurate. Then only does the presumption become fact.

At the start the first sketch explains $F = G \dfrac{M_1 M_2}{r^2}$. The rest is an unknown and putting facts aside and fiction in place. When we bring in intellect and we put stupidity aside then not even Newton can fill in the blank spots. It is all make believe and it becomes so clear that the academics in physics are keeping the minds of students busy with the biggest cover up the world has ever witnessed. Your local academic is not a wise old man having all the answers he wishes to share with you…instead he is a shrewd criminal that wants to deceit you with lies in order to carry on with the biggest cover up ever produced.

The comet performs all the other manoeuvres as the sketches indicate except returning to the Sun, and it is all the manoeuvres that the comet does except running into the Sun, which Newton's formula totally ignores. Yet on this very principle is every aspect of modern physics based. If the formula forms the basis of all physics used by science, then it is the basics, which are around for hundreds of years that suddenly are trash and simply does not perform as it is supposed to. If there is out there one professor in physics that will tell me the profession didn't know about the comet not falling into the Sun then I can show you one liar as I cannot show you in playing poker games. For many hundreds of years every person in physics were aware of this flaw but did nothing about it. I call that criminal and I call that deception because they took money from others to betray the innocent, the young and the trusting vulnerable by spreading untruths to the young and the trusting vulnerable and forcing young minds to believe what they well know is not remotely true.

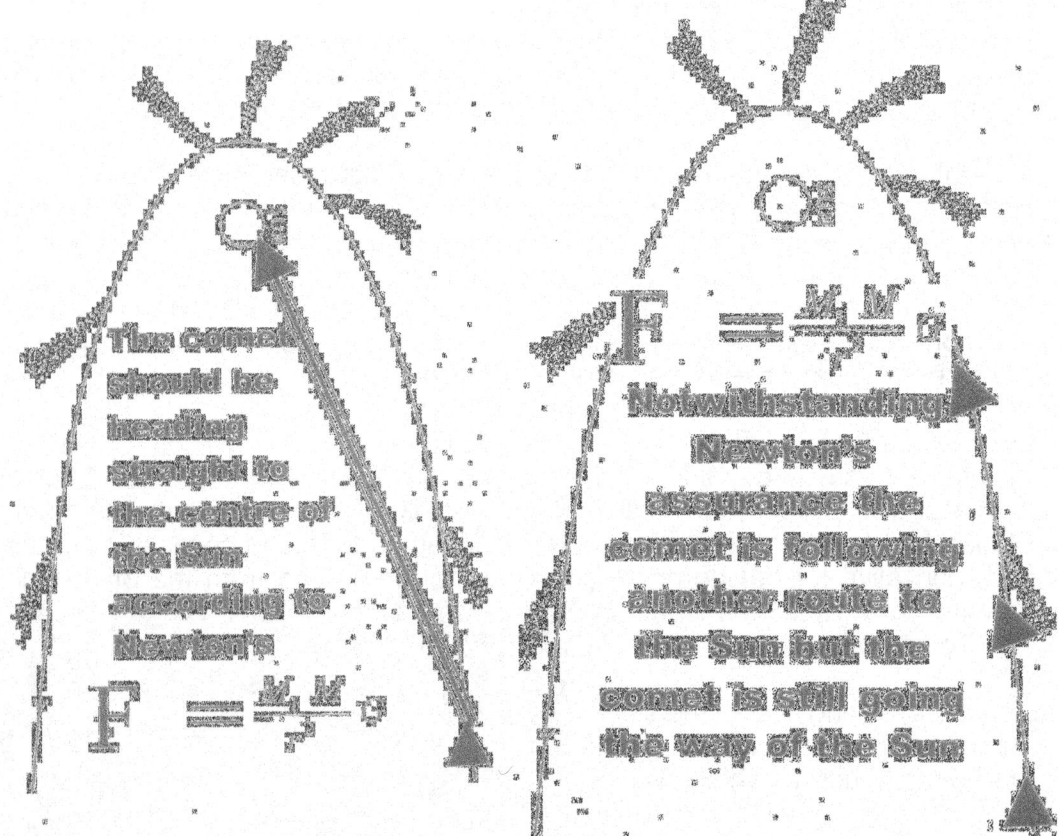

If it is true that it is the mass that is pulling the comet to demolish the radius, then ask you're most wise amongst all those that are wise Professors what is pushing in comet into the dark beyond. What is driving the comet away if it is mass that is pulling the comet towards the Sun. Where does Newton's formula $F = G \dfrac{M_1 M_2}{r^2}$ allow for that the phenomenon of attraction to turn around and become the complete opposite of what it was before? If it is mass that pulls then have them clearly state how and what brings about that the comet escape this destruction so easily. The normal deception that they use is that it is momentum, but don't let them fool you with more garbage that thy swindle because if it was momentum, then why is it not Jupiter coming that fast and the comet being so small that it can hardly be pulled with the mass it has. In this you would be able to see who the fraudsters are and who is telling you about the scandalous brainwashing they put in place to keep your thoughts under their control and that would keep you as a student in your place.

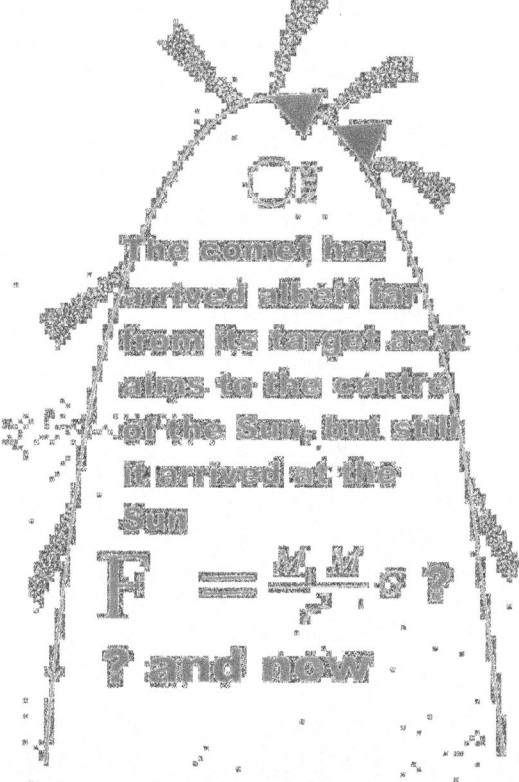

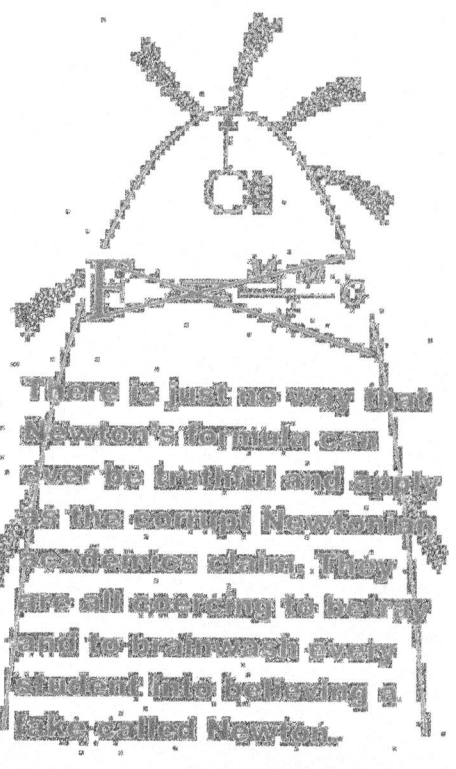

Comets are as regular as the sun and the moon's orbit but the cycle just takes extensively longer to complete its cycle.

The moon and the sun both eventually fill a cycle also off coming and going but coming and going will be in relation to our solar position we have.

Yet they never had the insight to see what would push the comet if mass pulls the comet because one can't ever put all the blame on mass alone going in two directions?
Don't you get the feeling of being suckered and fooled? With that they prove we are weak at mind.

Don't you also like I do feel silly that it was possible to be cheated by the Brainy – Bunch that easily?

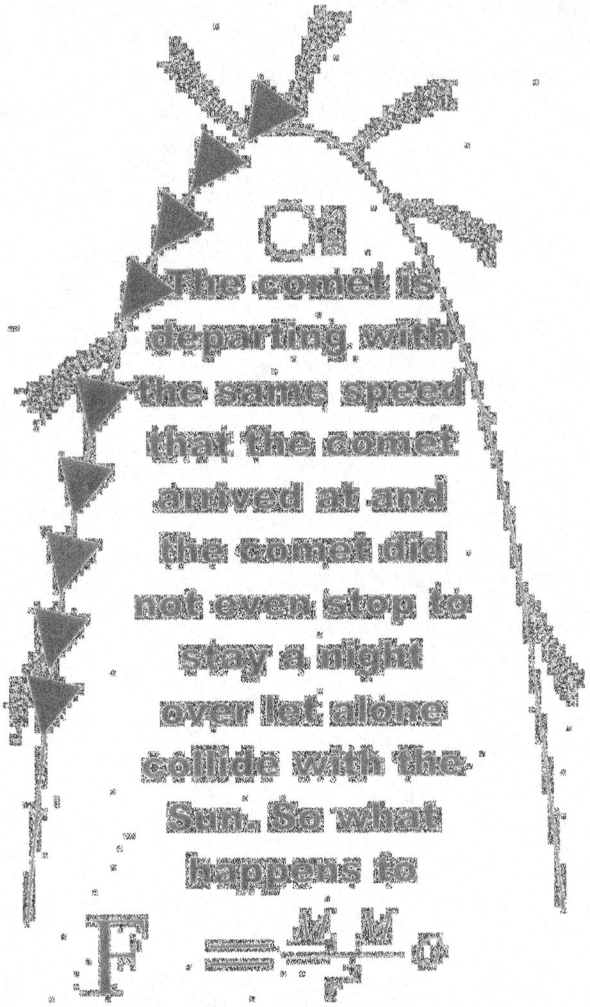

01

The comet is departing with the same speed that the comet arrived at and the comet did not even stop to stay a night over let alone collide with the Sun. So what happens to

$$F = \frac{M_1 M_2}{r^2} G$$

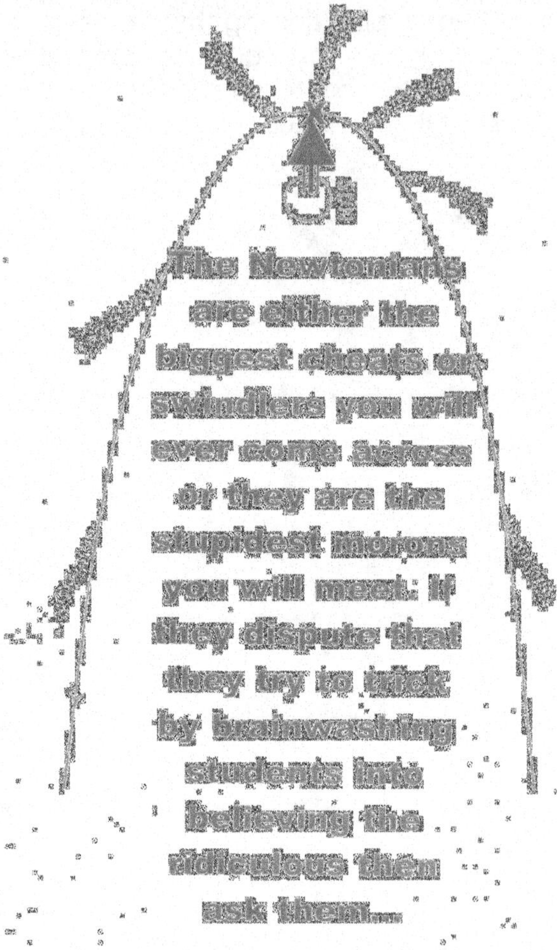

02

The Newtonians are either the biggest cheats or swindlers you will ever come across or they are the stupidest morons you will meet. If they dispute that they try to trick by brainwashing students into believing the ridiculous then ask them...

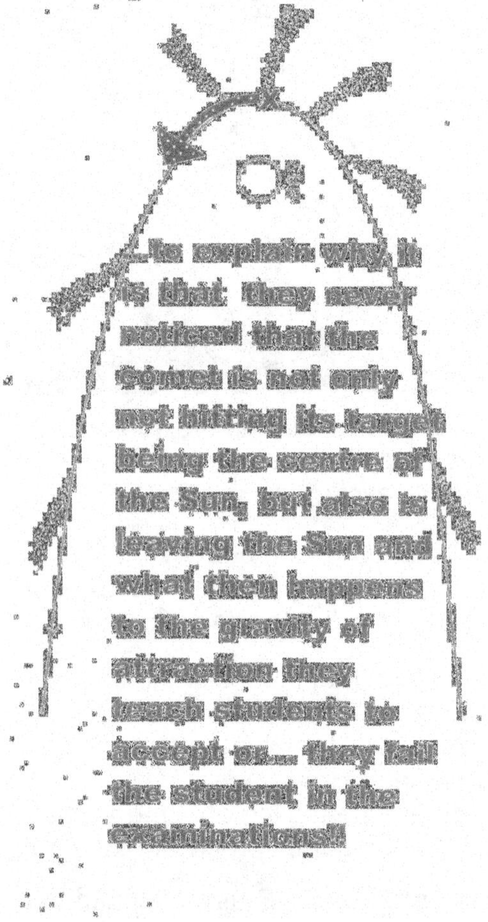

03

...to explain why it is that they never noticed that the comet is not only not hitting its target being the centre of the Sun, but also is leaving the Sun and what then happens to the gravity of attraction they teach students to accept or... they fail the student in the examinations!

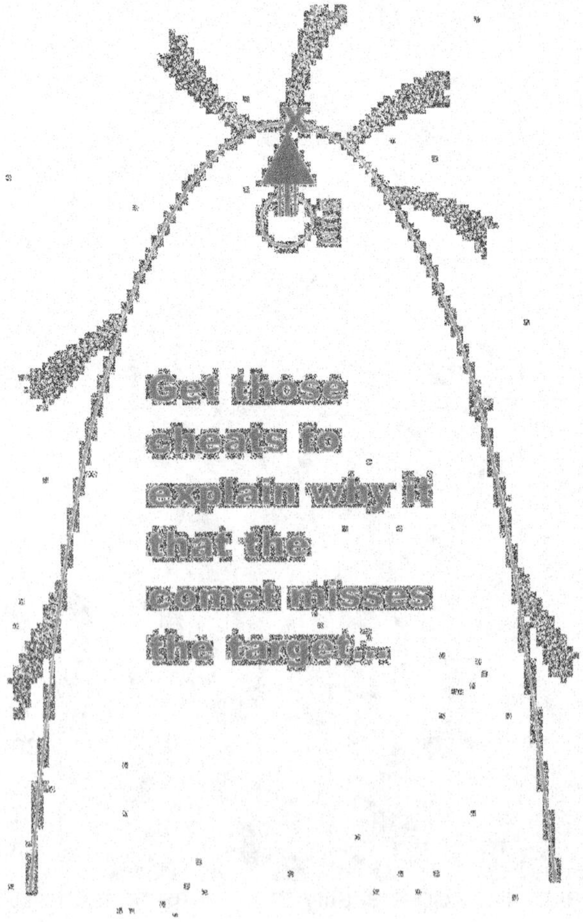

04

Get those cheats to explain why it that the comet misses the target...

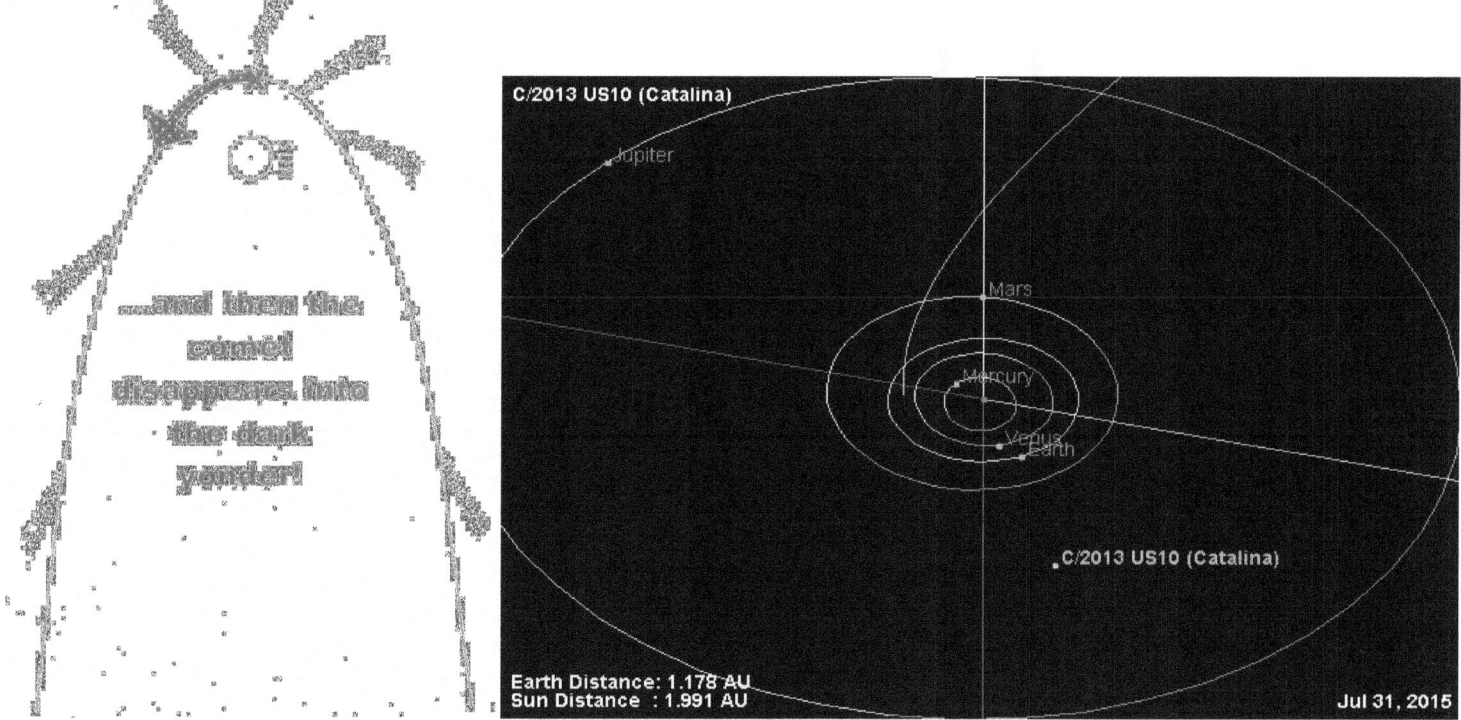

All evidence fills books about the comet's coming and then the "going". If "mass" brings on the "coming" what produces the "going"? If they can calculate and measure what it would take to sidestep time by passing through a wormhole and forecast the end of the sun's life span how is it that they never thought of looking at what makes the comet disappear into the darkness after "mass" pulled the comet to the sun?

That is an act of a criminal mind notwithstanding what motivated the person to commit the crime. Defining a criminal is to have a person that is prepared to place himself or herself in the centre of the Universe as to deprive all others of truth and possession in order to further the needs and wealth of the criminal while the criminal acts by never thinking about the rights of the victim or the harm done by the criminal's action in the matter. Spreading untruths and forcing payment for such actions is most certainly criminal! You students in physics go on and confront you Professors about the truth. Insist on the truth for once and all. Mass has nothing to do with gravity but to prevent gravity from being in place.

If it is mass pulling the comet as Newtonians declare then they better inform you being the paying students that keep their wallets filled what is pushing the comet away. The comet is leaving as fast as the comet was coming. Confront them because you have the right to insist not to be lied to about Newton and his shambles and criminal fraud. Go on and ask them to explain these Newtonian inconsistencies and see how they try to carry on with the criminal deception and covering of the truth that has been going on since Newton first thought up this scam.

This is the basis of physics. $F = G \dfrac{M_1 M_2}{r^2}$

It is not a concept or a suggestion or an idea but it is a formula used as a mathematical expression in terms of a mathematical formula and therefore it is expressed in terms of mathematical accuracy. It has to be put to the test in relation to its mathematical accuracy or Newton is a fake. That is as simple as it is!

This is but a small part of a big picture uncovering the scam Newton came up with and which all the academics are knowingly still participating in… and read on for I am about to inform you of more criminality the academics came up with and which they force feed you. You will see fraud as you never saw fraud before.

Module SIX
Glued or Not Glued

If you are a student in the science of physics, then ask your Educated Masters to please explain the following abnormalities you are about to read in this book and insist on a clear explanation about the inconsistencies they promote while tutoring physics as if the physics they present are the most flawless and accurate institution there has ever been. Ask those academics supporting Newton about the following flaws that no one mentions ...ever... except me in this book you are about to read and get them to explain the inconsistencies never talked about, which I present in this book and then after confronting those charged with tutoring physics and seeing who should be believed, then get wise instead of brainwashed. Let them mathematically show how one would go about and use Newton's visionary

formula $F = G \dfrac{M_1 M_2}{r^2}$ to calculate the force of gravity by replacing the symbols with the actual values

in mass that the items referred to have. Put in the Earth's mass in place where it belongs and put in your mass in place where it should be and then divide that with the distance between your soles and the Earth measured in micro millimetres by the square thereof!

Do you know what gravity is?

This is the question I asked myself about thirty years ago. In the late 1970's when, one day, I came into a room with a TV on and this may sound unusual in today's society, but back then in South Africa things were different. Today my saying that I entered a room with a TV on is unusual but at the time the unusual part of saying unusual was for me being able to enter a room with a TV on in South Africa, was unusual. In South Africa we just got TV and the concept was rather new to everyone. On TV I saw this man going on about an issue I knew little to nothing about. On TV there was some person going by the name of Carl Sagan or something having a program called Cosmos or something that was explaining things my mind found boggling. What he said was more than what I could readily take in. This concept in itself was as new as everything I was hearing. I took a few minutes to get the information sorted in my mind but after those few minutes of sorting, I was never yet that excited except in a racing car or on a fast motorbike. I sat down and stared at the box and that alone was extremely unusual for me to do. I started listening and the more I listened, the more I knew that I knew nothing of what this bloke was going on about but what he was going on about, was the most fascinating information I heard in all my life. I knew I had to know more because I have never yet encountered such astonishing concepts in my entire life. My fortune was good for that the program was broadcasted for the first time that night making the broadcast being the first episode. That night the broadcast was program one, which meant the series just started and the program itself just started very shortly before I entered the room and that meant I missed very little. My life could have been totally different if I did not enter the room at that particular point and began to listen to the TV as the program started. I was surprised at my interest because I am not a TV person and the coincident for me to be in the right place at the right time was extraordinary. I started to take in and grasp things and for the first time in my life I was hearing things that was at the edge of my mental development. What I took in, I realised, was the limit that my mental faculties were able to cope with and that experience by itself was a very first for me. It was something I never encountered in any way or form before with anything or in any discussion I ever had. I normally know what is discussed and often can add one or two aspects to whatever is discussed. The very next day I then went into town going from bookstore to bookstore in search of the book going by the title Cosmos. My searching paid off as I found the book Cosmos and I started reading. It was a new world that opened and the information that was presented was more than I could manage at first. What I encountered was a new Universe filled with information and the idea of so much new things opening up was most thrilling.

These made me ponder on every aspect there was to ponder upon and I started asking questions more to myself than any other person. Every question I asked myself I found a person going by the name of

Albert Einstein had an answer for. I have heard of Einstein and knew about such a person but never encountered his work on a personal basis before. I had thousands of questions, all looking for answers. Today I can't even recollect one of the questions I had at the time, but at the time they were plenty and they were quite intense in relation to the mental development I had back then. Compared to what I now have figured out, the questions were mundane, but then even something as simple as space-time and a Black Hole were on the edge of my understanding. I thought about time and went in search of an explanation and found Einstein gave the explanation that brought an answer to my nagging question about time…but not entirely… I pondered on space and went in search of the specific issue I was puzzled about and guess what… a person by the name of Albert Einstein did explain the specific issue I was confused about…but not entirely… Everything I thought about, this person Einstein answered…but not entirely… I vividly remember thinking one day while standing in front of a class of pupils just after completing a lesson on the subject I was presenting at school at the time, (what ever the lesson was is of no importance) that with all my investigations it seems as if there is nothing going on that this Einstein fellow did not have an answer for. He came to conclude whatever there was in need of some conclusion. He speculated on almost every conceivable aspect there was to speculate on in physics an ponder on most thoughts there was…except one… he never said what gravity is…and that part was the part I realised was the part that I could reply "Einstein did not entirely understood physics…" and that part makes Einstein just another Newtonian. He did not know what gravity is…and that made me ask myself if I knew what gravity is. I was looking for the meaning of gravity but not in some intellectual reference about a force not known where or why and without being specific about anything, but what are the specifics forming gravity.

Do you know what gravity is?

I ask this question because Einstein couldn't answer it. I ask this question because even the so presumed inventor of gravity, which everyone thinks is Newton, couldn't answer it. I ask this question because I have serious doubts about the fact that there is gravity anywhere. I have doubts as to the realisation in physics about the authenticity there is formulating the concept applying as a force called gravity. Hearing this, everyone in hearing range steps back in shock and disgust because everybody suddenly suspects my mental status. Everyone thinks my state of mind could be infectious and contagious. But please hear me out because there is some sense in the seemingly apparent madness I present. In every book I read and in everything I studied I never could find any trace of gravity. How does mass bring about gravity? What is mass that it finds the ability to employ something as vague as gravity to pull and pull? Those well to do Masters of physics already gave the undiscovered particle that does not truly exist a name. It is going by the name of graviton but that too is matching the same principle as the Phantom (and his white dog or is it a white horse?), Tarzan, Superman and a host of aliens not yet introduced to man, but in retrospect when all are considered the idea is foolish, which by any standard is not entirely scientific.

I have serious doubts about gravity being present. To my thinking I concluded that there is no such a thing as gravity. If gravity was present we should find gravity and we should be able to define it much better than it being a Neanderthal concept of forces presenting little understood dynamics. That gave me a goal and a direction in which to search. Ten years later I was still convinced there was no gravity because no one had found gravity. Then twenty or so years after my introduction to Cosmos and Carl Sagan I was reading one night about Einstein being of the opinion that he thought if he fell out of the patent office in Austria which was a multi story building, that he would then experience the feeling as if he (Einstein) would be without weight while descending to the ground. He realised that if he, the pen he had, the table he was next to and the chair he was sitting on would fall through the window, then the lot would fall together. While descending, all were then in a free fall and it would seem as if the lot that was falling would have the same mass because they all were falling in the same manner. For the first time in my life I had an issue about something that Einstein said. If he felt that he was without weight it could only be because he was truly physically without weight and not because his imagination was running wild. He couldn't imagine things that weren't real. If he and his desk and the pen were falling at the same rate, then they were having the same mass and the same weight, and his imagination played no part in his physical experience. His desk is somewhat more massive than he is. His pen is somewhat less massive that he is. They all travel at the same rate and will hit the ground at the same instant. They're travelling together and having no mass, was not a feeling Einstein was dreaming up, but was a reality. If they travelled together while descending, they then had the same mass while falling or they had no mass at all. There was no middle ground of having mass and imagining being without mass. That means I am questioning gravity. If they are falling equal then there is no gravity pulling them differently. That means the gravity Newton saw is in Newton's imagination and not in Einstein's imagination. However, that is not the way I am perceived, because since the first time I uttered the notion I was considered by friend and foe as fulfilling the role of the village fool. Still I am of the opinion that there is no gravity…mad as that may sound.

If I say there is no gravity I am not trying to convince you that you're being on the ground is entirely just your imagination holding you on the ground. Your being on the ground is not due to deeply rooted physiological issues following you in the form of a concealed depression since your childhood and is reminisce of a dark period from your early days that you never got to terms with. It is also not due to your willpower or maybe your lack thereof that is preventing you from flying to Mars. Hey, I am not that mad… I wish to give you a test to judge your intellect seeing that you (and almost everyone else on Earth) think of me as filling the role as the village fool since I am in doubt about gravity and the being of gravity presented as a reality.

Do you know what gravity is?

When hearing the question I put to you, you immediately jump to the conclusion that I refer to that which holds you steadfast on Earth. You don't think another thought about what I might be referring to because you already know what I am referring to. I am referring to gravity and you know what gravity is. It is gravity that holds you on the surface of the Earth, where you've been stuck ever since you can remember and the sticking goes on relentlessly. It is what you have been fighting since birth and if not for gravity, you would be a Superman. The only thing Superman has that you seemingly, obviously and definitely don't have is his Superman ability not to be restrained by gravity and confined to the ground. He can fly as he pleases while you are tied down by gravity. If it was not for the effect gravity had on you, then you could have been the local superhero along with six billion other Superheroes that had no restraining from gravity. By your suffering from gravity during all your life, that puts you in terms of being an expert on the subject of gravity and no one knows gravity better than you do. Then I have the audacity to ask an expert such as yourself if you are the expert on what you know better than anything else you might know about. The only aspect of your life you are unable to change is your attachment to the restraining of gravity.

To your knowledge there is only one form of gravity and you know better than most that you are standing in aide of that gravity, where it is that gravity that is keeping you on Earth, so by experiencing the restraining it brings with the concept for your entire life and during your entire life you very well know what gravity is. Your expertise on gravity had you fighting gravity more than anything you have ever fought or had any other fight with, including you Mother-in- law. From this on-going relentless battle every second of your entire life you have gained the experience only coming by being in a continuous fight and it is this fighting that makes you the ultimate expert on gravity! You know gravity is what prevents you from having the ability to jump over the Moon or run the hundred meters in three seconds flat. If it wasn't for gravity, you could out accelerate a fighter jet in mid air. You know it is gravity that is going to get you old and it is the very same gravity that that is telling you that you now desperately have to lose weight or over strain your heart and die young … then I come and accuse you of not knowing what gravity is by asking you what gravity is! You are so sure about gravity that you are no more familiar with any other subject and little else has your expertise as gravity does.

Are you sure that gravity is what you might think gravity is and that you and Newton have the same concept about what gravity is?

Have you ever thought that which you think of, as being gravity, is not that which science presents as gravity? The idea you have about gravity is not the gravity science says is gravity. You might think that your being stuck on Earth is that which you think of as gravity…that is the concept you formed ever since your first attempt to sit up straight and that was just after you formed an opinion about milk. Since then you needed no one to inform you about gravity because since then you have a very clear idea of what you envisage or what you think personifies gravity. I say that if what you think of gravity as that which is keeping you glued to the Earth then it is not the gravity science defined as gravity. That which Science defines as gravity and that which you are thinking of as gravity is very much not equal but also it serves the Masters in physics well to leave you with having that idea about your gravity and their definition about gravity being the same because now they don't have to inform you what gravity really is. They leave you with what you think of as gravity. You now are so well prepared as to become a candidate to be brainwashed in believing your way of believing is their way of believing what gravity is and in believing you share their view about gravity, you are in the best prepared state to become another one of their millions they have mind control over. What Science says gravity is, is far from your view because the Newton's approach to gravity is the same as a magnet hooking onto a metal. The grip coming from that is gravity, according to science. They say there is a force between you and the Earth and this force is pulling you as much as it is pulling the Earth, but it is pulling the Earth much harder than it is pulling you because the Earth has much more mass than you have. That means there is a force within you and there is a force within the Earth and these forces coming from within pull together that, which should be apart. That form of gravity will have you locked onto to the Earth with no release but when you do have release then the release will allow you to escape. The closer you are to the Earth the more significant and powerful the force then must be. The worst part of breaking the force is the first millimetre and from then on the rest is child's play. That is what they suggest when they say mass is pulling mass to reduce the

radius in $F \; \alpha \; \dfrac{M_1 M_2}{r^2}$, which is $F = \dfrac{r^2}{M_1 M_2}$, which later was changed to $F = \dfrac{r^2}{M_1 M_2}$. What this

means is if your mass is hundred kilograms and the Earth has a mass of 5.974×10^{24} then when you are standing on the ground with an infinitely small radius between you and the Earth, then the force keeping you attached to the earth is eternally big. With the radius of say one billionth of one millimetre you have no chance to be released from the Earth while when you are say one kilometre way from the Earth the force is one thousand million billion times weaker. That is shit at its best because notwithstanding height,

the gravity remains the same. $F = G \dfrac{M_1 M_2}{r^2}$, $= M_1 = 5.974 \times 10^{24}$ and $M_2 = 100$ kg divided by the

incredibly small radius of 1×10^{-15} m then the force gets to be $1 \times 10^{+15}$ making the mass more proficient by a margin of $1 \times 10^{+15}$ and that is totally ridiculous in reality but if the formula is correct that is what should mathematically be true. With a magnet, where this concept is true, the last millimetre makes the magnetic pull so strong it becomes almost humanly impossible to control the distance between the two magnets whereas when there is a meter distance between the two magnets there is no pulling power to

speak of. The same should apply in the formula if the formula $F = \dfrac{r^2}{M_1 M_2}$ did apply. Once you are free

from the ground, your first stop should be the moon, because then you overcame gravity at its worst. We know that is not true because jumping the first meter is the easy part. Getting higher than one meter becomes more tiring and from two to three meters a person needs other devises aiding the jump. The

further the jump is the harder the task is but contrasting this formula $F = \dfrac{r^2}{M_1 M_2}$ would suggest that the

smaller the radius is the harder the effort must be to break the gravity strangle hold.

$$F \; \alpha \; \frac{M_1 M_2}{r^2}$$

When the radius is insignificant the mass becomes enormous. The opposite also applies because when the radius becomes enormous the mass and therefore the force become insignificant. That is just the way it works when there is such a directly suggested relevance applying between the mass bringing on the force in relation to the radius influencing the force.

$$F \; \alpha \; \frac{M_1 M_2}{r^2}$$

With the radius big, the influence of mass and the force is weak. This would then suggest that if we are able to break the first meter between the Earth and us, it is then the end of confinement as we would be able to fly to the Moon for weekend shopping on a Saturday morning. We know that is not true because if that were true walking on Earth would be desperately strenuous. It is getting into the air at an increasingly higher distance that presents the problem. What we seem to experience is like a blanket of something pushing us down and the first lift is the easiest. It seems as if the strenuous pushing starts when we try to lift the blanket of air way up into the air. Partially lifting is not the problem but breaking the cover altogether serves as the real problem.

Now, you might say Newton did some damage control when he changed the formula from $F = \dfrac{r^2}{M_1 M_2}$

to become $F = G \dfrac{M_1 M_2}{r^2}$ because Newton saw some controversy in the way I explained the working principles of the formula. Don't you believe that one because Newton made the principle even less applying and much more complicated!

The changing of the formula from $F = \dfrac{r^2}{M_1 M_2}$ to become the formula $F = G \dfrac{M_1 M_2}{r^2}$ was done in order to give the concept a cosmic significance. When using the formula in terms of $F = \dfrac{r^2}{M_1 M_2}$, it only refers to an apple falling from a tree to the ground. There is no sense of dignifying the true nature of the event by supporting the refulgence that would reflect upon the appreciation of the spectacle Newton witnessed the day from which such splendid scientific demiurge eventuality arose, where $F = G \dfrac{M_1 M_2}{r^2}$ indicates gravitational implications of a cosmic nature. Using the formula $F = G \dfrac{M_1 M_2}{r^2}$ places Newton's lustrous occasion in appropriate perspective. What the grounds was for making it a cosmic notion still eludes me to this day…because after all was said and done, it was only an apple that fell from a tree as the formula $F = \dfrac{r^2}{M_1 M_2}$ would suggest, and not the Moon coming from space and landing near Newton as the formula $F = G \dfrac{M_1 M_2}{r^2}$ would suggest, but I am getting to that later on in the book. The minute the radius r disappears, the mass is overbearing and the minute you find the means to break the strangle hold of mass keeping you on Earth; it is like a magnet releasing with no effort available to secure your position again. In the case of gravity being the same as a magnate, it will mean that an object will fall faster and indefinitely increase the descending rate as the object falls, since the object's mass is increasing in force by the diminishing of the reducing radius. Such reducing of the radius will increase the strength of the mass. An object will fall while descending onwards with limitless acceleration and in a fall the object should even be able to break the sound barrier, if the object is massive enough or if the object was dropped from far enough. This same story they try to tell when they tell the story of the person that fell from the balloon that took him almost into outer space. They try to tell the story that he fell faster than the speed of sound. If that did happen, he would be in pieces because his body would not take the strain of breaking the sound barrier. These Newtonians struggle for most of a half century not to get to grips with the sound barrier because they try to look at the principle's tonsils while staring up the arse. Claiming that the person reached the sound barrier is principally an indication of just how poorly they really understand physics.

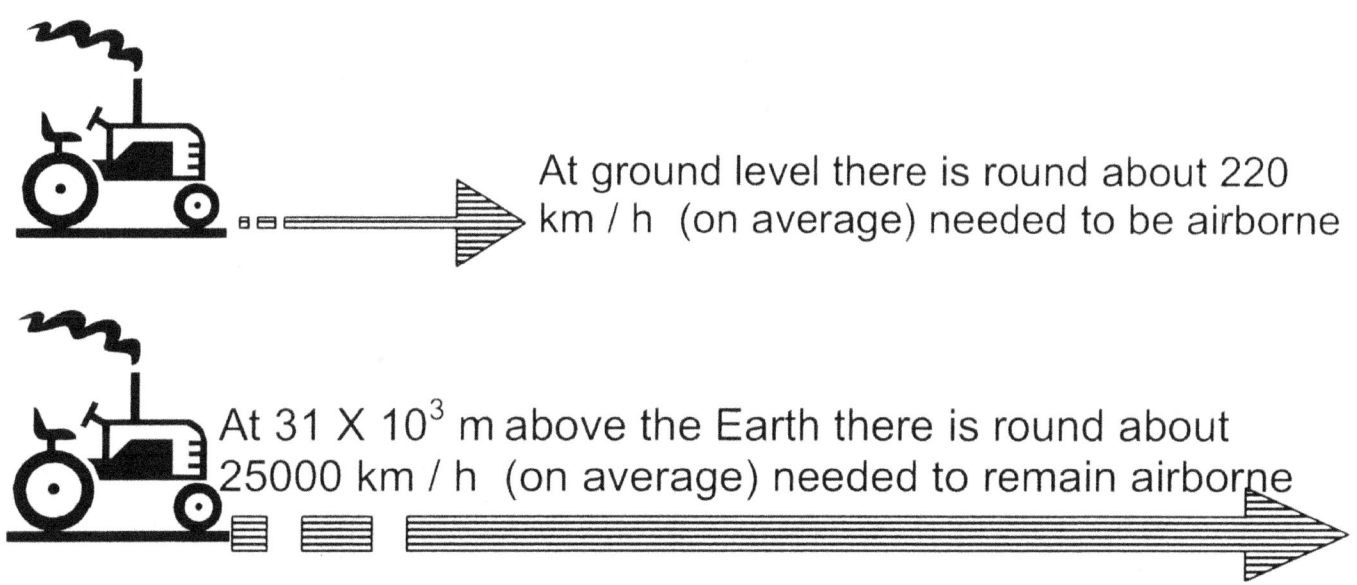

At ground level there is round about 220 km / h (on average) needed to be airborne

At 31×10^3 m above the Earth there is round about 25000 km / h (on average) needed to remain airborne

The length of a kilometre is not a constant but is depending on the density that makes up the material forming the kilometre. The density allows the aircraft to be better fuel consumption efficient at high altitudes and helps reduce flying time. Newtonians are unable to notice the concept in hand. Newtonians want mass and mass related arguments. For that reason more than anything else, science passed them by. Newtonians have no concept about the sound barrier but always fly at the speed of the sound barrier notwithstanding their inability to even hear sound at that altitude. That too has gone past our brilliant Newtonians because about that, Newton never said anything and what Newton did not reflect on does not exist! The fact that there is any relevance in the situation and the entire argument depends on relevancies and not the fabrication of mass, is far too complex for the Newtonian to comprehend. Newtonians can master mass and there it stops.

At 31×10^3 m above the Earth there is round about 25000 km /h (on average) of pure rocket thrust needed to remain airborne

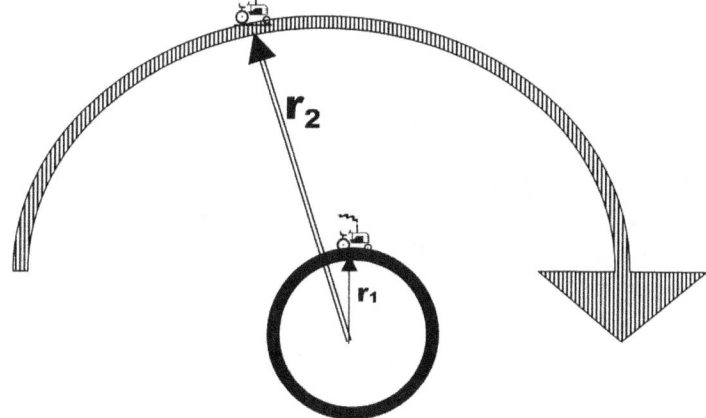

At ground level there is round about 220 km / h (on average and depending on wingspan) needed for any aircraft to get airborne

When an aircraft reaches lift-off speed, the relatively required speed would be around two hundred and two hundred and fifty km / h. To reach to the sound barrier, the speed at that level has to be 100 to 1200 km / h. At a height of 31 km straight up into the sky a rocket driven jet propelled supersonic aircraft requires the trust of 2500 km / h just to remain airborne. Clearly the two thousand five hundred kilometres per hour is meaningless, because one can see that propelling a craft at 200 km / h on the Earth is the equivalent of 2500 km / h in space. It has to do with the density of the space through which the object travels and not the man made suggested norms that become meaningless as the situation changes. This means up there at 31 km into the sky the sound barrier (if there can be one which I most strongly doubt) will be $2500 \div 200 = 12.5 \times 1200$ km / h (sped of sound barrier) = 15000 km / h. It is a question of air density and not distance travelled that determines flying. But since the word mass has never come into the argument the concept goes beyond what Newtonians can appreciate since Newton only told them about mass and nothing more. Now they can either appreciate mass or nothing. Up high, the travelling distance being in relation to time travelled will seem more but as density increases lower down, the speed "will reduce" which is utter nonsense because the relevant density of air displacement travelled to time spent during the exercise will remain the same. Again I repeat: without the word mass uttered once, no Newtonian is expected to understand or appreciate physics because going beyond mass kills all Newtonian participation in understanding and debating.

The man fell from his balloon at a super altitude and has a declining speed never exceeding 218 km / h. In terms of the density the relevance might indicate differences, but in space used for material distribution, the descending was no more than 218 km / h at best. When he descended at a high altitude, the air he went through was lacking density. This made the kilometre air that he went through very thin and very long when placed in comparison with air we use at ground altitude The fact that the density up there in the thin atmosphere at 31 kilometres above the Earth would not permit any living being the luxury of breathing, never occurred in the mindset of the Newtonian. The fact that the person required space protection in clothing and life-aid went past their understanding the term density because a kilometre is a kilometre as far as the Newtonian understanding permits. An aircraft has to travel at two and a half times the speed of sound just to remain airborne at that altitude. This tells a Newtonian nothing because there is no mass used in the concept and when excluding mass from any concept the concept dissolves as far as our brilliant Newtonians' intellect goes. The fact that the distance stretches ten times further at that altitude because the density decreases ten times eludes our Newtonian brainpower completely. In relevancy his deceleration at one point was in relevance still 218 km / h but at that height the kilometre in compressed space is far less than what it is on Earth. That means if the density factor reduced ten fold up there, the falling person then travelled 2180 km / h just to be able to maintain the gravity descending at 218 km/h. The kilometre up there does no hold the same amount of meter density than that which applies on Earth and the kilometre is no kilometre up there. But science hasn't got the wit to figure that one out. If the person was not slowed down by a parachute and had landed on the earth, then the person would have landed at a rate of no more that 218 km / h and most likely at a speed of 207 km / h. He could never break the sound barrier or exceed 218 km / h because that is only possible in the stupidity of the Newtonian mind. The magnet idea is not gravity and having that idea is proving a rather backwards mentality that only the mentally brilliant Super-Educated Newtonian-Masters in thought can achieve. They give you one concept while their approach has quite another. Using the formula $F = \dfrac{r^2}{M_1 M_2}$ or in its cosmic interpretation being $F = G \dfrac{M_1 M_2}{r^2}$ connect your position on Earth in gravity in the same manner as that which the magnet is applying. That means when your foot is clutched to the ground you are very much unable to shift it, but once the grip is released even by a millimetre, the gravity will decrease exponentially. That is not the gravity you have in mind, is it? What you have in mind is that the further you jump the more difficult the jumping gets. The radius extending increases the mass by multiplication and not by division. It takes a lot more effort to fly at 31 kilometres above the Earth than it does flying at 310 meter. Jumping 2 meters can be achieved by human power but jumping 250 meter requires machine-aiding power. This means jumping 250 meter is not dividing the effort but is multiplying the required effort. When translating this scenario into their formula the force that mass would require will diminish by 250 times when the radius becomes 250. The effort required to keep the body afloat will be 250 times less at 250 meters than it will be when the person is on the ground. Your concept about their gravity is not nearly that which they say is gravity, when they say the force of gravity is the product of the mass multiplied by another mass in conjunction with the multiplying of the gravitational constant and this product is reduced by the radius between the two objects holding mass. That is not the way you are walking on Earth but they wish you to be confused with the truth and their lie so that they can get way with murder.

But they would love you to have that concept about gravity and they would hate it if you thought further about the issue because then you would exceed their knowledge limits about what they know as gravity. It fits their comfort zone to have you believing you are an expert on gravity. If they can keep you thinking you are very well informed on the subject of gravity, then you will never question their authority on the matter of gravity and you will never learn that they know nothing about gravity whereas they will have you thinking there is no further point in discussing gravity since we are all experts on the subject of gravity. They will put up a fight to keep you fixed in the idea that gravity is a one-word concept and by thinking in terms of gravity you then entirely define the whole concept involving gravity. You then can define everything that gravity defines by using one word and that is gravity. It is not what they present as a scientifically defined explanation of gravity, but who cares. To them you and I and the rest are cow fodder, there to be used when needed and then to be left in the dark and to their mentality, they could care less whether we have any opinion of sorts, for what we think is of little importance to those mighty academics occupying towering intellectual heights. They could care less about what you think than they care about the opinion your dog has on the human civil war crises in Africa (and this statement will never go out of use because there is always a human civil war crises in Africa). The crisis will be there because the western nations that care so much about the disasters and the prolonged human suffering while the genocide is ongoing everyday. That is all the same to them because they are the ones lining their pockets with money they receive from weapons sold to the warring factions on both sides of the confrontation. Never once was there ever a genuine independent commission launched by the United Nations as to investigate the weapon transactions before the genocide even began or to find who the parties were that provided that arms used in the genocide. It is all a scam to confuse and depress the truth, just as they depress the truth about gravity. It is a culture that drives the powerful to subdue the weak. While one thing is happening they whish you to think another thing is in progress. It is a state of mind the controllers such as Industrialists, Academics of all sorts, Hoggenheimers, Mammonists and politicians use to commit the true crimes against humanity and they are those whom are all criminals with criminal intent to defraud and corrupt the truth. They use many things to defraud the public such as the fraises using the words Human rights, human crises, democracy, racism, political correctness and so many more meaningless expressions just to defraud and confuse. It is everyone that has a personal opinion that such a person is with some authority that allows the person to jump on the wagon of authority and everyone that thinks of himself in terms of being a so-called expert see his role as taking over from God Almighty and start to rule in His stead. This goes from fox hunting to what children eat to what and how parents should discipline their children and criminalize the ordinary citizen while decriminalising the true criminals. They give a concept a meaning as in using an idea while behind the scenes they use an entirely different definition they share very sparsely.

This is mind control just like UFO's are used to culture the very same principle, which all the institutions from the Government to the press and all the Academics at NASA and in agents at NSA benefiting from the misinformation because that suits their use in mind control on you being the victim. They use it to hide behind while they dictate the terms by brainwashing the people into reluctance. The entirety is very deeply rooted in society but in this book I only aim to tackle gravity and therefore we return to gravity.

We have to analyse what gravity is. There is a defining to do by dissecting the factors we see as gravity. Gravity is not many things to lots of people, but is very specific one thing. To investigate the identity of gravity there are questions one must first answer. Is gravity the part that has me standing on the Earth or is gravity the part that has me glued to the Earth. It is surprisingly not the same thing. While I am standing on Earth I am standing still as far as my perception carries but there remains the intention to move, even with me standing still. I can't stand still and at the same time intend to move by the same driving effort. My standing on Earth but remaining in intention to move further can't be motivated by a similar or the same principle. The concept is not the same because when standing still the Earth does the moving on my behalf and there is no independent motion of my body. But when I jump, I am moving away from the Earth and when I walk, I walk along the surface of the Earth giving me independent movement on top and in addition to the movement of the Earth. My movement is stopped by mass and not committed by mass when in mass. When in mass, my intention is to be motionless. There are two occurrences where one is immobility through mass and the other is being without mass while moving. Mainstream science wants you to believe it is the same thing, but it is not. Gravity is the inclination that my body has to move further towards the centre of the Earth while mass is, that, which frustrates my body, and prevent it from moving towards the Earth. Science tells you it is the same thing but it is not. The very second that that which prevents me from moving falls away, the inclination of moving, which is gravity kicks in and gravity lets me move freely again. I will continue to move towards the centre of the Earth and that process has a very well defined and scientific term which we use to describe the event: We call it falling in English but every other language devised another word to describe the same concept. When I stop falling that which give me mass is what ends my moving. What gives me mass changes my gravity from movement to being inclined to move. As soon as that which prevents me from moving allows me to continue to move to the centre of the Earth, my gravity will start giving me full motion and then that which gives me mass by frustrating my moving will not stop my progress towards the centre of the Earth. When my gravity is stopped by mass blocking my movement and that, which gives me mass, then has the ability to turn my descending towards the Earth over to a frustrating of gravity or my freedom to move and change such a freedom to move into a tendency or an attempt to move. The price I pay for the loss of free movement is the gain of mass in mass preventing further movement. It is not as simple as being glued or not being glued…

This is what I have tried to convey to academics throughout the world and in person to so many in South Africa. But conveying this message is criticising Newton and not once did I mange to get one of the esteemed to listen to my criticism of Newton. If they hear it is about gravity implicating Newton, the debate stops with them defending Newton and not giving any more attention to my argument. When trying to convey my message about gravity I firstly have to start to show that gravity is actually the moving of the body and when the body has mass the body isn't moving independently as a cosmic independent object but becomes in mass meaning normal concept that all persons and the body moving is the with the body moving is that the culture that Mainstream and fifty years never got this don't believe me discriminate between test your nearest thoughts and you will mixed up and confused the concept behind gravity what involves all the aspects that factors forming the concept proper categories.

it forms a unit with the Earth as the Earth. To the have, the thinking is that the body not moving same thing and the body not moving is gravity which is formed by having mass because science promotes the past three hundred past mass. If you, the person that is reading that Mainstream science does not mass and gravity in the correct way, then physics lecturer. Ask him by testing his see it suits them well to let everyone get all about the two concepts. To get to the bottom of one needs to analyse and dissect and then realise covers the concepts entirely. We have to go splitting the we hold as gravity and defining the two concepts into

To the average people, the idea of expanding a mental view that is normally achieved by promoting thought becomes overwhelming overloaded which leads to confusion and they experience it as me aggressively trying to harm their sanity by dangerously attacking their mental stability. Everyone uses slogans that bring comfort to all person as the slogan covers the concept and dividing the concept is taking out the comfort of the advertising intellectual age.

If I say gravity is the part not connected to mass that brings on the intention of motion to carry on moving downwards notwithstanding the blocking action which comes from intervention of space occupying by a controlling body while mass is not connected to gravity since it is stopping the motion of moving downwards, the concept starts to override the normal commonsense. Mass is having a much more demanding space filled with material in a position that will intervene further movement of descending to a centre of the body having gravity and therefore performing the descending motion. But the descending is part of a rotating motion as well and the complexity of the combined effort to move, forms gravity, while having mass is rendering the control of the movement to a larger body that takes hold of the moving altogether. This is far too complex to cover al aspects in one slogan and by crashing the slogan, people feel I am trying to short circuit their wits and understanding. By getting more detailed about gravity being or not being what the slogan refers to, people hold the view that I attempt to or I dare to try and bust their slogan mentality and break the confines of what the Powers that control society invested in so heavily and what the advertising industry depends on so much, and politicians feed on frantically by promoting the culture so that it is much nurtured by all and which everyone is addicted to. Their comfort zone becomes threatened. Then my threatening to burst this super intellectual "two-word in one idea"-slogan mentality making everyone a genius, most people see as culture threatening and in the defending of that superior feeling, they only see that I wish to destroy them and challenge me to protect their comfort zone. By me attacking their senses of genius which overrides all of their modern nature, is more than what they will tolerate because the easy life the slogan mentality sponsors brings a feeling of intellectual superiority to all concerned and my effort to remove that is most uncomfortable and no one will accept that.

They immediately feel threatened because suddenly the person facing the question also faces the demand on the ability to start thinking, and not just by conveying a slogan to express an encyclopaedia of understanding. When the slogan mentality is forced out and there is a demand to think, the comfort levels drop considerably. That is the last thing anyone wishes to accomplish because in all sincerity that society can offer, we have to realise that thinking is what the TV age and Video culture tries to avoid most. By my asking questions that surpass the slogan stage, I succeed to make that person to feel stupid and threatened and the person goes on the attack. They are not attacking my argument but they are attacking me for making them feel insecure and stupid. I suddenly take away their expertise and that makes them feel naked. That takes them to a point they have to attack me by being without reason because it is their reason I attack and it is the reason to understand what they find lacking. I came to reach this conclusion about human mentality, which I just shared with you on the grounds that not once did one person ever ask me to explain what I just said, when I suggested to a person for the first time that I challenge the authenticity of the concept portraying gravity. Never did one person first consult with me as to what it is exactly that I am saying. Everyone presumed I say they are not glued to the Earth and with me being that mad I am clearly more stupid than they are and for once they find someone with a lesser mind which they can demolish.

When I state that there is no gravity they all respond in a manner that is putting me under suspicion and not one person responds positively by asking me why I put gravity under suspicion. Now I give the reasons why I challenge the concept of gravity and I would love to see a person challenge my challenge. This is why I discard the fact of gravity as promoted by the Brainy Bunch.

By definition gravity is defined as being:
Gravitation is the force of attraction that operates between all bodies. The size of the attraction depends on the masses of the bodies and the distance between them: the gravitational force diminishes with the square of the distance apart according to the inverse square law. Gravitation is the weakest of the four forces. Newton formulated the law of gravitational attraction and showed that gravitationally a body behaves as though all its mass were concentrated at its centre. Hence the gravitational acts along a line joining the centres of the gravity of the two masses.

It is not you being glued or not being glued to the Earth that I discard. It is the definition holding this whole idea that I do not share in the least. What the definition describes is magnets pulling and it is the total opposite of what I experience. Breaking the first millimetre of gravity clampdown is the easiest and not the most difficult. The difficulty increases as the radius grows and not as the radius decreases. When I say there is no gravity everyone thinks I say we all are going to fall off the Earth at random and with me thinking that way, then it is obvious that I must be a nut. Everyone thinks of me as the clown acting mad when I say gravity is not to be found in nature. But I do not say we are not standing on the Earth. I do not say there is nothing that is keeping me glued to the earth. I say there is no attraction between two bodies by the force of the mass that in such doing is diminishing the radius parting the bodies by the inverse square law. I say there are a connection by motion between the centre of the body and the material surrounding the centre. This is what I say when I say there is no gravity.

I dispute Newton and so do all students at first when students are forced to learn about Newtonian physics because Newton's arguments are an onslaught on human intellect. Think of the resentment that students have towards Newton under normal conditions when they have to cope with understanding the Newton principles Mainstream science says are applying and how that confusion of what is possible and what Newton suggests is possible clashes with their intellect which makes them feel stupid. Students hate Newton because they don't understand Newton and for that they are accused of not having the intellectual capacity to follow Newton. Every student from the past going into the present and even including those forming a future generation of students will purchase a book that is showing that Newton's legitimacy is cracking up when exposed to some vivid scrutiny.

In short I will now explain what I explain throughout the book you are about to receive and which is named *Newton's Fraud* or whatever it will be named as. The Newtonian formula $F = G \dfrac{M_1 M_2}{r^2}$ is the formula used by science to explain and define gravity. It says the that the ($M_1 \times M_2$) mass of the one object pulls the mass of another object and this process in relation with a gravitational constant (**G**) (a supposed force keeping the Universe attached) and the pulling subsequently destroys the radius (**r²**) being between the objects. That says that objects **ALWAYS** <u>**MOVE CLOSER**</u> *BY FORCE* in relation to <u>***MASS***</u>. Newton submitted the suggestion that objects fall as MASS provides the force that will cause the falling by the inducing of a force he named gravity which he subsequently only proposed was the acting supposititious force. I disprove this formula in so many ways in this book and I show that this formula and the ideas Newton introduced just don't stand up to even the smallest tests. Then, if Newton's idea on gravity has validity and mass is responsible for objects falling, then all objects that are in a process of falling must be subject to mass and in that idea rests differentiation and discrimination in size and compactness producing speed variations. If any and all falling is subject to the variation mass introduces and the influences coming about is the result of mass interfering in the gravity force being generated, this then must bring different speeds to cause substantial variation in the falling of different objects holding different mass factors. There can't be conformity in the falling of all objects while such falling is the result of the discrepancy that mass has to inflict due to variations that result in mass differentiations. This is a vital issue that science eludes and has all clever ways to avoid direct questioning. This part science just run around and never addresses and avoids confronting the issue. This avoidance of confronting the issue whish will disprove the validity of Newton is done with such cunning as you will not believe. The fact that objects fall due to conformity in the falling, science accepts but portrays a picture of deceit that mass brings falling distinction and therefore equal falling doesn't happen, while they at the same time admit to Galileo's presentation that falling of all objects are equal in tempo, irrespective of size or any form of differentiation. While they promote the obscurity that Newton and Galileo is in harmony the truth about their deceit is that the two can never have the same issues. That I prove is a fact and also I show how big a part this is in the overall covering up of Newton's initial fraud.

I have written several books in which I challenge the thought process of Mainstream physics and especially Sir Isaac Newton's arguments about physics. I am of the opinion that even though everyone thinks of Sir Isaac Newton as the genius who established every aspect that is used in modern physics today, but in spite of every other person hailing Newton, I remain of the opinion that the man did not have a foggy clue about any of the principles driving the concept that he named as gravity, or what brought about gravity according to his explaining of what forms gravity. I am able to explain gravity but it doesn't even vaguely resemble Newton's version of gravity. I can explain gravity by proving my explaining with the use of simple mathematics. I use Johannes Kepler's formula to back up my statements. By using Johannes Kepler's formula I found a way to prove there are four phenomena found in the cosmos. There are the four phenomena applying in tandem that together forms gravity. They are: The Titius Bode law; The Roche Limit; The Lagrangian Point System and; The Coanda effect. As the phenomena don't support Newton's vision on cosmology, the phenomena has no support amongst Mainstream science although they did apply it with a positive results in locating the missing planets at the time of their discovery. When they located unknown and undetected planets in the past, the existing of the phenomena was never disputed but when the argument of proving them comes to mind, then they are dismissed as some coincidental abnormality occurring. But since it holds no similarity to Newton's view on science, Mainstream science rather disclaimed the validity of the phenomena than they would find fault with Newton's ideas. In the mind of science the cosmos can be wrong and God can be wrong but Newton can never be wrong. In using the four correct principles correctly, which I back up with the correct mathematical interpretation thereof in support of the function that each phenomena has in forming gravity, I did a far better job than what Sir Isaac Newton did and what I achieved is of a far more acceptable level as well as being mathematically far more correct than what Sir Isaac Newton did achieve with his guessing about issues he couldn't explain. To be successful in my quest to find an explanation for gravity, I had to redirect all my concepts I previously had and also alter all the otherwise normally accepted thinking on physics. I had to find the phenomena and I had to dissect the function of each phenomenon as well as mathematically valuate the phenomena. In this process I realised that to come to realise what gravity is, I had to realise that gravity is not what Newton saw forms gravity.

Newton devised a formula $F = \dfrac{r^2}{M_1 M_2}$ that represented gravity. Newton thought the mass of the apple

that fell drew the Earth as much as the Earth drew the apple. The one mass factor represented the mass of the apple while the other mass factor represented the Earth and the radius was in place of the distance that the apple had to travel as the apple fell from the tree in view of Newton. This falling he saw as the gravity that the Earth's mass and that apple's mass were achieving. Let us have a look at the force F that Newton introduced

What is F and what worth has F while we find out what role F plays. Let's place F in $F = \dfrac{r^2}{M_1M_2}$ and find what F really has in a mathematical sense.

$F = \dfrac{r^2}{M_1M_2}$ can be replaced by $F = \dfrac{a^2}{a_1 \times a_2}$ which then would leave $F = \dfrac{a^2}{a^2}$ that leaves **F = a^0** and that outs the factor of gravity without value or worth being a factor of **F = 1**. This doesn't make much sense but Newton never saw this imperfect outcome to his otherwise perfect formula. But calculating $F = \dfrac{r^2}{M_1M_2}$ in terms of real factor worth it makes no sense in another sense.

Replace all the factor values in terms of $F = \dfrac{r^2}{M_1M_2}$ and with the mass of the Earth multiplied by the mass of the falling body, which was the apple, the force is exceptionally small. If I calculate the force in terms of my view the force that comes about in this formula the result we find in calculating Earth with the apple divided by the distance between the two is something in the region of less than what the mass of one atom would be. This left our genius with some headache and a large problem (or is it a very small force of gravity) to solve. The force coming from this equation is less than microscopic small! Then Newton improvised masterly by cheating the wits out of all mathematical logic.

Newton changed his initial formula that was $F = \dfrac{r^2}{M_1M_2}$ to $F \, \alpha \, \dfrac{M_1M_2}{r_2}$ and the entire world still to this day think this move is brilliant. This is coming from the best mathematical minds found on earth. They applaud Newton in his brilliance by saying Newton never made a mistake. If you think that way then answer the following argument. Newton placed he value F would have in terms of the formula $F = \dfrac{r^2}{M_1M_2}$ as being equal in context to $\left\{\dfrac{F}{1} = \dfrac{m_1m_2}{r^2}\right\}$ and by changing the formula by only changing one symbol α the entire outcome of the formula changed without changing anything. Newton saw it fit to replace ▮ with ▮ and the formula was reborn in value while staying the very same. There is an applying rule or law in mathematics that says when one change a formula from $F = \dfrac{r^2}{M_1M_2}$ to $\left\{\dfrac{1}{F} = \dfrac{m_1m_2}{r^2}\right\}$ then F being F ÷ 1 must also remove a position to become 1 ÷ F making F the fraction value. All those that know even the least about mathematics and of which Newton and his followers not part of knows very well that if any part on the one side changes dynamics from being on top of the dividing line then the very same must apply on the other side. One can't just say that to change a formula $F = \dfrac{r^2}{M_1M_2} = \left\{F \, \alpha \, \dfrac{m_1m_2}{r^2}\right\}$ would not translate in ultimately change the outcome of the formula because the truth about mathematics is that $\left\{F = \dfrac{r^2}{m_1m_2}\right\} \neq \left\{F \, \alpha \, \dfrac{m_1m_2}{r^2}\right\}$ but when it changes the ratio of what is divided and what divides there is a principle in mathematics whereby one then changes every aspect in terms of such a change to alter the ratio on both sides of the equitation as to maintain coherency in mathematical logic and the equation changes become $\left\{F = \dfrac{r^2}{m_1m_2}\right\} = \left\{\dfrac{1}{F} = \dfrac{m_1m_2}{r^2}\right\}$.

Newton had this idea that because he was Newton. The Great (Cheat) normal rules did not apply and with him being Newton even mathematic laws was below his status. He could replace symbols $=$ with α used in changing the formula $F = \dfrac{r^2}{M_1 M_2} = \left\{ F \ \alpha \ \dfrac{m_1 m_2}{r^2} \right\} = \left\{ \dfrac{F}{1} = \dfrac{m_1 m_2}{r^2} \right\}$ and that will change mathematics forever. It never dawned on him or his followers that came after him that $\left\{ F = \dfrac{r^2}{m_1 m_2} \right\} \neq \left\{ F \ \alpha \ \dfrac{m_1 m_2}{r^2} \right\}$ but the correct application is in fact $\left\{ F = \dfrac{r^2}{m_1 m_2} \right\} = \left\{ \dfrac{1}{F} = \dfrac{m_1 m_2}{r^2} \right\}$. Let's find out why the changing that Newton did $F = \dfrac{r^2}{M_1 M_2}$ $= \left\{ F \ \alpha \ \dfrac{m_1 m_2}{r^2} \right\} = \left\{ \dfrac{F}{1} = \dfrac{m_1 m_2}{r^2} \right\}$ in order to improvise for his theoretical shortfall is total mathematical corruption on the highest level.

In $F = \dfrac{r^2}{M_1 M_2}$ the factor of force is with the mass multiplication as the mass presents the radius by the square with the diminishing or increasing value. A large mass will produce a small radius and the mass reduces the radius by determining the reducing of the radius. The force will reduce the radius.

Then in the improvised version $\left\{ F \ \alpha \ \dfrac{m_1 m_2}{r^2} \right\}$ that actually that with the factor manipulating inexplicably becomes $\left\{ \dfrac{F}{1} = \dfrac{m_1 m_2}{r^2} \right\}$ such changes brings total factor revaluation to the entire prominence of the force changes all together. When using the ratio as $\left\{ \dfrac{F}{1} = \dfrac{m_1 m_2}{r^2} \right\}$ the radius becomes the force carrying factor and the radius will determine the mass value in ratio to the formula. A large radius will provide a large force of gravity and not as with $F = \dfrac{r^2}{M_1 M_2}$ where a large mass will influence the ratio as to establish a large force and the result will be to produce a small radius. In the one $F = \dfrac{r^2}{M_1 M_2}$ we the force lies with mass and in $\left\{ \dfrac{F}{1} = \dfrac{m_1 m_2}{r^2} \right\}$ we find the force being in relation to the diameter size. This Newtonians missed all the time during the past (about) 300 years. The Force of gravity changed from initially being in the mss $F = \dfrac{r^2}{M_1 M_2}$ with the mass dividing (driving the force) to being in the radius between the points holding mass $\left\{ F \ \alpha \ \dfrac{m_1 m_2}{r^2} \right\}$ where it is the radius that will determine the influence the mass has on the Force. This is a small anomaly but it shows how little did Newton consider the impact of the stage he brought about and this shows Newton was inspired by one motivation and that was to defraud science.

But then he went much further and cheated the cheated by introducing $F = \dfrac{r^2}{M_1 M_2} = F = G \dfrac{M_1 M_2}{r^2}$. There was never one Newtonian that even hinted that the Newtonian could explain how did the initial thought of $F = \dfrac{r^2}{M_1 M_2}$ than mathematically changed to

$\left\{ F \; \alpha \; \dfrac{m_1 m_2}{r^2} \right\}$ which was intended to become $\left\{ \dfrac{F}{1} = \dfrac{m_1 m_2}{r^2} \right\}$ and then with normal,

mathematical principles still applying change this lot to $F = G\dfrac{M_1 M_2}{r^2}$ Furthermore, how could academics in mathematical physics teach children or students in physics this as the truth! How could any

mathematician explain a process of following logic maintain that $F = \dfrac{r^2}{M_1 M_2} = F = G\dfrac{M_1 M_2}{r^2}$

...explaining it is preposterous.

Let any academic mathematically show how one would go about and use Newton's visionary formula $F = G\dfrac{M_1 M_2}{r^2}$ to calculate the force of gravity by replacing the symbols with the actual values in mass that the symbols should have. Put in the Earth's mass in place where it belongs and put in your mass in place where it should be and then divide that with the distance between your soles and the Earth measured in micro millimetres by the square thereof! If it can't be done, then that is proof of Newton

committing fraud when he introduced the formula $F = G\dfrac{M_1 M_2}{r^2}$ being able to calculate the force applying as gravity. Take any formula used in daily physics and show where they use the mass of the Earth as a factor in calculating anything. Never, not once, do any formula used by physics hint that the Earth's mass has any influence on any part of physics when any one calculates factors to determine whatever they wish to determine. If the Earth's mass is never used in any calculation, then the Earth's mass has no part presented as a factor and then the Earth has no mass that influences any aspect of physics. That means the Earth's mass doesn't produce gravity because if it did, the calculating formulae used in physics must use the Earth mass as a factor in all calculations! Newton cheated to bring in the Earth as a factor that has mass that produces gravity and never does the mass of the Earth contribute to any part in any of the many calculations that form part of physics. The Earth has no mass because the Earth's mass never plays a part in any formula. It is as simple as that! The formula Newton first devised

has not even a ring of truth to it. If it is true then show how the formula reading $F = \dfrac{r^2}{M_1 M_2}$ is used to

indicate that this brings about gravity without cheating it to become $F \; \alpha \; \dfrac{M_1 M_2}{r_2}$ and then

committing blatant fraud in changing the formula to able $F = G\dfrac{M_1 M_2}{r^2}$ while even in this form it still doesn't apply.

If you think I am going on about academics then think how much they tormented me by ignoring me in eight years. With the clear evidence I show they still dismiss me as the one that is mindless because I am unable to "understand" Newton. What is there to understand when everything I am supposed to understand is tainted ands flawed! If it seems I am going into rhetoric about academics then it is because I wish to describe their deceiving methods in dismissing me.

The point I wish to make is that they say gravity is $F = G\dfrac{M_1 M_2}{r^2}$ while they also say that the value of "F" as in gravity "g" is $F = g = 9.81$ and further more they say that $F = mv$ while they first said

$$F = \frac{r^2}{M_1M_2} = F \; \alpha \; \frac{M_1M_2}{r_2} = F = G \frac{M_1M_2}{r^2}$$ Now get this lot married mathematically…that is a challenge they can never manage and yet they say it is true because Newton said it is true.

Let all the physicists show how they manage mathematically to get $F = G \dfrac{M_1M_2}{r^2}$ equal to the measured value that they say gravity has being the "**g**" value and not the "**F**" value at g = 9.81 Nm/s². They advocate that gravity is another symbol that somehow replaces F with g but also is gravity with a totally new value than that which Newton had in mind and then as "**g**" apart from "**F**" has a measured and physically determined value of g = 9.81 Nm/s² So let them do the calculating of the Earth mass and any person's mass multiplied by the gravitational constant and get this lot divided by the distance between my feet and the Earth when I stand on the ground by the square thereof and to top this, they then get gravity to be g = 9.81 Nm/s²

I'd love to see them accomplish that!

When they use another formula that also uses the symbol F in the formula F = mv I still have to find one academic that can show me whereto did the mass of the Earth disappear while taking with it the gravitational constant as well as the diameter parting the mass m from the other disappeared factors. This is one of the many small issues they never think of because they can't explain it while upholding the correctness of Newton at the same time. Let one of them with the many doctoral degrees, show how they

come from $F = G \dfrac{M_1M_2}{r^2} = F \; \alpha \; \dfrac{M_1M_2}{r_2} = F = \dfrac{r^2}{M_1M_2}$ to eventually reappear on the

surface as the formula F = mv. If you thought gravity was an act of magic try this magic. Where did all the factors (**M₁, G and r²**) go while being on route to change in appearance to become F = mv. The mass of the Earth that academics in physics claim is there and that is supposedly is doing the gravity pulling, is a relevance that the object has with the Earth having a factor of 1 and this relation is effective viable only when the object having this mass is resting on the surface of the Earth or having some direct contact through another medium connecting the object to the Earth. The object rests on the link by a link or otherwise is resting directly on the Earth, but the condition of mass of any object has is that the object is standing still or moving while being in direct contact with the Earth. But all action the object has is relevant to the position the object has in relation to an allocated relevance with the Earth and relating to the movement that the Earth has. The object in mass has to move directly with the Earth or slightly more than the Earth. The object only shows having mass when connected and when accepting the movement the Earth has but the mass the Earth should placed into the calculation alongside the mass the object has, that as a complimenting factor is totally absent in normally used physics because the Earth has no mass. The Earth's mass is lacking all visible presence in influencing physics by lending support or increase any calculation in physics. This proves my statement that it is because the Earth and all other planets do not have mass and therefore can't be used as a calculating factor.

Planets have no mass and neither has the Sun got mass except the mass Newtonians wish to credit planets with. Bigger planets don't move faster because they have more mass and smaller planets are not further from the Sun because they have lesser mass. All planets big and small spin at the same speed around the Sun and in relation to the Sun and all planets are scattered going around the Sun while being big and small where all sizes are well mixed. This is because planets have no mass except in the imagination of Newton and his devoted followers. The mass of the Earth never plays a role in physics and the mass of planets do not draw any of the planets closer to the Sun and let one physics professor bring proof that the planets do draw nearer to the Sun!

They just can't because planets do not have mass that can produce a pulling gravity! If and when the mass of the Earth do not feature as a factor in any formula that is used in physics, then the mass of the Earth is no factor playing part in gravity. This then can only indicate that the Earth has no mass. If there is an absence of mass as a factor that influences physics, this can only be as the result that the Earth mass has no gravitational presence in any physics formula. Gravity does have the value of g = 9.81 Nm/s² but that I explain and the value $g = 9.81\ Nm/s^2$ I prove as well. With that evidence being that clear, then the mass that the Earth should supposedly have, does not produce gravity as Newton suggested. Prove me wrong by getting gravity at $g = 9.81\ Nm/s^2$ from using either any of Newton's formulas being

$$F = G\frac{M_1 M_2}{r^2}\ \text{or}\ F\ \alpha\ \frac{M_1 M_2}{r_2}\ \text{and}\ F = \frac{r^2}{M_1 M_2}.$$ Let me see Newtonians do that and I will become a believer in Newton! The Earth has no mass because physics can't show the Earth's mass playing part in calculating formulas and if there is no mass that plays a part that should produce gravity, and then mass can't be responsible for the producing of gravity as Newton declared. That makes Newton's suppositions total rubbish and that makes Newton responsible for a crime of defrauding and falsifying the science of physics. If you, the reader is able to get academics in physics as far as even reading this argument I make, then you are more influential than I can ever be. They plainly dismiss all these arguments with arrogance by discrediting my credentials!

What Newton saw as gravity can't withstand even the slightest test of proof and I showed that it is not possible to use Newton's formula as Newton suggested it applies to mathematically calculate gravity. I come back to this issue later on. I have tested Newton's thinking and the book I offer to you for investigation serves as the testimony to all the testing I did on Newton. This any body who can see, will see when reading this book, I tested Newton from all the angles to see if he possibly could be correct but found his thinking wanting every time. The truth about Sir Isaac Newton's concepts I came to conclude, was that the reality is that it is not in any way overstated to declare that Newton conspired to defraud science and moreover that he committed blatant mathematical corruption in trying to prove the concept he had about what he thought forms gravity. There is no backing for Newton's ideas and even the ideas which are in use are not in the form that Newton said it applies where physics in daily use serves as the best discredit to Newton bringing no proof about any of the claims that Newton made on matters concerning science in cosmic gravity.

I show that every thought Newton introduced that later proved useful and was correct, was what he stole from another far better cosmologist called Johannes Kepler. Not one of his laws are directly relating to any concept Newton ever introduced at any stage but is the result of academic theft he committed against a much larger figure that preceded him by almost a century. But he stole, he lied and he raped the work of a predecessor in order to defraud the world of science in his time. Newton brought no original input into science except that he gave a concept the name "gravity" and even that is inappropriate. Newton made suggestions that break every mathematical principle he could think of. That, Newton did in his attempt to win over the prevailing academic thinking of the day in his time as to lay some sort of groundwork to form backing for his ideas on physics and to attempt to explain gravity or what he thought gravity is. If this is shocking and sounds outrageous, then a lot more shocking detail awaits the reader in this book.

Newton's claims about the principles he declared as being responsible for guiding physics carry no proof and after I realised that, I was able to start forming another line of thought on gravity. After formulating my concept about how gravity was truly formed, I had to introduce my ideas to academics in physics. In my quest to find the method how gravity formed I used the four phenomena and the principles of these phenomena as well as determining in which way each phenomenon applied. Then I placed each one in the way that were known how they work and then implicated that specific formula's function mathematically in forming gravity in the cosmos. This was no easy task but I did it and by formula shows that my argument is logic and the mathematics prove that it works well.

The phenomena that I use is still to this day unexplained by Mainstream science because it shows no sign of using mass and without mass the Newtonian mind understands nothing!. Newtonians don't understand the four phenomena due to the fact that science up to the present date has no means or method to explain the four mentioned phenomena while I can explain the working of each independently and how they work in a combination to produce gravity. I found a way to put those four phenomena in a perspective and put the four in a mathematical sequence that from there I could explain gravity in detail. When I first approached academics, I had the opinion that all academics were knowledgeable about the lack in the correctness we find in Newton's views and that every one in physics would be rejoicing in finding what gravity consists of. I was under the impression that I would be embraced by those in physics for finding a solution to Newton's errors. I was in for a nasty shock with such naivety.

I met with such rejection that no one even cared to look at my work because they were of the opinion that looking at my work would be sacrilegious to Newton. I was told on occasions that Newton has never been proven incorrect and therefore any attempt on my part in doing so is a waste of time. At first I was not confrontational towards Academics in physics and avoided any indication about disagreeing with Newton, but academics always threw Newton at me and eventually for self protection I had to start to confront them and confront Newton, with which I was in disagreement from the beginning although at first I was reluctant to voice any opinion about the matter. But slowly it dawned on me that if I had any serious plans to introduce my ideas I had to dispute Newton's gravity principles and show the inconsistencies and dishonesty in Newton's approach to physics. I came to realise that his flaws are there and the mistakes are present whether I avoid it or attack it; the inconsistencies are part of forming the basis for modern accepted science. It is that strangle hold I had to break before I could even think of finding acceptance about change.

Then slowly I concluded that only and after I can get people to see how incorrect Newton is, do I stand any chance to introduce my line of thought on gravity and I am so sure of my ideas being correct that I dare any one to disprove any part or the entirety that forms gravity as I see gravity! But that can only come about when I can get an audience to see how I expose Newton for what Newton was and in that is where I find no luck. I can't find one academic with influence that is brave enough to stand up and face my attack on Newton and argue me down or prove me wrong in a sound debate. The moment any academic realises he or she is reading my condemnation about Newton's correctness, their minds shut down! No other thought can penetrate their mind but to think in terms of Newton being correct even when confronted with facts proving Newton incorrect. They stop reading my work. They do not get confrontational but defensive and in defending Newton they refuse to read further!

I realise that every one has the view that my finding fault with Newton shows signs of madness and progressive signs of dementia on my part and in my thinking to even regard any possibility that I am the only person on Earth that is correct and all others that ever studied physics are wrong is pure foolishness but mad as it seems, if that is what I have to say to be correct in what I say, then that is what I say...Newton is wrong about gravity. I don't say this lightly or without understanding the enormity of what I suggest is going on, but be that as it may seem, it is the truth without question that Newton went on for three hundred and fifty years defrauding science with no one testing his claims.

Detecting Newton's misconduct is possible because I saw a way to break away from the invalid concepts Mainstream physics holds. I saw where Newton went wrong and correcting the major mathematical error is so small…if only any one would listen! …And of course one has to admit that the Earth doesn't pull or push by mass or any other way just like the physics formula they use to formulate indicates. Notwithstanding the pose Mainstream physics tries to uphold, the entirety of physics still uses the idea of magical forces intervening in nature and they still base concepts on unexplained novelties. Think of how they found four unexplained forces going around and influencing persons in an unexplainable manner except that they can see that it is through the inexplicable magic of gravity keeping people attracted to the Earth. To say the least, the concepts physics use in terms of Newton would not even be acceptable to children in the modern informed era we live in. I challenge any person to prove Newton, not to accept Newton but to undoubtedly prove Newton correct! I recognised the impossible double standards Mainstream physics apply to promote their much shady explaining. In short I tested Newton's principles and found the principles to be wanting on all levels of consideration.

The statements that Newton introduced have inconsistencies and to cover these holes science has in their understanding of cosmic principles, they have to apply standards, which are symptomatic of double vision. To compensate for these bogus truths that were supporting their incredible theories, they simplify issues to such a level where what they embark on is equal to and the same as witchcraft and soothsaying. They admit the cosmos is expanding but the expanding is not here in our neighbourhood because in our neighbourhood Newton rules and therefore notwithstanding a Big Bang still applying and a Hubble expanding going on, in our neighbourhood we contract because Newton said so and Newton just can't be wrong. Their pitiful explaining of the fundamental working of physics is meaningless. In spite of finding evidence that the Moon and the Earth is growing apart in distance they still uphold Newton's view on contracting because it suits their work and leave everyone under the impression that the contracting is valid as it should be if Newton is correct. The Earth and the Moon is growing apart at the same rate as human hair and human nails grow. Because they lack true basic understanding they have to accept the unproven and it remains unproven that the cosmos is coming together by the power of mass that is inflicting gravity. They proclaim to understand what flows out from what they understand but such concepts become meaningless because of many inconsistencies. To name but one such an example is the explanation they put forward in the Tunguska event. To claim that a mini Black Hole went through the Earth is demeaning just going on the basis that they claim there can be such a thing as a mini Black Hole. Such statements are beyond the ridiculous and to achieve some degree of believability from the public they create scenarios, which use arguments that are entangled with deception, such as what is obvious in the case I mentioned. What they declare as unwavering facts can't even be supported in the least form when tested. Even the least degree of verification of correctness is absent and Newton lacks all evidence of authentication in any investigation of even the simplest terms. It is as if they never read with interest that which they explain and they never scrutinise that which they advocate. They give values that are senseless and make that which they say meaningless. In all this they use billions of tax dollars to prove what they have no idea of. They try to commit matter to fusion while they have no idea why matter would fuse at all!

Do you think of astrophysics as being the department that is run by the wise and the level minded, the honest and pure at heart, the nobility of well-to-do academics and the sober thinking standing in front of the world as the absolute trustworthy? If you are a student, there is no other choice you have but to trust them while they feed you absolute hogwash! They force students in believing gravity is the result of mass and to see that students comply with the unequivocal acceptance of the brainwashing, they subject students to various tests and examinations. In those test they determine the degree the students' have developed by brainwashing and the levels they test goes by employing examination standards. They never prove this concept that it is a fact that it is the mass of the Earth that pulls the body down while all formulae used in science prove otherwise by not using the mass of the Earth. While knowing this all to well Newtonians insist on students accepting Newton and still students have to acknowledge these concepts as truthful facts. If you think those in charge of astrophysics are the pillars of trust, then get wise by reading the following. What you are about to read is simply mystifyingly simple and yet to this day I have not had the privilege to challenge one academic any where that had the honesty to admit to the fact of Newton being wrong. After you have considered the following you might agree with me that even small children can reach a higher level of clear-minded logic and find more sensibility than what those scientists promoting astrophysics have because science lives in a make believe fool's paradise. One such example is to put space travel and extra terrestrial life forward as even a remote scientific possibility. That is departing from any possible sane minded thinking. Even to consider space travel as an option shows the level of not even understanding the most basic principle behind gravity as we find the comet proves.

The scientific presumption is that gravity is established when one object holding mass is pulling another object having mass and forces the two abject to move toward each other. The entire basis of all physics rests on this formula where init it is believed that mass produces all gravity by distinction of differentiation in density as well as size and if physics is anything to go by, then what ever is proven, such proof must stem from and be in support of well as being supported by this formula $F = G \dfrac{M_1 M_2}{r^2}$. It is the formula that keeps the entire Universe in place and all of Newton's accuracy solely depends on $F = G \dfrac{M_1 M_2}{r^2}$ as a formula that has to be truthful and unquestionably accurate. The mass is the crucial factor because the mass is in a position where the mass destroys the distance of the radius from both ends equally. The mass generates a force and the force produces the gravity and the gravity produces the pulling and the pulling is what the time depends on that we have left to enjoy a Universe. We have to appreciate Newton's finding of mass until Kingdom comes because if not for mass, Kingdom is coming either tomorrow or never. Then we also accept the formula $F = G \dfrac{M_1 M_2}{r^2}$ has been tested and proven so many times by science that there is no other formula on Earth that has endured the testing that Newton's gravitational formula $F = G \dfrac{M_1 M_2}{r^2}$ has under gone. The force of gravity has the mass that would generate the gravity whereby the pulling of the other object orbiting and in also generating by mass the other object will also force gravity onto the first object $F = G \dfrac{M_1 M_2}{r^2}$. What goes up must come down. We fight our mass because we fight gravity the entire time during one life span we live through. When I jump the force of my mass that generates the gravity by which I pull the Earth and by which the Earth pulls me back and the pulling is the result of the mass of the Earth that pulls me down again while at that moment I am pulling the earth up again, thus the square value coming about in the radius factor. When I fall my mass kicks into action and by mass I hit the ground at a rate my mass will determine. Newton is a genius because Newton realised all these wonderful happenings. Newton saw that a planet pulls another planet by the gravity that the mass of the planet charges. But consider that if mass is what brings about falling, it then implies that objects just cannot fall equal but have to fall differently and according to their mass. If mass has nothing to do with the falling then objects must fall equal and in equanimity through out the entire distance of travel while in the process of falling and that fact goes without argument. Mass brings variation and conformity is the result of mass not applying! The Universe is in a state of contracting

$$F = G \dfrac{M_1 M_2}{r^2}$$

as Newton's formula must indicate. The objects are drawing closer to each other all the time.

Now marry that thought with the ever expanding Big Bang beginning and the Newton's concept of a Universe shrinking which totally contradicts the reality that Hubble found to be true and that there is a Universe out there of which we are part of that is exploding in expansion. To the world they declare openly that Newton's contracting Universe and Hubble's expanding is the same thing and we must wait for the Universe to admit being incorrect and start to employ Newton's contracting. They gave this blaming of the Universe going the wrong way on the Universe being the incorrect party because they are looking for mistakes in the Universe and wait to find out when the Universe will start to comply with Newton and start shrinking because the Universe has to stop this ridiculous expanding since Newton said the Universe is contracting. Since Newton just can't be wrong, therefore the blame of such silly contradicting of Newton has to be found at the door of the Universe. This blame game and detecting how far and why the Universe went wrong in disobeying Newton they named the Critical Density Theory and is the biggest scam and covering of fraud ever invented by any group of persons any time during the history of man. If the Big Bang is true (and it is true), then Newton just doesn't fit! In my book I show how this led to the biggest criminal cover up man has ever devised and was initiated by a person called Albert Einstein. The entire philosophy behind the Critical Density Theory is a scam and is even as ridiculous as what the rest of Newton is. You are about to read how far Newtonians will employ criminal cover up to form a blanket of deception!

The Newtonian formula $F = G \dfrac{M_1 M_2}{r^2}$ explains the comet arriving at the Sun, drawn by the mass of the Sun, pulling the mass of the comet as the comet comes closer to the Sun, but then if Newton's $F = G \dfrac{M_1 M_2}{r^2}$ has any validity the comet has to crash into the Sun after arriving. If gravity by mass was pulling the comet towards the Sun in the manner as Newton insisted in the Newtonian formula $F = G \dfrac{M_1 M_2}{r^2}$, then try and get any academic to explain why and how the comet moves away from the Sun and into the black yonder. After reaching the comet, the comet avoids colliding with the Sun as the formula $F = G \dfrac{M_1 M_2}{r^2}$ would suggest and head into the darkness of outer space. The comet then is moving directly in the opposite direction of what Newton's formula $F = G \dfrac{M_1 M_2}{r^2}$ would have us believe as the comet is not suppose to be pulling away because it is the mass pulling that was in place when the comet was drawn by mass as Newton stated. Does mass then start pushing mass to get the comet floating away from the Sun? Mass establishing gravity by pulling of a force is a gimmick Newton suggested but is unproven and it is nothing less than foolhardy to believe that mass does the pulling of the comet. Try and get those academics in physics to sensibly admit this reality and then in the explaining

be sensible by using their Newton formula $F = G \dfrac{M_1 M_2}{r^2}$ as Newton's formula presents the law to show how this going away happens when mass is doing all the pulling at first. Try and get any Newtonian academic to explain this escaping of the comet from the mass of the Sun in the face of mass pulling mass. Some try to use the idea that the momentum drags the comet around the Sun but the mass will pull

the comet into the Sun if $F = G \dfrac{M_1 M_2}{r^2}$ applies. Newton never created a detour as the mass pulling mass forms a linking straight line running from the centre of the Sun to the centre of the comet. Newtonians always bring more deceit to cover up Newton's fraud. Students otherwise never ask these questions I address because students are brainwashed to accept and not think about asking questions. In presenting my work I can and I do answer the questions raised above but my answers do not fit the Newtonian visions of mass doing the pulling and because it contradicts Newton, I am ignored. In my following describing Newtonians is not to moan and grumble but it is to show the means and the manners they use to fight and when using such utter arrogance, despicable high and mighty autocracy with plain bullying tactics and megalomania. They have this attitude that only they are wise enough to think and the rest is mindless dehumanised animals walking on hind legs. If they fought fair and used intelligence it would not be that bad but to use dirty tactics when confronting me by just dismissing my views from a position of having authority is coward ness. By bullying me from holding a position of being able to ignore me and I can do nothing about it doesn't frighten me, it angers me!

Only when and after proving that a student has totally lost all ability to think for him or her self may a student be promoted into the ranks of their sublime intellectual group. This form of accepting someone into their league they gave the name as being a postgraduate. The sifting process they named examinations. You write on paper what they tell you and never question their opinion and after passing that examination, only then will you ever enter their sphere of intellectual brotherhood. If this was not true, then how could all the misconception I show in this book remain on paper and be taught in Universities for all these centuries? Are there so many misconceptions as I claim there are. Does this sound far fetched? Then you better read on and I will remove your blindfold and show you what a world of deception the Academic Physicist force on us. Read the following and see how they, the high and the mighty, those that think they can replace God and those who think they can think on our behalf and think what to tell us what to think, read how much they are clowns and the jokers in society. Read how little are they, the Academic Physicists, able to understand concepts about Creation while they think they are able to replace God by using their superior intellect.

If you are a student in the science of physics, then ask your Educated Masters to please explain the following abnormalities you are about to read in this book and insist on a clear explanation about the inconsistencies they promote while tutoring physics as if the physics they present are the most flawless and accurate institution there has ever been. Ask those academics supporting Newton about the following flaws that no one, except me, ever mention. Get them to explain the inconsistencies they never talk about. Wise up and confront those charged with tutoring physics and see who should you believe. Then get informed instead of brainwashed.

One very simple example, which I mention now at this point but I do not elaborate on this matter any other place in the book since in this book I wish to limit space, used, is mentioning the gravitational constant. If any one wished to bring in an explanation by employing the gravitational constant also introduced in the Newtonian formula $F = G\dfrac{M_1 M_2}{r^2}$ then using this gravitational constant is one of the ultimate bogus ploys Academics use to confuse the public.

Newton first envisaged the idea that it is mass standing in relation to mass that is destroying the radius found between the two objects forming gravity as presented by the formula $F = \dfrac{r^2}{M_1 M_2}$ but subsequently the notion as well as the formula used changed to $F \; \alpha \; \dfrac{M_1 M_2}{r^2}$. To get Newton's miscalculation $F = G\dfrac{M_1 M_2}{r^2}$ to work with some dignifying crookedness' they devised a constant of sorts going by the title as the gravitational constant and is this constant holding the symbol **G** in $F = G\dfrac{M_1 M_2}{r^2}$ It is put in place as being the same as all the gravity but is apparently that gravity that fills the space between the Earth and the Moon. Now comes the Newtonian part… This same space filling ingredient called the gravitational constant and holds a measured value of 6.67 X 10^{-11} where it is using this value while it is playing its part in filling all the space we find between the Sun and the Earth as well as the Sun and Pluto and everywhere there is space in outer the gravitational constant is the space-filler to have in that space being filled. If you think of space then we have such space filled with a gravitational constant at a value of 6.67 X 10^{-11}. This was the case in the days when it was accepted that ether was filling the space the gravitational constant filled and therefore ether might have had the value of 6.67 X 10^{-11}. Then after finding no evidence of ether, the ether that was not filling the gravitational constant was miraculously and by a stroke of Newtonian magic removed and replaced with…nothing…yes, nothing is now filling the space ether filled before they realised ether was not filling the space but the marvellous part is that nothing that the replaced ether took from the ether that is not there the value given to the gravitational constant and now while space is filled with nothing it still holds the measured value of 6.67 $X10^{-11}$.

It is not my manner to speak ill of the brain dead or the dead by other means, but in the case of the Newtonian academics I am left with no option. Their forces haunt me to death and it is their forces and ghosts and witchcraft I have to fight. The lifting of a body comes quite natural when a certain speed is exceeded. By exceeding $7(3\Pi^2)$ the body will start to lift no matter what the mass is. A 747 Boeing of multi tonnage lifts off spontaneously at the excess of that speed.

Newtonians are forever concerned with middle ages and with forces they can't explain but such forces and witches there are not, therefore they do not have to fear and can sleep well at night. In the sketch the circle portrays a glass and the arrow portrays running water. The Coanda effect is the water that does not drop straight down but follows the

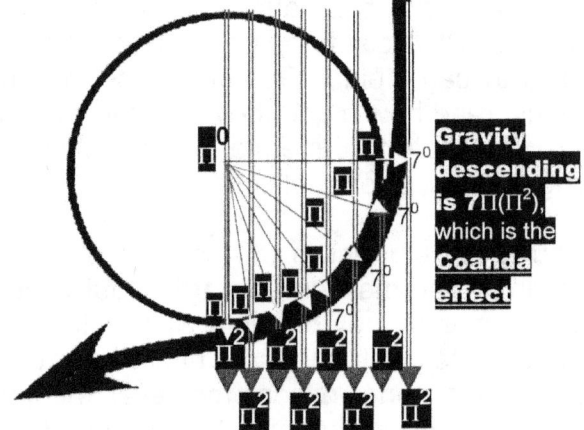

Force on glass

Force on water

Gravity descending is $7\Pi(\Pi^2)$, which is the Coanda effect

curvature of the glass.

The Coanda effect is gravity and my explaining this statement is part of many other books. The Coanda effect shows how liquid attach to the solid by $7(3\Pi^2)$ and the solid attach to the liquid by a relevance value of $7(\Pi\Pi^2)$. That is gravity.

Should anyone require more or better explaining, I would advise that person to purchase any of my books holding the title as an Open Letter. A flying object is under this gravity control of movement and it is this that has crafts fly and cars requiring down force by the aid of aerodynamic devices.

 Perceived natural direction of travel

When viewing any object travelling, we have a perception of the vehicle heading straight ahead in a straight line whether it is a donkey cart or an aircraft, to our perception it is all the same.

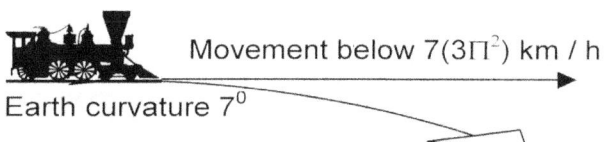 Movement below $7(3\Pi^2)$ km / h

Earth curvature 7^0

Our perception is letting us down because it is incorrect. The moving object is following the curvature of the Earth and that is turning by 7^0 as we travel.

 Movement below $7(3\Pi^2)$ km / h

Natural gravitational displacement or natural direction of travel

In truth we are dropping by **7^0** as we follow the curvature of the Earth. This will be valid as long as we go at less than $7(3\Pi^2) = 207$ km / h. It is gravity (Π^2) multiplied by time (3) multiplied by the space material holds (7)

When exceeding gravitational lines in singularity running at $7(3\Pi^2)$ we see the vehicle lifting into the air as we think the vehicle is getting airborne. Again it is our perception letting us down because it is incorrect. The moving object is instead following the curvature of the Earth and that is turning by 7^0 as we travel, now it is breaking free or releasing from gravity by following a diversion of 7^0.

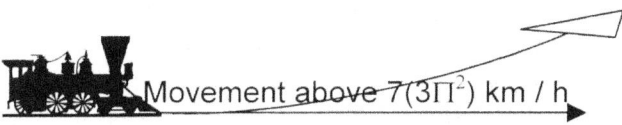 Movement above $7(3\Pi^2)$ km / h

Perceived natural direction of travel

 Movement above $7(3\Pi^2)$ km / h

Natural gravitational displacement or natural direction of travel

The object is rejecting the gravity the Earth enforces by applying individual gravity in the excess of $7(3\Pi^2)$ and in that it is relieved from the gravity the Earth applies. It has nothing to do with air lifting a car. By the way, the sound barrier is just adding one phenomenon further, being the Roche limit in double at $(\Pi^2/2)$ bringing the total speed required to match the sound barrier at the gravity velocity of $7(3\Pi^2)(\Pi^2/2) = 1022.795$ km / h.

As soon as the idea of gravity connects with the idea vested in thought of mass by sharing the split concept in a unifying bout, then I say that gravity there is not. I say, what is defined as gravity is not gravity and then what is acting as gravity is not named correctly.

Gravity is defined as a force that is present in mass pulling mass and it is that entire idea that there is no evidence of. When I refer to gravity everyone grabs on a cultural notion of a concept they formed and in that concept they link the smallest part of the concept to become and represent the overall gigantic principle and by knowing one line, everyone has the opinion that anyone then is the absolute master on the idea of gravity. When I freeze any substance, the substance contracts to a liquid and with more cooling it contracts to a frozen state of ice. The gas expanded more than what the solid did because the gas is hotter than the solid is. When we form the opinion that the outer space expanded to the limits, the idea springs to mind that outer space is freezing cold. When I say the Sun freezes hydrogen to a liquid because my eyes see the liquid squirting from the Sun, I am dangerously mentally impaired, since the Sun is blistering hot. Then through this culture my effort to say gravity is motion and motion is the cooling of an overheating and thus expanding Universe goes wasted. Everyone has the opinion that where gravity is the strongest such as the case is on the Sun or at the centre of the Earth, such a place is extremely hot and where gravity is least that place is unbearably cold.

Students tell your Professors to stop deceiving and stop trying to control your minds with their fraud. Those Academics tutoring you are telling facts about gravity that have never been proven. That is mind control.
They wish for you to accept facts on gravity that they hold as the truth. Should you question that mass produces gravity, they will expel you from University by letting you fail your examinations and it was never proven. Academics do put mind control to work on unsuspecting students by forcing students never to question the legality of statements they offer as being sound and correct.

For the first time ever and this statement takes time back as time runs further back than the time when and since the time Newton introduced gravity there now is a logical and simple explanation as to what gravity is and why there is gravity. Before I achieved that discovery, I firstly had to find the centre of the Universe because it is there that I could locate gravity. I can now show how gravity forms because I have detected the centre of the Universe. Academics guarding physics will never allow an outsider to enter their domain without the intruder paying a heavy price and in this matter I was the intruder. I argue that if

it is the correct practise to use $F = G \dfrac{M_1 M_2}{r^2}$ to calculate gravity, then the radius holding the

gravitational constant must lead one to the centre of the Universe. If the Sun for instance has mass that is apart from the Earth and the Earth also has mass and there is a gravitational constant between the Sun's mass and the Earth's mass, we have the radius in that location. It then must be the gravitational constant that fills the space that the radius holds. Through my venture I discovered one person that knows what gravity is!

Ask yourself the following: If gravity pulls towards a centre and gravity holds the Universe attached, the question arising from that simplistic answer is then … where is the centre of the universe?

Science sees to it that Kepler stays the least appreciated Cosmologist where as in truth Kepler proved gravity, proved singularity, proved space-time, proved the Big Bang, proved every dynamic most of the wise persons afterwards thought about.

Yet, no one gave Kepler any recognition up to now because science denies Kepler his limelight.

Module Two

Students Beware of the Mistake

I DISCOVERED A MISTAKE IN SCIENCE.

This is a crucial mistake since it touches every aspect of the foundation of physics and rocks physics like no Earthquake ever rocked any part of the Earth ever…it puts science on a cross road that will change cosmology forever.

It reverses the essence of cosmology.

It changes the most fundamental foundation of physics.

I contacted the custodians and arch fathers of physics and enlightened them to my discovery.

From them came no reaction.
After seven years of constant knocking on doors not one reply was ever returned by a single academic professor.

Not to disprove me.

Not to point out my mistake and my error.

I went to see countless academics wherever I could and whenever I could and that I did with much hope riding on their sincerity and it was done with much cost on my part.

What a fat lot of good all that did me by spending all that time and money going into all that effort. If they were sincere it was only to save their work by not recognizing mine.

 I have contacted by Internet more than 1500 academics in physics in numerous countries all over the Globe and conveyed this unfortunate occurrence in physics to them with no reaction on their part. Nothing came back in any form being positive negative or neutral.

Then to some I sent letters addressing the problem that I saw in physics.

The letters I sent to professors through out the world became so numerous that I started forming books from the letters I sent explaining my point of view about the mistakes and in finding answers and corrections about the mistake there is in physics.

No reply came back…not to discuss my point, not to discuss their point, not to elaborate on the mistake I so clearly saw, not to defend their views about the point I so clearly saw, not to set my mind at ease with some explanation about facts I am unaware of, not to say they are aware of the facts I am aware of, not to thank me for contacting them confirming that I must not contact them again in future for they shall contact me in further corresponding (at least my Bank manager wrote me this reply when I approached him for an overdraft although I know that on his part he did so just out of common courtesy), not one said they were aware or unaware of the issue I pointed out and not even once came one acknowledgement about receiving any documentation carrying the information on paper of the facts that I pointed out. I found that odd but still I went on trying to find some response somewhere about the mistake I discovered and which I was trying to uncover in science.

I could for my effort wrote them from the grave for I was not even a spook they could not notice although I had the ability to offer the solution that could correct the occurring flaw.

I wrote and printed many books showing my concerns I have about the mistakes. At one point on one day I posted eighty 80 books in one shot to most Grande Universities that were exceptional in their Astronomy Departments and to this day I have received no word in response. In the books I addressed the issues in detail and the work I named An Open Letter to Selected Academics in which I proposed the full remedy to the mistakes I indicated.

It drained my funding totally and I got nothing back for all the effort and the money I spent in trying to charge a reaction. The only response was just a vacant silence; they did not even E-mail me in response acknowledging that they received such a book that I sent their way.

At this point so far after all my numerous attempts in trying to establish some contact with academics world wide I wrote seven books in a combination I titled "Matters Time In Space: The Thesis" covering the entire issue of my work plus four books wherein I combine all the various letters I wrote to academics through out the six years of ardent trying to establish some line of communication. The last letter I addressed to academics I include as part of the content of this web page for your insight and which forms part of one of my books where I join and elaborate on the letters that I combine to form a unit as a book.

The Books holding the letters are entitled

1) Newton's Mythology
2) Newton's Fraud
3) Sir Newton's Fraud
4) Sir Isaac Newton : A Conspiracy to Defraud Science
5) An open letter On Gravity Part 1 Volume 1 + 2
6) An open letter On Gravity Part Volume 1 + 2
7) An open letter Announcing Gravity's Recipe
8) An open letter Addressing Gravity's Formula
9) An open letter About Gravity's Prescription
10) An open letter Explaining Gravity's Rules
11) An open letter To Selected Academics
12) A Cosmic Birth Dismissing Nothing

But in my work I do some things no one should do. I break rules never broken by man in three hundred and fifty years or more. I cross a line that is forbidden to cross by any man not dead or insane. I go into the darkness of the foreboding chambers of insane madness and mental instability.

In those letters mentioned above I call Creation by name and prove with science that we are in Creation. I employ science to prove that that which resembles the Biblical view of how Creation started. I prove that what controls Creation, is also which is not in the Universe yet is noticeably because it is not in the Universe. In the light of all proof and when facing evidence I bring I dare an atheist to prove me wrong about Creation. In mentioning this word Creation by name in a science book I break a ground rule enforced by the atheistic dominated world of science. I overstep all boundaries because I prove mathematically that Creation (the entire Universe) came about in the manner exactly and precisely as the Bible states…to the letter).

Then I go further and do the unspeakable, the act that proves insanity, the dead that vanquishes the force by distorting the gods presence in physics, becoming the reason why the cause of the future Earth will be destroyed… why what would have been will be no more because I unleashed the wroth of God onto man bringing punishment forth that man cannot endure, I sent my sole to physics hell just because I criticize Newton!

I challenge any person to bring proof about any part where any of my theory might be incorrect and furthermore I challenge any Academic in physics to prove that Newton's mass pulling mass is anything other than fraud. I charge any one to bring proof that the cosmos is contracting by the force of mass and that mass produce gravity as Newton advocated when he committed the biggest fraud of all times. If you are of the opinion it is ridiculous that I say that Newton committed fraud then answer the allegations I make and which I prove. I know every Newtonian is shouting for my scalp on hearing this allegation and I did consider the penalties I possibly face that goes with such allegations before I made the allegations.

Since the death penalty was abandoned in most countries I contemplated the fact that I could not receive the death penalty for my attitude towards Newton. The time has past where I could be charged by the elite in physics on showing much insane behaviour on my part and brought with charges hanging over me to the effect of hierarchy. I could no longer be held accountable for blasphemy against god Newton and that I had to answer the charges of hierarchy and address the evilness of my ways in front of the elders from which the punishment I would receive would be a justified torture to death as William Wallace received. I could no longer get whipped until death brings relief to my enduring agony. I knew with some comfort that I will not be arrested and brought in front of the International Tribunal for Human justice in Den Hague where my case will be heard in front of a panel of justices that was paid beforehand to convict me while excluding all chances of me not meeting my destiny and death. I had to be on guard and vigilant in every step my family or I took but that was a small price concerning that even in such an event of assassination by bullet from a distance using a high power rifle meant that death I receive in such a case was quick and simple. They could no longer hang me from any lamp pole of their choice without giving me a fair trial and then for further punishment after my death to leave my body to decay while everyone smelling my degenerating corpse will have a life long remembrance never to commit such a heinous crime, such an unspeakable sin, such a disregard for the holy and the precious and show such an ostentatious example of mental vexing. However I realise regardless of what civil law allow to the academics in physics world wide I still deserve such punishment in the eyes of God and of man and all life forms on Earth for I do what no man may dare do…

I challenge the correctness of Newton's view about gravity.

I even improve by showing where, how and why it is incorrect if only I could find an audience prepared to listen.

I show that Newton is incoherent and that Newton broke mathematical principles to validate his corrupt view about physics. He stole the work of Kepler, raped the content there of, vilified the meaning and corrupted the truth in Kepler's work. Then he pinned all the blame on Kepler by naming this corrupt lot after Kepler, as being Kepler's laws giving the lasting impression it was Kepler showing the insanity. Newton disgraced the work of Kepler by destroying the correctness thereof and even in this web site I prove this statement where my proof can leave no doubt as to its authenticity…and yet I am sure that again there will be not one Newtonian amongst the lot of them that will even glance at this page just because I criticize Newton.

As soon as they find I discuss Newton in a negative light the Academics stop reading notwithstanding what the content reads. I think they are the one showing stupidity…and they realize it. I think in the end I am the only one that truly understand Newton and can therefore see the mistakes he made in the formulising he did on the subject of gravity coming from mass, while they wouldn't dare read what I have to say about Newton because they are the ones not understanding a word about Newton. If they read what I say then they would be in a position where they had to agree or disagree on what I say about Newton and give reasons for agreeing or disagreeing on what I have to say about Newton. It is not possible to understand Newton and not also understand the mistakes he made. It is not possible to understand Newton and see the laws in physics as well as in mathematics that he broke to justify his incoherent claims.

If only there were one incident one Academic turned around and showed me where my mistakes in my assessment about Newton's reasoning is and where I go astray in stead of saying I don't understand Newton and it is not for every one to understand Newton because Newton requires insight and years of study. I am sick to my sole of hearing that trash because there is nothing to understand after one realizes that it is Newton himself that was the one that didn't understand cosmology. Newton did not understand Kepler! Or the other defence that usually comes to bring an end to an embarrassing situation on the part of the academic I visit by appointment is ending the conversation and the meeting abruptly with conversing the words to the affect: Mr. Schutte, Newton explains the situation very clearly where he can leave no doubt and there is no point in debating the issue any further. Then in the same breath without reading one word of my work further they add that Newton's explanations are accepted across the world for many centuries and there can therefore be no doubt about the correctness of Newton's work and no one ever could prove him wrong.

Some even add that because no one to this day was ever able even once prove Newton wrong and therefore any attempt in doing so is fruitless and foolhardy. Only the insane would attempt it... and if they read one paragraph on they would see that I bring the evidence I put in front of you now. They are at that time minutes if not seconds into an hour appointment. With that my appointment ends a few seconds into my appointment and just starting an hour appointment with whichever professor I made the appointment notwithstanding the time and effort as well as cost on my part to come to honour the appointment. On some occasions at some Universities I was told to leave my book at the main gate where security checks passes and prevent unlawful entry as security will see that professor so-and-so will receive my book. I am told to leave the book I wrote on the matter of Newton at the main gate before entering the institution because the professor could see no reason for meeting me notwithstanding the money I paid and the effort I made to travel a great distance to meet the person. This takes frequently place when I tell the academic and academics (plural) I had written a book where I highlight the errors Newton made.

One of my books I named Xepted Newton mistakes and not once did one academic see it fit to give me an interview on the content of the book after hearing the title I chose for the book. I was eventually forced to change the name to Xepted Science Mistakes, which brought me just as little joy in the end. On so many occasions I was told they see no point in reading my book about Newton's mistake since there are no mistakes and therefore there is no point in meeting me on the matter. I came to the conclusion they are on the offensive because they are on the defensive about Newton and that they do because they make no sense of Newton in even a small percentage of his work. It is easy to refer the blame onto me where I have the disposition of having no academic standing or authority while the truth is that it is not I that don't understand Newton because I am the only one that does understand Newton. The behaviour they show is both not rational or sober for the guard they produce to shield them from me is not responsible behaviour of men with clear reason.

...And since I still show no remorse as much as I show no regret for my insanity I am the one that can gain no grounds or benefit from at least one case where there is one academic somewhere showing one bit of doubt about Newton and therefore are prepared just to read my books from start to finish without the Newtonian bias interfering. Not once could I find one academic that would sit with one of my books from start to finish and read past the page I start to show Newton's defects although I am so obviously correct on every matter that I state. I am washed off the Earth for I show little regard and even less respect for the consecrated sacrosanct hierarchy of Newtonian wisdom and for that attitude I find no ear in the world of science prepared to listen to my insanity.

Later on I was to blame and I am sure of that but not at the start. In the beginning I showed much respect and that got me nowhere very quickly and it was expensive. Later on I started attacking their religiosity with venom. Now I go into detail and prove what fools those honourable Academics in physics are and how they corrupt the young mind to brainwash the young and vulnerable in accepting the detestable criminality of lies and deception that they call astrophysics. I challenge any one to show the correctness of Newton's view about gravity in the face of the evidence I am about to bring. I charge students to challenge the academics with the evidence that the academics present to portrait what they advocate as being religiously correct. In order to reach the heart of gravity one must discover the heart of gravity. To find the truth then answer to your person in privacy and in all honesty the following questions that I put to you. It will gauge your state of brainwashing and show the amount of damage that you have suffered this far in your particular and specific case.

Do you know what gravity is?

If you knew would you tell me what gravity is...yes beside it being a force...well we all know that gravity is a force but other than being a force that acts on behalf of mass and pulling unsuspected objects all over, what is gravity, which is besides being Newton's pet force.

If you don't know what gravity is besides knowing that gravity is one of the four forces and it is Newton's original force, then do you know someone that knows what gravity is?

 If you still answer in the negative and you still don't know what is that which is behind what is causing of gravity, then have you heard of any one that knows what gravity is?

Maybe there is one Academic professor or a NASA Scientist that knows about some person that knows one who knows what gravity is, other than knowing the fact that gravity is presumed to be one of the four forces and being Newton's personal pet force.

There has to be an Academic professor or a NASA Scientist known to someone somewhere that knows about some person that knows one who knows what gravity is, because there are so many Academic professors and even more numerous in numbers amongst the many NASA Scientists with super human mathematical abilities in the art of physics acting as if they know. There are even those going around with ideas to renovate the cosmos by assembling space whirls. They apparently plan great voyages that will take man over great distances while they are so informed on the matter of gravity and space-travel amongst gravity. With their absolute phenomenal calculations as they present physics they are showing such abilities in a class which no one can imitate, which must represent a picture about having super knowledge on gravity in the most precise detail.

 When reading what they say they are able to accomplish in terms of astrophysics it stands to reason that they know gravity to the smallest detail there is to know and know everything anyone can ever consider knowing or hope to know about gravity…after all they can present the cosmos as they are able to explain the cosmos with gravity taking centre stage in the past, present and future of the cosmos. This they accomplish by presenting a few mathematical formulas in which all of the Universe then are defined. They can calculate all the matter throughout the entire cosmos, adding every atom by mass into the conclusion of their calculations when they determine the critical density that the cosmos has, which is responsible for the entirety that is providing all the mass that provides all the gravity. They have gone as far as even calculating the explicit required quantities of atoms that forms mass which should be available, which they then find to be short falling in the mass availability through out the vastness of the Universe and the missing mass is not establishing the matter density throughout the vastness of the Universe required to bring Newton's vision on contracting to reality.

Their calculation ability is so vast that they have the opinion about the mass they measure that is forming the gravity they require which is probably less than the requirements needed to substantiate the cosmos' effort in rendering a constant supply of gravity that will eventually secure the returning of everything to where the cosmos came from. This eventuality they named The Big Crunch even before locating the Big Crunch. It is like naming a baby even long before knowing how the procreating is taking place that will lead to impregnating of some member of the specie (which member it will be is still unclear) where it later on will lead to conceiving the baby … that is the manner in which science dogma is enunciated but that is how clever those are that knows everything there is to know on gravity, or so they pretend to have in their promotions.

 They know gravity to split detail where the detail goes to such precise extend that they are able to calculate how much gravity the missing dark matter in all the Black stars must be to provide a force allowing the cosmos in experiencing the next big implosion that is coming somewhere in the future. That the implosion must come even in the face of insufficient gravity is a certainty otherwise Newtonian physics is completely inadequate in their cosmic vision about the Universal future! With their having this qualified virtue of intellectual splendour spawning such phenomenal abilities they then would have to know what gravity is!

 Well…if you don't know any one that knows anybody that knows someone somewhere that is familiar with the ins and outs of what is causing gravity, I then can assure you I know about someone that knew all there is to know about gravity. I am the person that knows someone and that someone knew gravity…but he is dead now...died a premature death long before his time (have you ever heard of any person that died spot on at the second that was his time where he was suppose to die). He sadly passed away and is no longer with us. Still I would like to introduce him to you…and about his work of course, if you would page on.

Where would you, the person that is reading this page, place the centre of the Universe? Whatever your insight into physics might be, it will be unfulfilled because of your inability to accurately place the position of the centre of the Universe. Without such critical knowledge to your disposal, you then have no idea where gravity comes from or where gravity is taking you this very minute. You have no insight into your immediate future and therefore have no control over your life or your destiny... Test your thoughts about the following: If gravity pulls towards a centre and gravity holds the Universe attached it has to be pulling to a centre, therefore then the next question arising from that simplistic question must then be… **where is the centre of the universe?**

$$F = G\,\frac{M_1 M_2}{r^2}$$

If that is the formula the cosmos abide to and that is the formula controlling the cosmos the radius of all things should lead gravity on an eternal quest to voyage towards and finally into the centre of the Universe. If gravity is taking whatever there are to where all things will end and all things are forced by gravity in relation to move according to mass to such a point that will unite the entirety of what is into what will be that outcome can only be where the centre of the Universe are allocated at this moment. The first thing to do is to find where r^2 ends because where r^2 ends it will be where we will locate the centre of the Universe.

The factor r^2 should eventually lead everything to the centre of the Universe in the end…then where is the centre of the Universe and where is the factor r^2 going and conclude in the end at the final destiny?

Module SEVEN

Brainwashing and Mind Control is Everyday Practice In Physics

If you are a student in physics then this information you are about to read is most important. In the classes you attend in physics, has any one confirmed a location where one might find the centre of the Universe? Have you been told precisely what causes gravity to pull? Have you as a student in terms of the fact that you are being a student been informed how mass confirms gravity? What evokes the force that establishes the pulling that confirms the mass that produces the gravity? If no one went to the trouble to tell you this, is it not about time that someone exerts himself and do the honours? On the other hand have you been asking what evokes the force that establishes the pulling that confirms the mass that produces the gravity? If not, why have you not gone to the trouble and just ask this simple question? It would be most interesting to hear the answers those lecturers will come up with since these questions have not found answers, up to now that is. I wrote a book in which I found a means to define gravity. I did accomplish this for the first time ever since the time Newton introduced gravity. This is more than Newton achieved and it is more than the whole lot of Newtonians achieved in three hundred and fifty years. I could do that by accomplishing one thing all others thought not to be possible! Before I achieved finding what gravity is, I first had to find the centre of the Universe because it is there that anyone and I could locate gravity. I can now show how gravity forms because I have detected the centre of the Universe.

Is there any Newtonian applauding my effort and congratulating me in my achievement? If there should be one such a Newtonian, that Newtonian still awaits birth. I couldn't find one Newtonian even being prepared to read what I have to say about what they have nothing to say about. I could therefore not locate one publisher that was prepared to publish my work because before publishing they first have to read my work and no one was prepared to even glance at my work let alone read it intensively with publishing in mind. But I need to get the information out to everyone to get anyone to read my work. In achieving that I had to resort to private publishing because from the nature of my work I take Mainstream science head on and am confrontational on most aspects of astronomy including astrophysics and the founding principle guarding the authenticity of physics. To have a publisher backing me in order to publish my book the publisher had to find an academic prepared to back up my statements that Newton is a criminal that committed extensive scientific and mathematical fraud! In that sense there does not seem to be any publisher that wants to go head bashing with the Physics Custodian establishment of science on official science principles, which I have to do to convey my message in no uncertain language. I argue that if it is the correct practise to use to calculate gravity then the radius holding the gravitational constant must lead one to the centre of the Universe. This fact Newton nullified by using the argument that the rotation nullifies work done because the rotation is in repeat of the process and through that the radius between the centre and the point rotating is nullified. If you don't believe me then explain what he says in this statement.

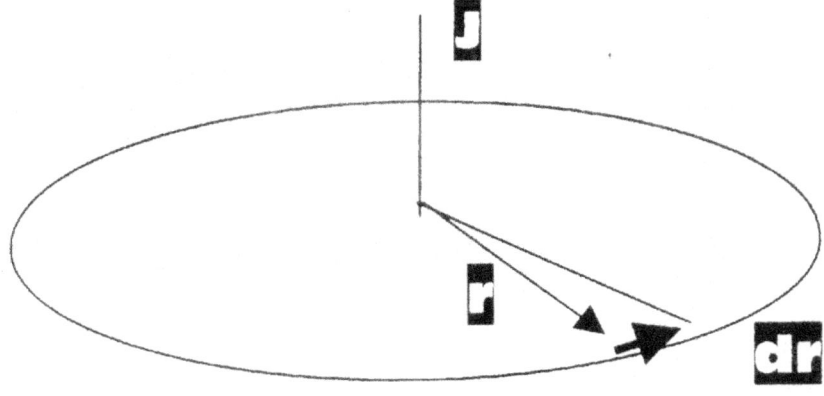

This picture in its entirety condemns Newton's statement of the rotating distance nullifying the motion totally. $\frac{dJ}{dt} = 0$ Newton, and science, made one enormous blunder, from taking this stance. It is as if they took the idea that when a wheel spins the radius of a wheel has not to any influence on the wheel. In doing that, they removed the very fact that keeps the wheel at a radius and size and cosmologically the universal attachment together. They put two objects in an attaching relevancy and then announced that there just is no relevancy applying between the two. When one divides into another there is an irremovable ratio in place. Removing that ratio is breaking the most fundamental mathematical principle. $\frac{dJ}{0} = dt$ or $\frac{0}{dt} = dJ$ This disputes mathematics. DJ / DT can have any number except the only number not possible to have is zero. This is mathematical and physics fraud and there is no other way to put it but to put it into its correct context. I challenge any academic to show how the radius does vanish by the circle running a full rotation or how the radius removes the circle after it ran a full circle…and make no further judgement error when mathematically expressing that $\frac{dJ}{dt} = 0$ then the statement must directly translate to also saying that $\frac{dJ}{0} = dt$ or $\frac{0}{dt} = dJ$ placing the following is true $\frac{dJ}{0} = dt = 0$ or $\frac{0}{dt} = dJ = 0$ $0 = dt$ or $0 = dJ$. Now match this statement with the sketch on the previous page and reconcile that with the disappearing of either the radius (becoming nil as in disappearing altogether because that is what nil is expressing) or the circle (also becoming nil as in disappearing altogether because that is what nil is expressing in the case of becoming 0) while the other is still remaining and see how mathematically corrupt that is. Then tell you Professor to explain how the supposedly best mathematical brain ever could nilly-willy come to this conclusion without corrupting mathematics and physics. Tell the professor to convince you why he as the academic is backing up such corruption in mathematical law while he remains as innocent as a newborn baby in the matter of promoting corruption and deceit. What the statement expresses is that the radius between the Earth and the Sun disappears and the Sun and the Earth combines as a unit. There is a distance between the Sun and the Earth notwithstanding Newton's fraud. But let's return firstly to my quest in avoiding the consequences of Newton's corruption by locating the true gravity that is out there waiting since the arrival of the birth of the cosmos to be discovered. First let's go in search of the centre of the Universe where the birth took place.

If the Sun for instance has mass that is apart from the Earth that also has mass and there is a gravitational constant in between the Sun's mass and the Earth's mass, we then have the radius in that location. It then must be the gravitational constant that fills the space that the radius holds. It is rather obvious that while the radius is filling the vacant space between the Sun and the Earth, it is the only place left where the gravitational constant can hide. To find the centre of the Universe I had only to find the gravitational constant that holds the centre. That then would be where gravity is or so should logic prescribe.

If you think scientists know what gravity is do not be duped that easily because no one in science remotely knows what gravity is…not even Newton knew what gravity is except Kepler… and because of what Kepler introduced, now I know I can prove what gravity is. Gravity is precisely what Kepler said gravity is and only Kepler knew where to find the centre of the Universe because only Kepler knew what gravity is all about. I used what Kepler brought into the world to locate the centre of the Universe and therefore the point where gravity starts. But I did not use Kepler in the way that Newton corrupted and raped Kepler because one cannot commence when corruption destroyed the truth.

Try to get an answer from an academic in physics about where the centre of the Universe is, is like trying to touch the moon. Science can't see past Newton and Newton couldn't see past mass and mass does not exist so to Newtonians it is an endless cycle of getting no where as far as detecting the truth while it is taking them an eternity not to get there very fast. So Newtonians put it on an elusive undetectable and unexplainable force that can conceal their stupidity and Newton's fraud.

By merely putting gravity in the Universe by telling everyone that gravity is acting as a mysterious FORCE that is pulling towards a common point in an allocated general centre is rather avoiding the question with simplicity because the question about how and why remains unanswered. Not knowing the answer to where the centre of the Universe is, will leave you feeing empty and unfulfilled, because being a student and not knowing is the same as suicide on a mental level. That is why you are primarily a student. Being a student is being in search of information and knowing you might never achieve the prime information in physics must be devastating to eager minds such as yours. Ask yourself the following: If gravity pulls towards a centre and gravity holds the Universe attached, the question arising from that simplistic answer is then … where is the centre of the Universe? Newton was unable to find the answer. Newtonians took all of three hundred and fifty years not to find the answer. Do you wish to spend a lifetime searching and never find the answer? Then become the next generation of Newtonian Masters. However, if you discard the falsifying of facts that I charge Newton with, physics will present you with an answer as we follow Kepler's lead.

Should you decide to purchase and read this book, it will bring along a new perception about Kepler. Science sees to it that Kepler stays the least appreciated Cosmologist where as in truth Kepler proved gravity, proved singularity, proved space-time, proved the Big Bang, proved every dynamic most of the wise persons afterwards thought about. Yet, no one gave Kepler any recognition up to now because science denies Kepler his limelight. All they can see is the way Newton raped Kepler by falsifying everything Kepler introduced.

I came upon a mistake concerning physics.

This mistake is about the cosmic phenomena called gravity. Detecting the mistake is simple because it is uncomplicated to understand. Academics in Science say that a feather will fall with the same speed as that of a large rock.

That is according to Galileo and that is accepted as a principle in physics. For the first time ever since the time Newton introduced gravity, I seem to be the person that questions this interpretation.

How does mass pulling mass and falling by the gravity power exerted by mass then fit into this interpretation, because a feather has much less mass than a large rock? If this statement is untrue Galileo is incorrect and the Pope does not need to apologise to Galileo as Physics insist the Pope has to do. Then the pendulum doesn't indicate time and mass does implement falling of objects as Newton protested it to be. But we know Galileo is correct and that makes Newton's suggestions what they are…merely suggestions that prove to be incorrect.

Objects of different sizes all fall equal as seen on television so many times. An army tank will fall as fast as the tank's corporal, while being either next to it or on top of it or in it. The corporal does not stay behind and watch his tank speed to the ground while he is following much slower and at a sizable distance behind. How can a large mass pull as equal as a small mass pulls to travel equal at the same speed over the same distance and still be driven by the power of mass creating gravity. Have you given this a thought?

When you disagree with any academic in any lecture hall about mass not forming a picture as responsible for pulling gravity and you come to a conclusion that you doubt the mass part that they bring into the picture as establishing gravity the academics wipe you from the table with a swipe because then they contemplate that you are so stupid you fail to see   all the facts that physics present as proven facts and they hold you as being too stupid and mentally underdeveloped to appreciate or to understand physics. They tell you that Newton is not for such stupid people that are unable to see how true Newton's arguments are!

I have been at odds with academics for years and only because of the superior positions they hold in office, are they able to bully me into silence but not into submission because that is what this book is about…to expose their corruption.

They can push me into abasement but never into abeyance. By the important Academic positions they hold in the huge academic institutions that give them sanctuary they might dictate the terms of our meeting and in terms of those advantages they can hide behind the criminal wall of deceit and suppress me into silence but they will never get me into submission.

Now I am taking my case to the members of the public so that the truth must be brought into the open. I have had the tour they give and then more came my way. I never got around swallowing their gravity that comes as result of mass creating gravity or any part they present as facts as to how it happens and where science is of the opinion that mass pulls as gravity is... Academics condemned my work and therefore me and for six years I could not get a publisher to come around and bother to read my work let alone seriously proposing a publishing contract. I had to finally go private with the publishing as all doors shut in my face as soon as the academics read the content of my work because from the nature of my work I take Mainstream science head on and as I am sure about my arguments I am confrontational on most aspects of astronomy. There does not seem to be any publisher that wants to go head bashing with the establishment of science on official science principles, which I have to do to convey my message in a no uncertain language. If you also have doubts about the academic's indisputable correctness please read on and confront either them or me on everything you read here.

After reading this book you will have to take sides because you will know the truth and you are free to decide who presents the truth...is it the Newtonians or I.

By that decision you then either become my partner in also recognising the crime I uncover or you become part of the crime syndicate as you cover the truth up. You either will be part of the truth by helping me confront them to acknowledge the truth or you will remain part of their cover-up by ignoring the evidence as everyone this far did for about four centuries in any case.

By not confronting the establishment, you give the establishment grounds to allure you into being sheepish. Because they see you, as just another stupid senseless student they have the opinion that they can brainwash you into accepting these fallacies that I am about to tell you. They will literally brainwash and condition your mind to accept what they never yet were able to prove. The truth is that this process of brainwashing is going on successfully for four centuries without any backlash to those committing the atrocities.

The academics wish to brainwash you by mind control in accepting that it is the mass that is responsible for the gravity and by mass pulling you down it is gravity that makes you fall. And if they will tell you flying pigs cross Siberia every seven years to land in Brazil you will believe that too. Where is the proof of mass that according to them is that which is producing gravity. They tell you Galileo said all things fall equal and we can see from the TV monitors how all things fall equal. Where is the mass that makes the gravity to let you fall if all things fall equally? They tell you that the truck has a mass of 15 tons and that mass is making the gravity that is having the truck fall while the truck is falling at the same speed and distance than the frog does.

If you take that as proof then they got you. Then they brainwashed you into a zombie. They take young minds as students have and they use your eagerness to become something bright and that they use against you. They never offer proof as to how mass is responsible for gravity and they will admit they have no idea how mass is responsible for gravity but they expect you to believe it. They take the fact that you are new to life and that you trust them as much as you trust your father and then they turn that trust against you and use that trust you have for your father which you hand onto them as they use it against you like a sword. They put the dagger into your mind and force you to accept that it is mass that charges gravity to let things fall while they know all too well that Galileo said that all things fall equal and that too they say is also true. So do you believe that mass pulls you down and then all things fall the same because mass pulls all thing down except that all things do not have equal mass but they fall equally. Then if you don't repeat after them and echo every word test after test and exam after exam, they will fail your papers and kick you from campus. That is mind control, better than what even the KGB is able to implement. You repeat after them and you live an academic life or you disagree and you go home to play with your toes. They have no one to report to. They can do as they jolly well please. They can ruin your life if you don't play their mind games. If they say there is mass and can't prove the mass influencing the outcome then you can say by the same rules that snakes run faster than hippo's can fly. If they can add what they wish and not be responsible to any one who then is working with facts and facts alone! Who then is making and fabricating at will while presenting no proof? If mass is in the picture, then mass must be represented by a factor of more than just one because if mass is not part of the overall picture then mass has a factor of one which proves that mass is not part of the equation since mass can't change the results. With all the objects falling equal mass has no role and if mass has no role then for my money academics in physics can't just go and put everything in as their heart's desire. If it is Galileo that is correct and if all things fall equal, then mass has no part in gravity. If mass is the inspiration behind gravity the truck must fall a million times faster than the frog and in fact the frog should almost land in another country because that is how slow it falls.

The fact of the matter is that I don't wish to be near when any of this lot hits the ground because the truck will cause a quarry and the dancer will be a splash of red fluid while the frog might not be that worse for wear if the truck or the dancer don't land on the frog. But that is mass. The differentiation of having mass or having equality and then not having mass and between individual differences in mass by each component that enters the equation when the objects touch the ground. Then every one gets the mass it has. Only when they touch the ground and land on the soil is mass as a factor awarded. While they fall they all fall equal and there is no distinction between the falling at all. What then is gravity? The gravity is the falling. The gravity is the motion. While the object is in a state of mass it is not moving. The tendency to move and apply gravity is the part that the mass restrains. The mass is preventing the falling from continuing. There is always an asserted effort to move by all objects. It is the role of mass to prevent further falling and independent motion to continue. Why that is, is not difficult to explain but there is so much to explain that I leave that part to my books. This brings us back to your tutors and their dishonesty because if you read the books you know more in a short while than they knew for one lifetime and in this I am not exaggerating in the least. Some of them might even still honestly believe it is mass that produces gravity because they were taught that it is mass that produces gravity and never thought about the matter again afterwards.

They were brainwashed by their tutors as their tutors were brainwashed before them. You don't need the brainwashing because you now can find out what the answer is to gravity. You are the first generation not to have to endure the lying that went on in physics for three hundred and fifty years or so. You are the first generation that can receive the light of knowledge about what gravity really is, or you can be the last generation that will live in the lie. You are in a position where you can teach your tutors the truth about gravity if you read what is in the books. The truth is there and the truth is out and the truth will be because the truth is written for all those that wish to read. The academics on the other hand have ignored my work and my being on Earth for the past six years while I was writing them letters about gravity. They ignore me as if I am a rattlesnake because to them I am a rattlesnake. I am to them as dangerous as an atomic war is to the Super powers. With what I say I will have them tumble down from their pedestals because by accepting my work they suddenly find their position equal to yours as students, and then they will have to learn my work in the same manner as you learn my work because to them everything is as new as it is to you. I can assure you very little remains the same and even less is unchanged. The Academics of the day have too much to lose to recognise my work and therefore have to protect their interest with all they can muster. For that reason if no other, they will rather go on lying to you and cover their corrupt fraud than face up to the truth and admit their work is lost. The truth will be whether it is recognised by them and they can become the first to admit and repent or they will be the last of the laughing stock that those in the future will refer to as the bunch that couldn't see when things fall equal they cannot have mass and when things do not fall by mass then one can know mass has nothing to do with the falling and the gravity.

It is up to you as students to rattle their cages and make them admit they've been lied to as they are lying to you. Or you can be the last of the fools that couldn't see that when things fall equally they have no mass by which they fall. It is up to you. My book is written and those that read it first will know what gravity is. If you prefer to be the zombies then those coming after you will treat you as the idiots. If you do not accept the role as being zombies who have been brainwashed, then confront these academics that treat you with disgust and betray your trust. They might tell you the mistake is not that serious and the damage is small but then how will they know how big or small the damage is if they don't even know what damage there is or what the damage is.

Science has strayed so far from the truth that they can't even see the truth any more. My explaining has to involve so many corrections but after being corrected the reading and understanding makes as much sense as this part that you have read up to now. But there is such a long reverse to return the untruths into the truth and that takes many pages to rewind. In the end cosmology is the simplest subject of the lot. That is after many corrections are made about crooked presumptions and misconception that are treated as gospel. If you carry on you will learn about some of it and when you read my books I will entertain you with many more than you ever believed possible. After you have completed my work, you will sit back and think: Now why didn't I think of that because it is so simple and clear to realise. It is like looking into the dark because no one showed you where to turn on the light but once you found the light switch then all you have to do is observe and everything becomes clear to you. My books will serve as the light switch that brings the light to you.

I charge your young minds to confront those fraudsters about the truth. See how they react when confronted by that which they cannot afford to admit to. If they lose the mass concept forming gravity then they have lost everything. I wrote to them in the last letter where I informed them that they protect the criminality of their corrupt teachings because when the corruption is removed then nothing remains because they have lived a lie for too long. It seems to me that society has spiralled down into deception. If you purchase a car and you buy something designed to give you trouble a year or so before you covered your last instalment to the bank so that everyone can make as much money in order for you to lose as much money as possible. It is the bank and the car dealer and the factory and the spares consortium that cleans you out just because they can. Buy a house and you are in debt for the rest of your natural life. Go and visit an attorney in law and he prepares the trial on the grounds that he can make the most profit out of your case. He is not in pursuit of the truth, but to see how much he will get from your case. Go and see a doctor and the doctor will see that don't get well but that you come back as many times as possible and as soon as your medicine he prescribes will permit. Judges see how soon they can release criminals to commit from incarceration in order to repeat the crime again. There just seem to be no honesty left in any profession and this even includes physics. In other books I show how much money and how many times these physicians' takes the taxpayer to the cleaners and show nothing in return afterwards. They spend billions of dollars to detect the mass that a Neutron has ands a neutron can't have mass!

When you do confront the academics in physics about their fallacies their reaction is going to be that they are not fraudsters but think about this and will even be surprised that any one can think of them in such terms. I wish to paint a picture in which you supposedly entrust you future in me where I am the professor and you are acting as students that entrusts their future blindly in the hands of academics. I pretend to be the academic that is suppose to be trustworthy and you take the role of the student that is trusting blindly. If I come to you with a proposal about something I wish to share with you knowledge in the pretence that it is meaningful wisdom but I am to share it with you on condition that you pay me an amount to share with you that knowledge what I know, then I am an academic wishing to teach you. When that which I tell you is not honestly the truth you might think that I am a liar and you will not be mistaken. When that which I tell you is a lie and I know it is a lie but still I tell you, then I am dishonest. When that which I tell you is a lie and I know it is a lie but still I ask you to pay me in exchange for me telling you what I know is not true, then I am a confidence trickster and a common criminal. When that which I tell you is a lie and I know it is a lie but still I force you to accept what I tell you while I also force you to pay me to brainwash you in accepting that which I know is untrue, because I have the ultimate authority and you either pay me and accept my brainwashing and mind control for the rest of your natural life or you die an academic death, then what am I? I have no name for such a person but have you got a name for a person as evil as that? Have you a name for such a person that will force another person to pay him to be brainwashed and be mind controlled because the tutor has absolute control over the life and death of the academic future of the brainwashed and therefore is willingly forcing this unfortunate creature in accepting what will never amount to the truth? I think they are called Physics professors and rule Universities as draconian authoritarian dictators bent on sadism.

They are of the opinion you will swallow any rubbish they throw your way just because every generation before you were mind controlled in the way they are about to control you. You may think these are big words, but read on and see after you come to know all the facts whether I exaggerate even in the least. They see you as slow-witted and mindless because they think they are the academics being superior making you the lesser and inferior party. If you are not aware of the facts beforehand they know you will follow their teaching without asking questions as it is going on for four centuries this far. Talk about the Catholic Church putting the fear of God into Copernicus and you will find their manner in disagreeing about statements not echoing their perceptions, is as bad and more ruthless than the Church was during the Dark ages. The Church just killed those confronting them on science dogma but did not make every student a brainwashed mind controlled Zombie!

They think that your naivety makes you a Zombie and with such a degree of mindlessness that state of mind will incapacitate you into their control. They don't want you to ask nosy questions about contradictions existing and they refuse to answer any uncomfortable questions asked in any confrontational manner. This process of brainwashing and mind controlling in physics has been in progress for hundreds of years. If you are surprised then control me by asking and just answering how it is possible to agree that a feather and a large hammer fall equally while also agreeing that Newton is most correct. How can any person thinking logically agree that Galileo is correct when denouncing objects falling under mass differentiation when Newton insist on mass driving gravity as a force. If you can't...well they can't either! Their task is not to explain but to mislead since they think you can't think while they think they know how to control you.

The motive behind this book is to promote my other books and by trying to be as least complicated as possible, this book aims to present Newton's deceit. This book does not aim to represent the full entirety of the original thesis as it represents the entirety of my theory in a single copy, for that is not possible but is reduced to aid any possible potential reader in the examining of the purpose of the information this book wish to present in order to uncover Sir Isaac Newton's fraud. Anybody and everybody are aware that all objects fall at an equal rate. If an object such as a car weighing one ton falls at the same pace as a person weighing fifty kg, how does mass come into the picture by committing a force to do the pulling? Mass has to pull because according to their teaching it is mass that establishes gravity. However, mass is a factor that produces differentiation whereas all objects show equality during their fall. If it is mass that is establishing the force gravity, all objects must fall at different speeds. That they do not do as they all fall equal. That means Sir Isaac Newton's physics is wrong from the start because mass cannot have any input in objects falling.

This is not the only untruth that the Paternity called Mainstream Science is keeping concealed as a cover up that is wrapped under an airtight blanket of deception. If you sit in class and listen while also experiencing the sinking feeling that the facts you hear are not adding to a total you are comfortable with while you disagree with what is said, then you better read on because this book has it at task to show all that will read this document how much discrepancies academics lay on unsuspecting students that trust Academics with their future and their life. Do you as students realize the inconsistencies that physic Academics teach you as the truth?

Students, tell your Professors to stop deceiving you and stop trying to control your minds with their fraud. Those Academics tutoring you are telling facts about gravity that has never been proven.

That is mind control.

They wish for you to accept facts on gravity that they hold as the truth. They claim those truths are beyond questioning, yet with the least examining those truths they stand by then prove to be totally void of substance because it was never corroborated by one single experiment.

Should you question that mass produce gravity they will expel you from University by letting you fail your examinations and it was never proven! They will expel you and have you fail tests should you question their authority on the matter of gravity while at the same time they can't for one second bring evidence in support of what they wish you to accept as the unquestionable truth.

That's brainwashing by mind control because if you don't accept their baseless facts as God given truths they dismiss your academic career.

It is either put up and shut up or be gone. Academics do put mind control to work on unsuspecting students by forcing students never to question the legality of statements they offer as being sound and correct.

What they present as correct I prove in this very book are openly laughably, totally incorrect and by just reading my evidence you will see how feebly easy it is to rubbish it. Take the evidence I am about to share with you and confront them with the fabrication of facts that they present. Go on and challenge those teaching you with the falsified facts as I challenge any one to prove me wrong.

What they maintain is gravity is total incompetent nonsense and can't be corroborated at all, but what they can't corroborate because they don't understand, I prove to be that which the Universe employs to form gravity. There are four phenomena they dismiss because they have no idea what they are. I studied each one and formed an explanation by implementing Kepler's formula as the Universe gave it to Kepler.

By understanding the formula and implementing the content into the four phenomena, I am able to prove what forms the motion we think is gravity and when reading it, only then the Universe makes sense. All the questions in these books I managed to answer while they can't … and in the books I answer a lot more questions than those I ask here in the rest of this book while Science fails to answer any...

The book I present has the dynamics to change science forever and that is not cheap exaggeration or promotional talk. I say that because I investigated Kepler and believe it or not but that investigation was the first one done since Newton explored Kepler the first time and that was four centuries ago. Science never went back to Kepler after Newton included the work of Kepler into his work but when one reads my work, one will find Newton compromised Kepler's work as he did with the work of Hook and many others. For my saying that about Newton not correctly analysing Kepler, the Newtonian Academics at various University Institutions bluntly ignore my work. Academics would not touch my work notwithstanding the twenty-seven years of research that went into my work. I have a very limited three South African Universities dealing with cosmology to turn to and their reviews of my work in the past left me in doubt about their sincerity in the performing of the reviews. The South African Academics do not attack my work because then I can defend my work…no they just ignore the work by sending reviews that totally miss the point about my work. Their reviews are about my linguistic capabilities or about my presentation of the books but they never refer to the work in detail that I represent. Their attitude spurred me on to present my work to Academics outside Africa. I have sent the books to various Universities but because of various factors not in my favour, the Universities in question treated the books outside Africa as junk mail. The review of the Academics in South Africa was always mainly (and more or less only) about my accompanying letter by which I introduced my work, such as the letter you are reading. Their review never went further than my introductory letter since the facts they mentioned were only aspects, which I mentioned in my letter. So the facts in my letter were the only response I received. I say this because the review presented no evidence that those Academics even read my work, much less understood my work. That forced me to write a letter where the letter is an introduction manuscript about my work and that letter of introduction I present as a manuscript on offer for publishing. After all the attempts I decided to chance the commercial press because I have absolute faith in the correctness of my work. I need to find one academic who will read my work with a degree of sincerity to establish the connection of academic acceptance.

The following presentation is as simple as gravity can be represented but in order to have the presentation as simple as it is we do surrender some part of the accuracy to achieve simplicity. Under a microscope one would find that in this explanation the explanations strays a little from the truth in order to make it comprehendible to everyone reading it. The truth is the explanation about gravity can be somewhat more complex than what the following presentation has to offer but then on the other hand it will never be as simple as dumping the entire concept on one thought about mass that produces gravity. That simple, gravity in explaining can never be, because it is so far from the truth as telling that fairies produce summer and witches bring on winter. The whole concept of mass being responsible for gravity is one big hoax and forms a scam. Putting mass as being responsible for creating gravity is the biggest fraudulent lie that ever hit the Earth on any scale ever.

Gravity is a rotating solid moving through a liquid space. The Earth is a solid that much is true. The atmosphere is regarded by physics to be a liquid and that much is also accepted as true. Gravity is the

Coanda effect and the Coanda effect is where a car tire spins through water and the spinning wheel gathers the water onto the surface of the tire. At speed the tire picks up the water and secures the water around the tire. The motion of the tire, which is the solid, contracts the water, which is the liquid onto the solid tire surface. The contracting of the liquid onto the solid by rotating motion produces the gravity that attracts the liquid to the solid. That is gravity. The solid tire can cover the surface of the tire by a layer of water where the water is as hard and as sturdy as the solid tire can be. But the water might be hard and sturdy, yet it remains a liquid with all the characteristics attached to liquid. That is why driving in the wet is so dangerous. The solid tire can surround the tire surface with as much as one inch of water. That is gravity when the solid tire spins and by spinning it contracts the liquid water. The water being a solid in the form of ice can't perform in the way the liquid does when the liquid is surrounding the tire. An inch of ice will never be strong enough to allow a car to drive over it but in the case of the Coanda effect the motion of the tire allows the water to be much stronger. While when being in the position where the water is surrounding the surface of the wheel through the spin of the wheel that is contracting gravity, it makes the water as strong as the tire, which enables an inch of water to support the entire car running on the water. The spin of the tire produces a gravity contracting by rotating motion, which turns the density of the liquid water to the same compactness as the tire surface being a solid. The spinning solid of the tire turns the density of the fluid water into a solid equal to that of the solid tire. The tire asserting a rotating motion does the producing of gravity by motion. The expanding of the rotating action produces a contracting of the liquid space it moves through. By expanding the solid that is rotating, takes away some of the space that the liquid holds.

The tire rotating is expanding the space it holds. That is called fleeting momentum. The matter tries to move away from the centre in an effort to gain more space. As the tire spins the tire tries to capture more space and the tire can thrust this so hard that the tire does go oval. The tire tries to hold more space than it has because it is capturing more space in an effort of expanding.

While the tire tries to capture more space, the tire also reduces space that the liquid water holds. By capturing the space that the liquid water holds, the tire is capturing water and in the effort the tire is trying to gain the space that the water has and in that the solid tire is making the liquid water solid. Therefore we can drive on an inch of solid water while the water is liquid. By the water contracting we find the density of the water changing where it meets the surface of the solid. The solid tire expands and in the expanding it makes the liquid water solid. Then where the liquid water finds itself being reduced and being made denser, the liquid water gets so dense it becomes a solid water area. The tire wall turns the water into a solid by the rotating action of the wheel. That is when motion applies to the wheel.

When considering the wheel at speed and the Earth at speed, we tend to think of the Earth being still and motionless. The Earth is spinning at a far greater speed that the tire would ever be capable of. The Earth is spinning so much it is concentrating air to become a liquid.

When objects fall, the object has no mass and this is in spite of all the claims the Academics in physics try to produce. Galileo said all objects would fall at an equal pace and hit the Earth at the same time when falling the same distance through the same air under the same conditions. Newton said mass is responsible for that which produces the gravity by which objects fall. That means the object being more massive must fall faster than the object being less massive. If mass brings on gravity mass must distinguish the amount of gravity by applying more or less falling pace. If that does not happen there is no evidence of mass applying because then all objects hold equal mass while descending to the Earth. We see frequently that the object can fall at a specific rate depending not on the size it has or the shape it has but the distance it travels through the air. We see so many times that a car drops from an aeroplane with a human falling next to the car. There is a car advertisement where the human that is falling has a parachute in a bag falling next to the person that is falling next to the car. The car is around twenty times more massive than the human and the human is about twenty times more massive that the bag containing the parachute. If the falling process depended on mass that is supposed to instigate the gravity action as the intellectuals wish to declare, then the lot cannot fall at the same rate. The car must fall twenty times faster than the person and the person must fall twenty times faster than the bag. We can see that the lot is falling at the same rate and the descending bares no implication to any mass that shows differences. That is what Galileo said when Galileo said all things fall equal at the same rate and land at the same instant and a massive object will land at the precise instant that a very light object will land on the condition that they are dropped equally and that they fall through the same space at the same time. That statement excludes mass from any part that gravity has.

From that one can see that the falling has no implications brought on by mass. Mass has nothing to do with gravity but to restrain any further gravity effort putting individuality to the item falling. The gravity is produced by the moving of the object and while mass restricts the moving, the moving still applies as gravity because the moving remains as a tendency to move when mass applies. Even where mass stops, the gravity moving the gravity remains as a tendency to move towards the centre of the Earth. It is not that the mass is pushing but the mass is stopping the moving of the item to the centre of the earth. Mass only applies when any further motion of objects are restricted. If the restriction of the larger object that inflicts the mass suddenly also start moving, the mass turns to gravity that very instant. While the larger object retains in position of preventing the independent object from any further individual moving, mass comes in as a factor that stops further motion. Mass counteracts gravity by stopping gravity. The object only obtains mass when gravity becomes no longer applying. When the objects all fell and they all hit the surface of the Earth at the same time, only then is there distinction about size and mass. That makes mass a factor of the Earth holding a restraining on the object and then mass is not part of gravity because gravity is part of the falling or the moving of the objects. When objects fall they have gravity and the gravity is equally applying to all because the gravity is the moving of the objects without restriction. When the objects hit the ground, these objects lose independent motion and with the accepting of mass the objects retain the motion that the Earth provides. The mass renders the objects the motion of the Earth and having the motion of the Earth they move at the same pace as the Earth. The mass then makes them having a relation to the Earth where the mass puts them in a restricted part of the Earth. They move at the rate that the Earth moves because they then are part of the Earth by the provision of mass. Mass shows how much the object that then landed on the earth and no longer moves independent of the Earth then became Earth.

The gravity is the Earth forming the solid that rotates and the air is the liquid through which the Earth rotates. The liquid is contracted onto the surface of the Earth by the solid of the Earth trying to expand into the space the liquid holds and this happens due to fleeting momentum or rotating motion.

Mass is the resistance that an moving object shows when stopped by a larger object that blocks the smaller object from moving further as an individual object where by showing mass, the smaller object resists the effort the larger object asserts on the smaller object to compromise the form the smaller object holds as a unit and not to accept the form the Earth imposes on the object. By having mass the object no longer holds independence but retains some form of individuality by not compromising the unit it forms as an independent structure. When the object lands on a larger object and the larger object is part of the Earth the larger object halts further motion. The larger object removes the individual characteristics that the moving object has.

Let us have a good look at mass. When there is a ship we see the painted waterline of the ship indicating the load in mass the ship can take. The ship is lighter than the water because the ship floats on top of the water. By loading the ship the ship can take in a lot more mass than what the water is because as long as the hull displaces more water than what the mass of the water is in terms of the area the ship claims, the ship will float. The ship is less dense than the water when taking the area it holds in relation to the density of the water it displaces.

When the ship has an equal density to that of the water the ship will float in the water while being buoyant. The ship has to accept a certain percentage of water to allow the ship to float inside the water and prevent the sinking of the ship to the bottom. This is an act of precise balancing. The ship then being somewhere inside the water has no mass but being equal to the mass that the water holds. The ship has to displace as much water as what the area it holds will be in mass when being only with water. Then the mass of the water is the same as the mass of the ship and therefore the ship can hold ground inside the water without sinking or floating. If the ship again wishes to float the ship will have to displace some water it has in its hull and exchange that water for air. Then having more air in the hull than water, the increase in buoyancy is what will enable the ship to have less density per volume of space and less density will render the ship in having less specific density than the water. When the ship sinks the ship has more mass than the water it displaces. The ship has a bigger specific density than what the water has and the mass of the water pushes the ship to the bottom of the water. At such a point when hitting the bottom the ship then has mass. It is the mass of the ship at the bottom of the water being more than what the mass of the water is that floats over the sunken ship that puts the ship on the bottom of the water. In water or in liquid the ship or the solid either has buoyancy or it has mass. It either floats in buoyancy or it sinks whereby it receives mass when the sinking ends. When it floats it has motion by buoyancy that the water provides.

The ship while floating has less mass than the water has. The ship, while submerged has the same mass as that of the water. The ship has more mass than what the water has when the ship sank. The mass is a relevant factor because of the fluid aspect. When floating, the ship has little mass and all floating objects has mass that is less than what the water has. When being submerged it does not matter if the ship is a big ship or a small ship because the ship has the same qualities as the water and therefore a big ship will be as buoyant as a small ship will be. This is very important to note. When in a liquid there is no mass factor. Only when the liquid suppresses the object and onto the solid and with no distinctive difference between the motion that the object has and the motion that the solid has does the body turn from being a liquid to being part of the solid does the mass factor enter the equation. A big ship will float submerged next to a fish and from the mass aspect the two would be equal. It is depending on motion.

Any object will float in outer space without showing any sign of mass. To cheat us as science always do about these matters, science tells us that when any body floats in outer space the body has mass but the gravity is micro. That is as much a fabrication and a distortion of the truth as the entire mass idea. The body has maximum gravity and micro mass. The body requires gravity or a movement of 11,7 km per second to counteract the gravity the Earth applies. The body has to move at 11,7 km per second to move faster than what the Earth has the atmosphere moving. The gravity or motion that the Earth provides while moving through space puts a motion requirement on any and all bodies of 9.81 km/ sec and to beat that motion, the body wishing to escape the Earth's atmosphere meaning to move faster through the air than what the Earth moves through the air, that body must move or have gravity at a rate of 11.71 kilometres in one seconds. That is movement of the body and any body has to move at a rate of 11.7 second whether the body is big or is small.

The faster the body moves, the further will the orbit of the circling body be from the centre of the Earth and that also depends on motion or gravity. Being big or being small is not part of the requirement and therefore mass has no influence as a requirement. When the body starts to move slower, the body starts to descend. Big or small does not matter and having mass or no mass does not matter. It is when the motion of the structure reduces to a speed that is below the rotating speed of the Earth that the body starts to plunge to the Earth.

The falling or not falling and the escaping or not escaping only depends on the speed of motion and mass as a factor is only required in relation to the imagination of the physicist. While falling, all objects show equal mass and having mass differences puts no extra load on any of the factors in any way. The buoyancy of all objects in the atmosphere depends on the speed in motion of the object. There are many other factors that come into play and all those factors forced me to write four different books on the subject where every book highlights different aspects that play a part. The only aspect that plays no part is having mass or not having mass.

This letter you are reading is my effort by which I hope to interest you in reading my manuscript in unrest because that is the least it deserves but that least it does not get from academics in control of Universities. I decided to offer the letter I wrote to academics, which is inexpensive. The book on offer has the title of <u>an open letter To Selected Academics</u> ISBN 0-9584410-9-X and is the actual letter I sent to various establishments that then developed into a book. It is about serious misgivings I have on the subject of gravity. The book now has a much simpler title being *Gravity' Recipe* with three more books on the same theme to follow. The book also has an accompanying web page. If you first study the accompanying web page that I include, such reading should help you to understand my ideas better by giving a platform to grasp the ideas in a shorter time. It would mainly help to introduce the very new ideas and the information, which I try to bring across. The accompanying web page is condensed to suit the reader and provide some shortened briefing of a book. I hope that by you're reading the short web page first and then follow that up by reading the book, it will supply a background about the content and firstly familiarise the reader with the information in the books.

Up to now every Academic at all levels in science are normally acting as if gravity is a commonly explained factor, proven in detail without the tiniest whim of uncertainty, where every one knows every aspect about all principles that are involved in gravity down to the smallest detail. In truth, no one in science anywhere remotely knows what brings gravity about and I used Kepler to unravel this mystery called gravity. But no one in science will admit this fact about Newton or any one else never being able to explain gravity in the least or admit that Kepler is the one who formulised gravity decades before Newton came and gave gravity the name. Newton did not underwrite or define gravity and even today the most informed in Science at best can only assert their suspicion on a rumour presumed about what causes gravity to perform as the part interlinking the cosmos but no one can go any further by explaining the concept. Newton started this realising of gravity but it had and still has no more substantial proof than a rumour has and Newton admitted to it being a concept he could not explain.

Still, to this day nobody in science at present will denounce the principle of gravity in a fashion by acknowledging that gravity has still as yet never been explained. All in science act in a manner as if Newton's gravity idea is the best-proven fact there ever was and only occasionally admit it to be just a rumour as Newton admitted it was when he introduced the name (not the concept). If Newton's concept was accurate it will by now have the moon much closer to the Earth than what it was during the time of Kepler's investigation. Yet we know from modern tests that came into place after man landed on the moon in the seventies, now know it is moving away instead of coming closer. Newton agreed that he could only declare gravity as a vague concept more in a suggestion and far from the manner of forming the proof as one would demand from a rumour when he was announcing a force that could be anything. Not once could one person in the past or present provide substantiating proof on gravity as a reality by defining the very principles. That includes Newton as well as Einstein and even Hawking. Scientists can declare that gravity was a factor at 10^{-43} seconds after the Big Bang but what brought gravity about or why gravity became or still remained as a presence, is still tightly concealed information which all are speculating on. Even to the best informed amongst the most educated do not know what gravity is because they all ignored Kepler and for ignoring Kepler the price they pay is not finding the principles bringing about gravity. Using Kepler makes the method to follow and understand even Einstein's discoveries shockingly simple. I started using Kepler when I was shocked by the lack of proof on the matter of gravity as it supposedly applies in Newton's manifesto. I found a lot of total nonsense thought to be the truth. That urged me to investigate the matter by using other sources than the mismanaged and incoherent nonsense Newton brought about when he raped every aspect of Kepler's findings.

By my applying Kepler's formula, I can define gravity precisely to the point where I now can explain why the proton and the electron form a mass difference of 1836 times between the two mass components. This achievement is obtainable only when one reads into Kepler and finds what Kepler did not say but meant to say when he said $a^3 = T^2k$. Kepler said it so nicely and simple, every body including Newton missed it because everyone is waiting for this mind-blowing discovery lurking in the subatomic structures. Science even gave the subatomic particle they are waiting for a name: it is called the graviton. The particle will supposedly be a part of the most inner part of the atomic formation and such a particle will produce all forms of gravity. This has been given a name before the discovery of such a particle ever took place. Can you believe the audacity these geniuses have when they get to their third or fourth doctoral degree? To go and name an undiscovered subatomic particle being a graviton, is evidence of how little those lot understand basic physics. I use Kepler to explain why this awe-provoking particle went undiscovered and the main reason for not discovering it is because Kepler said it is not there!

The graviton went undiscovered because there is no graviton to be discovered and no graviton produces gravity. I can explain why the strong forces are forty times more massive than the weak forces. Just by my studying of Kepler, this became possible. This book on offer is one of four books that I use for the sole purpose of linking the rest of the rather complex theory to the present way of thinking by using mainly Kepler as referring material. Of the four books, this one is the first. I still am trying to obtain funding from somewhere to have the other three published. The four are contributing to form the foundation of another set of books that I call Matter's Time In Space: The Thesis Vol. 1 – 7 ISBN 0-9584410-8-1. These nine are rather more complex and although being a Thesis it will be little less understood than the four less complicate books I present on introducing the working of gravity. There are in all four books the motive to lay the foundation for the presenting of one book consisting of seven volumes where the seven is forming one unit. This excludes the one I present being one of four from the unit forming the Thesis because it is made up of various letters I hoped would convince Academics to investigate my work and to realise their incorrect approach on the subject of gravity. In the letter I conveyed my work as a way of introducing the principles that the unit explores. The one I now present for your attention is one derived from my original letter that I named <u>an open letter To Selected Academics ISBN 0-9584410-9-X</u> to form part of the publishing of the four books dealing and exploring Kepler's formula. The basic letter was rewritten so many times whereby I tried to entice so many Academics just to read my work and if they found it is incorrect, they then indicate to me the incorrectness there are. To try and avoid confusion I will explain once more. The books started off as one book entitled <u>an open letter To Selected Academics ISBN 0-9584410-9-X,</u> from which the books

1) An open letter Announcing Gravity's Recipe

Where the author links gravity directly to the Coanda effect by applying the four yet unexplained phenomena going by name as
1) The Roche limit
2) The Lagrangian system
3) The Titius Bode law and how these all combine to form gravity
   as implemented by the
4) Coanda effect.

2) An open letter Addressing Gravity's Formula
Where the author explains how the Universe came about at the first instant the Universe came about in using evidence on the four cosmic pillars and it matches the Biblical explanation in explicit detail

3) An open letter About Gravity's Prescription
Where the author explains how the Universe came about at the Solar system, as we know it took place. It explains why there are four solid planets, four gas planets and one cold structure. It also explains mathematically why all the debris is encircling the planets and where they come from. This is one part of another book entitled the Seven Days Of Creation

4) An open letter explaining Gravity's Rules
As the Author goes into detail about a new cosmos theory where the four cosmic pillars produce a cosmos everyone ca understand. It puts time in relation to space and discovers what space is in relation to time. Never yet before was either time or space understood because everyone drooled on the misconception about mass and incorrectly interoperating that mass produces gravity.

This series forms as a unit with four individual titles forms a prologue to a Thesis that introduces a whole new concept about Creation.

Matter's Time In Space: The Thesis in seven parts
ISBN 0984410-8-1
<u>*Written By Peet Schutte*</u>

In this letter I call Creation by name and prove with science that we are in Creation. I employ science to prove, that which controls Creation, which is not in the Universe but is noticeable because it is not in the Universe. In the light of all proof and when facing evidence I bring I dare an atheist to prove me wrong about Creation.

In mentioning this word in a science book I break a ground rule enforced by the atheistic dominated world of science. I challenge any person to bring proof about any part where any of my theory might be incorrect and furthermore I challenge any Academic in physics to prove that Newton's mass pulling mass is anything other than fraud. I charge any one to bring proof that the cosmos is contracting by the force of mass and that mass produces gravity as Newton advocated when he committed the biggest fraud of all times.

Although it may not be academically as advanced as the other books forming the Thesis, the four that forms part of the series and is compiled with the intention to be read by most people with an slightly above normal interest, I also have to mention that the four on gravity is no light-hearted storybook and reading the books will require a degree of concentration. But the intense studying such as the Thesis books require is unnecessary. It aims to bring a link that connects my theory to Mainstream Science without hammering out all the mathematical detail as it does in my other books aiming to supporting evidence that conclude my theory details and therefore is the simplest of all of my work. All one needs to appreciate the information in this book, is the understanding of the mathematical expression $a^3 = T^2 k$ and what on earth can be simpler than that! T^2 is motion, a^3 is volumetric space = is equal and k is distance from a centre and this simple explanation serves the whole book. That is the formula Kepler founded that was too simple for Newton to accept and by Newton changing Kepler's formula, Newton destroyed the most correct scientific work that was ever devised by man. With the changing there of Newton did just such destroying because he (Newton) felt inspired to do so and that inspiration became my inspiration to write these open Academic letters, which I now wish to publish as a set of books.

The purpose of this book entitled Newton's fraud is my attempt to present to the public a much-abbreviated book with much less detail and explaining and studying the work in page detail then after reading the book it may bring more clarity about the what the other books contain, as it is similar to the actual skeleton from which **An Open Letter On Gravity Part 1 Vol. 1 and Vol. 2** as well as **An Open Letter On Gravity Part 1** were developed however the page numbers are drastically reduced. It should help you to understand my ideas and concepts better. The reading of all the work I am about to submit to you does require mathematical insight when one wishes to follow the content of the work but only a high school mathematical background is required. The information in this book, which I offer, is at the most on par with scholastic developed skills. The web site should provide a further platform to help one grasp the basic ideas in a shorter time. Of course the real information will be provided when the reader later explore the full manuscript. I hope that you're reading the web page will help you decide on the authenticity of my arguments and the merits thereof. When the publishing is done there will be another more detailed web sight that will accompany the book as it will also help any future reading of the book when published. I then aim to establish a web page that will supply a background about the content and will familiarise all future readers with the information in the book.

To be very blunt in explaining the main differences between my views on gravity and the views Mainstream science supports, as seen from my interpretation, is to compare gravity with a picture in which a dog is chasing a rabbit. Science sees this chasing as purely the rabbit pulling the dog around and the dog can do little else but pull back because the dog is just as much pulling the rabbit with some invisible cord or rope linking the two. The two in the pulling contest then will reduce the rope, but in three hundred years since Newton institutionalised the idea about gravity pulling material in all cosmic studies, this never happened. They claim there is some invisible link tying the dog to the rabbit that is pulling the dog that is pulling the rabbit where I dismiss the link they see. Then I come along and claim that the dog is dismissing the space between the rabbit and the dog while the rabbit is extending the space between the rabbit and the dog. While the rabbit is establishing new space the dog is all out attempting to diminish the space between the dog and the rabbit. It may seem as a slender differences and mostly the same issue but in this hides all the mysteries locked away behind the cover of not finding the ability in appreciating or understanding gravity. I admit one must find detailed reasons that will motivate the rabbit as well as the dog's running, but I believe I do just that.

I prove gravity is strongest where space is least (not where the Universe goes flat as Einstein promoted).

I prove there are relevancies that are all applying equally and without such relevancies in balance there is no gravity.

From the view the onlooker has, it may seem as if some rabbit is pulling some dog all over the area that is remaining in one place where the dog holds the running rabbit in one area circling around the dog. This has another angle where as from the dogs point of view the dog will love to dismiss all space and capture the rabbit.

The rabbit on the other hand would love to leave the dog at a distance where the rabbit will never again see the likes of the dog. While the space between the two is merely a common fact appreciating their differences, the space is the factor that has to resolve the issue but cannot because different relevancies sustain equilibriums. I prove there is no rope tied because with a rope there will be no characteristics while there are many characteristic or principals applying at both ends.

The angle at which science is looking at the issue, science either dismisses or cannot explain the characteristics or principals, which are there none the less. The explaining of the phenomena is quiet impossible when using the pulling rope magical attachment idea in the manner science tries to explain gravity. Therefore instead of dismissing the rope, they dismiss all other factors present about and surrounding gravity but not the rope idea.

I claim I can prove all their officially dismissed characteristics by just dismissing the rope idea, but the rope idea were by coincidence also Newton's idea. That makes the arguments becoming about Newton versus me and you may guess who was winning thus far? In my argument the space is no transmitter of forces applying between the dog and the rabbit but is merely one fact of three factors in the matter because there is no rope attachment. It is all about space being dismissed and established. There it is in a nutshell. I see space being depleted as much as space is introduced and where such introducing becomes too little, the dog will capture the rabbit and that is what the earth did to us and all other objects in space of the earth.

If the rabbit pulls the dog the action depends on the rope connecting the two with the only factor applying then would be the strength of the rope. But if it is the dog chasing the rabbit as much as it is the rabbit fleeing from the dog, the dog will execute certain methods and so will the rabbit execute certain methods where these methods then are not coincidental randomly occurrences happening by chance but acts deliberately executed to establish advantages on either end and explaining certain characteristics of both sides. But for all the life in me I cannot see that the rabbit is pulling the dog just as much as the dog is pulling the rabbit because then the rabbit will allow the dog to capture the rabbit if it is smaller or the rabbit will escape from the dog if the dog is not big enough. If there was this rope then the rope must under all circumstances remain in total equilibrium remaining at a specific length that never changes because all the focus latches on the rope and any reducing or enlarging of the rope length will become a permanent advantage and a permanent fact to be to the advantage of either side winning the tussle.

The only way the rope can reduce or enlarge in length is if the forces pulling become more or less growing or shrinking in mass and in the orbit the distance does change without any increase to the mass of either the sun or the planet. The very opposite is indicated as the rope never prove to remain at equal distances and when the rope flexes, equilibrium should become excluded as is the case with comets. The comet illustrates this best where the comet comes rushing towards the sun but in defiance the comets are never caught by the sun. The rope is reduced but never retrieved by the sun. It means when released and rushing off into outer space, the comet should never again return because the rope broke. However, research proved the attachment is still there because the returning by the comet is like clockwork and even cyclic. Again on the other side we have the opposite, when the comet returns to the sun the rope somehow strangely reduces in length. Thus it will allow the sun to capture the nearing comet and that also never happens because the comet escapes in a cyclic manner.

Gravity is not coming about from pulling and pushing but is a result of running. It is a balance of coming towards as much as going away. The running started with a process now called the Big Bang where particles are running. This theory I try to introduce explains why the dog is chasing and why the rabbit is fleeing and what their inspiration will involve to keep running and never tire. Remember, the dog is catching the rabbit since the rabbit is not getting away from the dog and that part is announcing the factor that indicates the presence of the Big Crunch as part of the Big Bang, which every one misses. I use Kepler to prove all that because thanks to Newton the world ignored Kepler for four centuries and Kepler is the cosmologist while Newton is only a mathematician.

Thanks to Newton, science has been mismatching the two sciences in favour of a mathematical view and as of late (post W.W.2 period) scientists have totally confused the two academic fields but such only proves their confusion and underlines their not understanding what to grasp. I do not say using maths is incorrect but maths is merely a language and when translations of languages go wrong the concept goes oval. That is when we read that Einstein said the Universe is going flat and popping out again every second while all of us can see it is not happening or space-time can stretch and shrink at will while light can bend. There is nothing wrong with Einstein's mathematical finding but the translation from mathematics to the verbal spoken language misses a few words and misinterpreted a few words while not following a few words. That made the message seems to go mad because the message does not make sense. Using Kepler all the prophecies of Einstein is again proven but then it makes a lot of sense in the normal verbal spoken language.

For instance Mainstream science has the theory that matter and antimatter developed and some matter formed as particles, which were nicely wrapped in containers we call atoms and stayed on as material or matter where the rest formed antimatter that disappeared. The anti material also formed atoms but then chose to just plainly vanish, as did singularity and other cosmic factors to now be nowhere. Singularity is a mathematical position falling outside the detection of the observable Universe in any case so in the case of singularity there is a possible excuse for the disappearing...or is there? We know that singularity produced space-time and space-time produced gravity but then space-time went away leaving us with gravity. Singularity is a single dimensional entity that is not material, holds no space, and holds no time and in our 3 D view can only be found outside the human visual spectrum. Mainstream science is of the opinion that singularity disappeared after the Big Bang process came about whereas I am of the opinion that everything must remain part of the cosmos once it was in and was part of the Universe simply because there is nowhere for it to go.

The same applies to space-time and antimatter. If it was part of the cosmos at the beginning, it has to be in the cosmos until the very end only and simply because there is no other place available to be or to move to. The Universe is the container that is the only container leaving no other container to contain whatever is in need of containing. If it is or if it was, it still is in the Universe! Our task is to find the place it went to or find what it changed into. While we then are on such a hunt, we might just as well find the cosmic principles not understood but which is there all the same. Take for instance the Bode principle where all nine planets show a relation with the sun in precisely the same manner and using the very same method of spacing, yet science brushes the Bode principle off as a coincidence. It might be a coincidence when the space between one or two planets shows these phenomena but when all nine planets plus even the fragmented structures adhere to the very same principle, no person can be of the opinion that it came about as a coincidence and still pretend to be serious or professional about cosmology. One then must simply find the proof lacking in our understanding.

Mainstream science knows about gravity, the Bode principal, the Roche limit, the Coanda affect, the Lagrangian system and the sound barrier, but cannot explain any of the phenomena although the presence of these phenomena is without dispute. It is the explanations about what causes the phenomena that should be part of the dispute, but in science, the way they defend Newton, scientists go overboard by disputing the phenomena and the phenomena as a principle existing in cosmology or not becomes disputed. Science fails to give acceptable explanations of such occurrences therefore disputes the validity of the phenomena and this failing to explain the presence becomes disputing the presence thereof. In such a light Scientists must somehow realise they are barking up the wrong tree with the information they have at hand to succeed in their explanations.

They cannot refuse the phenomena and not realise they must have the cat by the tail as far as cosmology goes. Please remember that with this I am referring to cosmology and not general physics. There is an Earth versus a Universe of difference between the two concepts but Newtonians fail to see that because Newtonians cannot appreciate the differences, thus blurring the understanding of gravity. If there are that many phenomena (it represents all there is in cosmology) to explain and such little ability to explain (science fails to explain even one) by using the information Mainstream science is using to explain the cosmos, then someone somewhere has to realise there is something drastically wrong in the way they present the knowledge they claim to have.

One cannot be serious about science but defend your view by dismissing the validity of all what is known as the unknown quantities that should be indicating factors presented as such. There then is some gross incorrectness in the way Mainstream science reasons. The Roche limit is there and no denouncing thereof can remove it from the cosmos. They may refer to evidence received from the Hubble telescope as "the star is blowing bubbles" for the lack of explaining what is occurring but occur it does. One cannot say it is some unknown gesture presented on occasions because not explaining the pictures presents the presence of certain foolishness.

For fifty years they lost many pilots but still have no idea what brings the sound barrier about, or find the link gravity holds in the process we call the sound barrier. Instead they try to interpret some effect established almost two centuries ago with steam trains back then travelling at the same speed that horses run. No further investigation with the science in hand brought them closer to new facts! It should be a sign telling them they are going about incorrectly, but it does not because Newton said so. It may sound as if I am anti Newton but I am not. But there has to be more than Newton with so many pieces of the cosmic puzzle still missing. Science should not serve only Newton but science should serve the seeking of the truth. When I first came upon the unknown it stirred a sense of disbelief and I decided to respond.

Some twenty-seven years ago I decided to start an investigating quest on my own to see where I could go with my private research. It came as a result of my frustration when I realised the discrepancies there are in theories presented about cosmology and all the unexplained factors no one ever made any effort to explain. Later I found that no-other person than Newton in person was to blame for the mistake that was made, but now I am jumping the gun. If you are a Newtonian and feel a repulsing urge to throw down the letter, you will do so at your peril. I say this because I have seen Newtonians get fits in the past when I say what I just said. Whenever I make this very claim the entire science community rejects me immediately without any reservation to any person. When I speak out against the incorrectness, I see science as a unit and an entire structure rejects all further statements that I make without excluding any body active in the field of science. To science Newton is reserved as a god and they placed Newton beyond criticism.

When they listen to me criticizing Newton they switch off their mental lights literally. They immediately go blank and I have witnessed it every time. I can visually see their eyes go dim. No Academic Newtonian priest will spend another second to listen to more of my views. Yet they remain unable to use Newton to explain the cosmic phenomenon as the Bode law, the Roche principle, the Coanda affect, the Lagrangian system or the sound barrier. But what they do not realise is the mistake was a result, not about what Newton presented because it came in what he admitted that he could not present. The mistake came as a later presentation of a concept he admitted he could not underwrite by scientific explanations. When he introduced gravity as a concept he admitted he did not know the origins of the force. What Newton admitted he could not explain and what he did not explain later became an institutionalised claim presented later as if he did explain it. The error is in the facts not yet ever explained and in that which Newton admitted he did not understand. It is Newton's incorrect suggestions that he made about gravity that were later accepted as explained and proven science that went on to become institutionalised facts. It absolutely came in the way Newton changed Kepler's formula. Newton admitted he did not know what gravity was and left it at that. He did not offer any more insight than reducing gravity as a force and a force it stayed all the time.

Three hundred and fifty years on science still does not know what gravity is and is still leaving it at being a force. Without knowing what gravity is, all other concepts in cosmology does not make sense because all the phenomena I mentioned a minute ago are sides of gravity and perform (each to its own but still) as another principle where the totality forms one concept we call gravity. If the World does not know what gravity is then the phenomena coming from gravity will remain unknown to all. With that in mind Mainstream science still takes a very dim view on my criticizing Newton! In all this time Kepler explained exactly, precisely and unequivocally what gravity is! The unbelievable part is that we all missed Kepler's announcing of gravity and in the "we", I use that 'we" to even include Isaac Newton and Albert Einstein in persons and by names.

All principles I use in the theory I introduce with the publishing of this book are part of nature. I base my theory on heat stabilizing through space using motion to produce cooling. That is gravity. But however this may sound basic, Mainstream science is also most guilty of their usual departing from this basic principle through the employing of terminology and such terminology has the tendency to cover many of the basic meaning behind the principles in nature. For example the one principle I do not applaud is a principle Mainstream science underwrites in the sense that matter in the beginning was coming about and anti matter came to destroy the matter by consuming it. This translates into a packman computer game that has no correlation to cosmology. It is moreover the disappearing from the Universe of that which came as the result between the two opposing materials that I strongly reject. Anything part of the cosmos at any stage before, during or after Creation, remains in and part of the cosmos and cannot leave the cosmos because there is simply no other place for what ever there was to go to. Leaving the cosmos just is no option.

In any test performed today by creating friction through motion the discrepancy there is between objects in motion will bring friction that will produce heat and the heat will result in space forming. In such destruction of matter, space and heat comes about and the net result eventually is space created where no space was before. The cracks showing is space created in the cooled material that was heated to a glowing red-hot. The cracks are space not filled but was filled during the heat coming about as a result from the overheating. After the cooling the cracks present new space where there was no space before the heating took place. The heating process started forming and filling space but afterwards when the cooling set in it reduced the filling of the space. The cooling did not destroy the newly formed space. The cracks represent the space not filled. The material in the cool state cannot fill the void that came as a result of the cold material contracting and reducing the space filled when the material was overheated. But the space remains although not filled any longer. That we can see as evidence with material having a heat building up when motion difference brings on friction and such friction brings on heat. I do not share the view Mainstream science has that when matter and antimatter came into conflict; the product that came from this just disappeared without a trace.

I believe the evidence is present and I think I know where that evidence is. I believe I can show that it is a motion discrepancy that produced matter and anti matter and we do not have to go and look for non-exiting positrons. A positron must produce a negative proton and such a performing sub-atomic structure cannot be functional. By changing legions it must then produce a product where it performs as gravity by rejecting material and pushing away other cosmic components. That will lead to an exploding Universe! I say that discarded material became heat that became space that became outer space in the Universe. I go to lengths to make people see that space cannot be "nothing". This is a factor that science has to accept if Mainstream physics have the will to find solutions about the Big Bang.

I say the motion between particles in a cramped space as the case was during the initiating of the Big Bang would have brought on friction in space between particles present that we couldn't even calculate. The result is that some of the matter particles produced a means of self-sustaining by applying gravity and the demise of the other particle that became destroyed resulted in plasma forming on the one side and material on the other side. I believe even to today and throughout the rest of the Universal motion through space in time the plasma is transforming to material forming particle growth in space through the motion we named gravity. This was how the cosmos came about and this is the manner in which the cosmos will conclude.

I believe some of Creation remained as some particles formed by applying gravity in motion and the lack of gravity turned the other particles of lesser motion into heat. In this is the destroying of singularity. Contact came as a product of light, which again I believe (within reason) I do prove. I believe heat is the destructed form of material and the atomic thermo explosions give this information to us. By releasing the heat that is sealed in an atom such release thereof produces heat, light that can liquefy the eye and most important the unexplained nuclear winds that destroy so much. Let me explain because this statement is at the heart of what gravity is.

A star (or a planet) is filled to the brim with atoms. The atom is filled to the brim with heat. The rotation of the star contains the heat in the star and when the star rotates to slow the star expands by heat releasing and heat forming space. A clever Newtonian name for this is the star is going Super Nova and the best intellectual explaining Newtonians can deliver is that " the gravity inside the star has gone mad!" can you believe that gravity with no intelligence whatsoever can go mad...and it is accepted science. That is a scientific explanation brought about by the most intellectual academics this planet can deliver. Now I wish to present you with what really happens.

The atom is a container that contains heat by movement at a speed, rotating equal to or on the inside higher than C or the speed of light. The rotation of the atom is the way in which an atom is cooled and spinning on the edge of the atom reduces the heat inside to a frozen solid. The cooling is done by the rotation at a speed of light or C. When the container we call the atom overheats all the heat inside expands as the circle of the electron then orbits a longer route and therefore in relevancy to the inside can no longer maintain C. Due the slower orbiting speed that comes as a result of the longer orbiting route that the electron has to follow, the atoms extends the space by heat becoming space.

Science also named this process a nuclear explosion or a thermo nuclear reaction but it is the release of atomic contained and atomic frozen heat into a liquid form. All that the atom bomb is, is that heat contained by spin that exceeds the speed of light and by expanding space due to overheating it moves slower and by moving slower it overheats and by overheating the movement inside the container called the atom as well as heat enclosing the container at the speed of light moves slower. When moving slower the solid heat inside that atom turns to liquid and the liquid requires substantially more space than what is available inside the atom.

Gravity is not a force, witchcraft, magic, and the unknown wroth of God that pulls you slowly to your grave. It is not the unknown flying from the region of despair pushing you to your final destiny; ultimately the grave. Gravity is heat being contained by movement just as Kepler said when Kepler said $a^3 = T^2k$ or said by using English space a^3 **is equal** = **to the movement T^2k thereof.**

Let's have a look at the set-up of the atom. The atom is solid heat. That we know because when that which fills the atom is released the heat released and escaping has the intensity that no other form of heat on Earth can equal. The electron circles at the speed of light, which carries the symbol, C. Inside is heat frozen into a solid as the heat spins at a speed beyond C. Because of the immense and fast movement, the movement freezes the heat into a substance that the atom is able to contain. Increasing heat is only done by slowing movement of some sorts. Decreasing space is only accomplished by moving material as fast as possible. Moving material is cooling material. Moving material at a rate of C freezes photons into a liquid form of solidity.

Increasing gravity is accelerating movement, which is reducing space by increasing solidity and freezing material into less space. This is what the picture above portrays. The higher the gravity the more movement will freeze material into a denser state by using less space. The more movement the colder the atoms would become. The colder the atoms would become the denser the atoms would become. The denser the material is inside the atom the more mass can be placed into space. The more mass that goes into a specific space will increase the mass to space ratio as the picture correctly presents gravity vs. weight or mass. Gravity is the containing of uncontrolled heat by moving the heat and gravity is the process in which heat moves from a more expanded region to a more confined region by condensing the heat into a smaller area…and that is hardly magic!

The faster an atom travels, the smaller the must the rotating circle of the atom be to compensate for the longer distance the atom travels per time unit spent to travel. The electron rotating is connected to the speed of light as the formula used to express time $t = \sqrt{1 - (V^2 \div C^2)}$ will indicate. The further V travels in distance the more the increase in distance must have a reducing effect on the circle the electron has orbiting at C.

Reversing the process of gravity is also releasing a hell of a lot of contained heat into an uncontained space. That gets a star going Super Nova!

By releasing the heat that is sealed in an atom, such release thereof produces heat, in the form of intense light that can liquefy the eye and most important produces the unexplained nuclear winds that destroy so much. It is heat going on to become space!

Light is the materializing of liquid heat into the smallest solid form. Releasing heat from a solid form such as the atom holds must involve heat as solid as light can ever be and therefore it can liquefy human flesh. The heat released burns shadows into cement and that is how intense the solidified light is in the form of heat when the nuclear thermal reaction releases heat from the destructing overheating atom. The winds are heat turning into space here the heat was confined into controlled space the heat expends into uncontrolled space. Newtonian call it shockwaves for the lack of understanding the process because of the absence of mass in this process. They can only understand mass and where mass is awarded.

This is a daunting process for Newtonians to understand because Newtonians can understand nothing…no that is coming out incorrectly. I should rather say Newtonians understand so much of nothing they are able to fill an entire Universe with nothing as only they can find nothing to form a substance that can be measured by distance!

But to realise that we must beforehand find what space is and accept that space is made of something. We have to see what forms space and why space can be the absolute basic container through which gravity can relay the influence it carries. We must come to realise that whatever forms space has to be that same ingredient that forms the lot of everything in the entire Universe.

When particles heat up the particles expand. Expanding is applying more space. The space that the particles hold amplifies. The particles claim more space when heated. The claim of heat on space creates more space and heat results in more space as the product of heat rising in material. Such expanding is one way of bringing about cooling. Heat produces more space and never reduces more space. The motion coming about from the expanding of space brings about cooling. When particles cool, motion applies in some form. At first the Universe was extremely hot and without space. Then space came about and heat levels declined. Compress space even today with a piston in a cylinder and such confining of space will increase the heat by the piston effort as the space reduces. The heat coming about has no relevance to particles colliding because compressor cylinders cool down with time and not necessarily with the loss or release of particles. It is not only the discharging of air that will reduce the temperatures inside the container. After the pumping of air increased the heat in the cylinder even to dangerous levels, the heat will reduce back to room temperature when further pumping seizes and the stopping of further air movement into the cylinder can bring about heat stabilizing. If it was the commotion of air particles colliding and rubbing of particles going on in the compressor, only the release of air could produce cooling. If this were the case, the temperatures would rise indefinitely because the action will increase in counter actions producing heat and spurring on more friction by collisions. The stopping of air pumping reduces motion and subsequently brings on cooling. The cooling process is not resulting from calming the excited particles within the container by whatever means. The process is natural and follows immediately when pumping stops. Stopping the pumping automatically leads to cooling coming into action.

This means it is not the particles in the cylinder that brings on the heat levels rising as much as it is not the particles that will eventually bring about the explosion that will follow, should the pumping continue regardless of danger. When the pumping stops, the heat immediately starts reducing. There is no further increasing of heat in any way except if the pumping continues again. The already over filled container does not continue with the friction of particles rubbing within the cylinder container. Afterwards when the compressor is left by itself and temperatures stabilize to the same levels of the temperature on the outside, a sudden releasing of the air under controlled conditions, such motion of the air relieved will bring cooling to the extent that pipes can freeze and block the releasing airflow. Two American Submarines were lost in this manner and yet to this day no person in science saw such a connection in the sinking of the subs. After a few days left undisturbed as far as heat distribution goes, the stabilizing will lead to conditions being equal on either side of the cylinder wall. The compacting of air molecules, which although still much higher on the inside than what the denseness is on the outside, will not produce the same heat levels inside the cylinder than those achieved during and immediately after the pumping operations. The temperature will only become affected with motion contributing to changes in the balance. The releasing of air will extract heat from the process to a point where it will lead to freezing coming about. We have to see the container for being a container and the Universe also being a container, but more important is that what ever comes about in the one container will be similar in the other container, since the containers contain what the cosmos is made of. It is heat flowing between materials in space. Heat will always flow from the highest value to the lesser value. This is the concept I use as the basis of my theory. I base my theory on gravity producing cooling and contraction while heating produces motion by expanding and creating more space. When an object is overheating, no amount of force can retain the container from becoming too little and with the heat coming about forming the expanding, the space produced with this action will destroy the form any container may have. Even a container as solid as an what atom proves to be is still not able to withstand the expanding produces by overheating. The Universe is a container and as all containers proves; the Universe also was unable to sustain the space that heat created when particles overheated. The Big Bang is evidence of this phenomenon. During Creation the compactness of the particles produced motion discrepancies bringing about friction where some particles overheated and formed heat. The heat that then formed space, we named the Big Bang where heat produced space and formed motion we see today in the Hubble constant. The Hubble constant is antigravity in its most splendid form.

I believe that my study of Kepler allowed me to achieve an all time breakthrough success because I can now explain what gravity is. Remember that not even Newton could explain what gravity is or where it came from, but Kepler did that without any person ever noticing. Kepler studied what kept the planets circling the sun and that is gravity. From such explaining, what Kepler said without Newton changing formulas on Kepler's behalf, I prove the Titius Bode principal also known just as the Bode principle. I prove that the Bode principle is forming the value of gravity when incorporating the Roche limit. These phenomena was never before explained or understood by Mainstream Science although they appear more than regularly in the cosmos. I explain how singularity forms the Roche limit and how singularity brings about the Coanda affect but most important of all Kepler showed me where to search for singularity. My achievements came from my effort where I separated Kepler's work from the opinion that Newton formed and gave to the world about Kepler's work. For instance from Kepler's work I can explain the operation of the Black Hole, which not even Prof. Stephen Hawking understands. That is because Hawking ignores Kepler. In my opinion my explaining of gravity makes much more sense than the accepted force of Dark Age proportions…and the best part is you do not have to be a genius to realise or understand it. Even a simple person such as I can see it clearly! From my view, a force is just motion applying and that is what Kepler said gravity is. Kepler said $a^3 = T^2 k$. I dissected k as a factor in the Coanda effect and found that the Coanda effect is proof of my view about gravity and singularity produces the Coanda effect. The Coanda effect is the establishing of individual space a^3 by applying motion T^2k. Where the Coanda effect is producing gravity and such producing is stronger in a small space than the gravity produced by the Earth in that spot, I use that principle to show that there was some manner in which the reducing of k brought about a stronger T^2 just as Kepler said. This was a crucial part during the Big Bang and therefore had to play a major part during the period of the Big Bang. Einstein came to this conclusion but failed to refer his view back to Kepler and by not referring to Kepler missed the point he wanted to make.

Presumptuous as it may be on my part of trying to disprove Mainstream Physics, such a presuming does not change the truth about Mainstream science being incorrect about gravity. After all, they admit they do not know what gravity is. I am not disproving anything because they agree they do not know, which paves the way for my showing what gravity is. By admitting not knowing what gravity is, they then also admit being incorrect about gravity but unfortunately mainstream physics do not see it that way (yet). The question in hand is finding what role gravity played when the Creation came about for the first time. I had to find a method that would allow me to explain why gravity played a role. In the book I present the analysis of Kepler's formula without Newton's interruption of Kepler's work. Please let me explain: Tycho Brahe and later Kepler made a study of outer space as never repeated afterwards. From this Kepler concluded that $a^3 = T^2 k$. We all know that a^3 is space and with the space indicated as being in the third dimension and the third dimension is unmistakably a cube that forms volume, which by definition is presenting space. We also know from the way calculations come about by using the formula of Kepler that T^2 is the duration of a specific period of time relating to a specific centre. On the one hand we have space a^3 and on the other hand in direct relation to the space, Kepler introduced motion coming from a centre that forms time $T^2 k$. Kepler gave us space-time a^3 / T^2 centuries before Einstein gave the concept a name but no one ever took any notice. In the formula is space a^3. In the formula the space a^3 has a direct relation to time T^2. If k is a^3 / T^2 it means that from the centre holding the gravity is space-time. Space is a^3 and the motion of space a^3 we accept as time $T^2 k$ and such an acceptance is part of our understanding for the past three hundred and fifty years. Kepler gave us gravity before Newton named it as a force. Kepler gave us space-time long before Einstein named the notion. With Newton's meddling, he missed Kepler's introduction of gravity as $k=a^3/T^2$ space / time.

The theme of the book is about the Universe not coming from nothing and therefore outer space cannot hold "nothing". By taking Kepler's $k = a^3 / T^2$ and using k as a line, I show through using the line as an example that the cosmic Universe holds everything and all concepts. However, the only thing it does not hold is also the only aspect not present in the Universe at all. That is the value of nothing or zero in as much as carrying the definition of the absolute absence of any value. This means the Universe is filled and not empty. The Universe is so full it is overflowing and we call that an overflowing expansion. With the line that light uses to flow, the lines eliminate any such a possibility of nothing being present. There has to be a contact that directs the line of light departing from one position and ending in another position.

Mathematics is a means of communication about matters concerning the cosmos. It forms one more language or expressive interpretation. As an intercultural language spanning across race and ethnicity or as a principle as such, mathematics cannot have zero because mathematics indicates lines, which is about not applying the numerical number or value of nothing. Everything came about from singularity and Einstein proved that. From singularity "nothing" never had the chance to enter space. I challenge any person that disagrees with this statement to show mathematically where nothing as a factor ever entered the mathematics of the Universe. Your attempt may either be before or after reading my work but my challenge will stand since mathematically nothing replaced ether when ether was removed from space. Then some party's in Mainstream Science brilliantly allowed the nothing they could offer to replace the ether theory as the substitute of ether where it produced a way for that nothing then to somehow become space. This achievement was never achieved by the use of mathematics. Kepler gave us the relation between cosmic objects as $k = a^3 / T^2$. From the formula k forms a connecting straight line filling the first dimension and not the single dimension because k in the single dimension is not zero. It is unproven how k can backtrack to become $k = 0$. I deliberately press this point and make it an issue because that removes all the theory of mainstream science from any logical base they have in support of their views that space is, holds and comprises of nothing.

Whenever Academics could not respond to my work with counter answers indicating where they see my possible misconception to be in and about my arguments and proving me wrong, they always came to a conclusion that space holds nothing and from that argument simply dismissed my work. I truly think that if you wish to invest in my work you should know why South African science denounces my work. They denounce my work largely on my standings on Newton but also on my being "incoherent" about outer space that I say can mathematically not be "nothing". They are in the righteous position that they can dismiss me without disproving me mathematically. But from such denouncing they never brought in logical arguments that could disprove my views and their dismissing left me defenceless and without grounds. The Professor can plainly say he does not agree with any one and only one of my statements and leave it at that. I found in the manner they treated me that he is not even obliged to say with which aspect he does not agree. On the contrary I came to a conclusion (I admit the conclusion must be totally incorrect) but still I have the impression those academics do not understand my work but I repeat that this impression must be totally wrong. This I cannot accept since the work I gave them is most basic in principle and academically on school level. In order to prove the Big Bang we have the reverse by reducing space and the line forming space to a point it was before the Big Bang started. By reducing space one can reach an ultimate reduction point where the reducing of any or all lines that form the universe, then can only confirm singularity, which was my first breakthrough. What is so difficult about that and moreover what is incorrect about that argument? Take the lines forming the cosmos back and see where it ends. Reduce k in the formula $k = a^3 / T^2$ and see where k ends, but one place it cannot end is at $k = 0$ because then k was removed from the cosmos totally and it will be very impossible to introduce k once it was removed.

Gravity is precisely what Kepler showed gravity is. Unrecognised it may be, but Kepler showed gravity before Newton named it. In Kepler's formula k equals space-time or a^3 / T^2. Gravity is singularity extending and forming space a^3 through gravity rotating space T^2 as well as containing expansion k. Gravity is not particles pulling one another in a tug of war. Gravity is about reducing space and maintaining different cosmic sides not sharing the same sort of space. It is how gravity does it and why gravity does it that produces the accepting of the Big Bang Theory, but it also includes the Plasma theory in detail. The Plasma theory is as correct as the Big Bang and Kepler indicated that. We only have to open our eyes and see it. At the beginning, before and after the Big Bang (yes before because there had to be a before) gravity was about bringing across heat that was in space to material that was in another space. That is how Mainstream science presents the forming of all particles. The only difference is that in the space where the space is filled with plasma or heat and not material, the heat was much denser than compared to now. Now they present that same space holding "nothing" but they never show where such a process stopped at a later stage. I say it never stopped and is still the result of gravity applying. Material is still growing but the growth now is hardly noticeable because the heat in space lacks density to provide such growth and space has much less heat density compared to what there was back when…

Mainstream science claims that many aspects of material (matter and antimatter) and singularity amongst others were present in the cosmos during the early phases but has vanished since then. Think about it clearly…What ever was in the Universe had nowhere to go but to remain in the Universe after the Big Bang. There is no place else to go! The Universe is everything there can be. That means if singularity was part of the cosmos during the Big Bang, singularity must still be in the cosmos. If antimatter was present during the Big Bang, it must still be present. There is nowhere to go except remain a part of the Universe! I am not disputing antimatter but I am disputing the vanishing claim. Space was little and heat was massive. Heat became space as the density of heat turned to form space through a process we named exploding, which is space expanding. The fact that heat turns to space was never recognised by science in the past. The fact that space compressed forms heat, notwithstanding all evidence supporting this knowledge was never accepted by science while all combustible engines work on heat coming from space compressing to bring about heat. If you push in a cylinder the space heats up but science does not recognise such obvious facts. While the whole world produces energy from a principle that is about converting heat to space mainstream science chooses to deny the fact. In the beginning singularity was present in the Big Bang. Mainstream Science promotes the idea that singularity and antimatter went on the disappearing by escaping from the Universe. But Mainstream Science never says where they went. Singularity and antimatter and all other Houdini acting stuff had no place to go. Everything that was in the Universe, is in the Universe and will forever be in the Universe until the very end of the Universe. Going somewhere is not part of any options of whatever is the content of the universe. If singularity was part of the cosmos in the beginning, it is still with us. We just have to find it by searching for it and use the characteristics it had during the Big Bang. It will still have the very same characteristics. However, by claiming that singularity vanished somehow since the Big Bang, is totally incorrect and proves a lack of understanding the cosmic concepts. Kepler and his formula also prove this fact.

Every aspect that was part of the Universe at the cosmic birth announced by the Big Bang has to be present up to this point. There is no room or place to which the singularity could dispose since the Universe is the only place there ever was or will be to be within. I prove that singularity is present within matter, a claim not yet made by any other person. By my finding such a location and producing as well as applying a very specific value to singularity, singularity makes the process, which the Universe adopted, much easier to follow. The connecting of singularity runs from the ordinary stars such as the sun and up to the Black Hole down to the ordinary material forming our every day living in the cosmos. We have to realise that singularity is a prerequisite for the Big Bang to have formed any and all material. If the presence of singularity is not in all material in the everyday cosmos, singularity was not present during the Big Bang and the Big Bang cannot be, because what is in the cosmos now, had to be in a pre Big Bang cosmos as much as it has to be in the cosmos of the present and the future as it is in all stars including Black Holes.

In the book an open letter To Selected Academics ISBN 0-9584410-9-X the book only and exclusively deals just with the fundamental basics of my theory. I do not elaborate or explain the broader aspects or form an overall view. I found if I do that before a solid understanding of the basic concept is established no concept becomes established. The most basic to explain is that the line cannot start at zero because then there can be no line to follow zero. The cosmos has lines forming cubes and lines forming circles, which in applying 3D manifests as spheres. Between the circles and the cubes runs lines, so the key to understanding the Universe, are through lines. The Big Bang was at a time when the Universe was incredibly small, making the running lines small. Understanding the Universe is taking the line connecting particles through space back to its limits where such limits were during the Big Bang. But the reducing cannot go to zero because zero removes the line all together. By reducing the line to where the line will not reduce any further, we will find at that point that all points land on the same spot. The spots all share one position because that is the only position there is to hold in the form singularity presents. That is singularity being one to all but it is not zero. Finding form in that point shared by all, will give a value of singularity. Extend that value received to a Universal centre and bring that value to align with Kepler's a^3 = kT^2 and understanding the Universe by finding the centre of the Universe makes the Universe simple as can be. The Universe becomes sensible making the entire different yet unexplained phenomenon as easy as children's schoolwork. There are suddenly no more mysteries in the Universe. It is only possible when we see gravity not as a grabbing force but instead seeing gravity reducing the space between particles. Gravity is not being some magic force found between particles grabbing onto everything. I mathematically explained the following phenomenon:

1) **The location, the position and the value of singularity as a factor forming space-time**
2) **Finding space-time** by dissecting Kepler's formula in relation to valuing singularity
3) **Finding space-time, proving space-time and aligning space-time with gravity**
4) **The working principals behind and manifesting of gravity as a cosmic occurrence.**
5) **The Roche limit,** and explaining the resulting of a law coming about from singularity.
6) **The Lagrangian system** and how and why that becomes the building form of the Universe.
7) **The Titius Bode law** and how gravity comes about from that
8) **The Coanda effect** and the producing of gravity through reproducing space-time.
9) **The sound barrier** by proving it is gravity generated by motion in space becoming independent where motion creates independence. Breaking the sound barrier is the motion in space duplicating space by crossing over gravity borders. It is $a^3 = k > T^2$ ($k > T^2$)

By using Newton one cannot even begin to explain any one of, or the combined effort of the above cosmic phenomena that are all over the cosmos and form all the laws in the cosmos. Newtonian definition cannot even recognise any of the principles but only Newtonian science are taught to students. No student can have the fortune to disagree with Newton and remain a student. If the student will dare to disagree with Newton, it is the end of such a student's academic career. By setting this firm condition, Newtonian science becomes institutionalised mind conditioning of the concepts of thought forming in physics. With my saying this I have not made one academic friend but neither has any one proved me wrong. Students are taught to accept Newton and to ignore Kepler and any student doing it the other way around will fail all examinations and other testing at Universities. Students accept Newton or they accept a ticket taking them home. Newton is an institution force fed to each following generation but saying that, reserves only resentment towards me amongst Academics. According to Newtonian science, space is simply nothing with no qualities but gravity separates space and space does not mingle, as one would expect if space was nothing because space does form borders. Disasters of unprecedented magnitude arise from such borders. The Challenger disaster of February 2003 is much testimony to those borders that was powerful enough to break the aircraft into pieces while the explaining contributed by Mainstream science is evidence of a shocking lack of understanding about what took place as cosmic laws were breached. I do not pretend to be of superior understanding and do not place myself on any pedestal. On the contrary, the information is so simple and so easy to understand that the lack of any Academic understanding frustrates me almost witless. But an academic taught culture demands all persons to miss the evidence, which is so clearly visible because academics demand researchers to look in other directions because students are forced to accept Newton's vision about Kepler's work. My denouncing Newton is about cosmology as his views of gravity do not extend into the cosmos and I denounce Newton on cosmology and only on cosmology. Students on cosmology are force fed to learn (study) or spending hours repeating concepts the logical mind cannot accept and thereby brainwashing their minds in the concepts of Newton which is another way of conditioning the mind to accept Newton. By the time they reach researcher's status, they too have tunnel vision that can only acknowledge Newton and ignore Kepler. Our not understanding laws, provide a platform for future disasters occurring because it will lead to us ignoring more of the applying principles that lead to space tragedies of magnitudes we have not thought of as yet. By not understanding the sound barrier, tragedies have and will again come about and will increase as misconceptions become more present in the future because demand on space travel increases.

The fact that space is the shape of a dome or a bubble is not a big coincidence but comes as a result from singularity bringing about dimensions. These dimensions are there whether science chooses to ignore them or recognise their powers. Space is a liquid and as all liquids do, space depends on specific densities. With the specific densities borders come about. The explaining I manage with very simple sketches indicating the principles. I explain with ongoing proof throughout the book indicating to the reader how I came to realise that gravity is not about particle pulling on each other with some inexplicable force of magic holding an effect on matter or matter pulling matter. Motion brings about the demise of space and that includes space holding material as well as space not holding material. This may sound simple in concept but understanding requires concentration and that makes this book not being a novel to read as it may come across.

What Newton missed in Kepler's work is Kepler's introducing of gravity as a formulated fact because Kepler said gravity is $a^3 = T^2k$, which is the space a^3, that forms through the moving T^2k thereof giving the space a^3 independence by a relevancy coming from T^2k releasing individual singularity from surrounding space. Gravity is space moving in a circle holding space including, that which is in space at a distance from a centre where that is applying the conditions of space in time. In outer space six dimensions secure and stabilize floating objects. Only when such conditions are broken, does space fall away and particles come crashing down to Earth. It is the breaking of the balance that leads to objects falling to Earth and not the Earth's gravity pulling. This falling comes from a lack of motion by the particle in space in relation to the distance from the centre. Again it is a^3 that becomes too little to support the floating of the object when the spin T^2 reduces in the presence of a certain value applying to k. A little science experiment such as the Coanda effect disproves the gravity-grabbing-on theory. Gravity is about matter concentrating space through the spin of the proton and the reducing space by accelerating the movement of space between the two objects. A good example is a blowing fan. By establishing motion of the air with the spinning fan, volumetric occupation reduces in favour of airspeed coming about from motion. By removing the space the particles come automatically closer, and that is the principle of gravity in operation. But the diminishing of the space goes about by applying very specific rules and in the adhering of those rules is applied to everything in the cosmos.

Gravity comes about as space a^3 applies motion T^2 and from singularity k gravity is as much part of space as the motion of space is part of gravity. Do not be fooled by the simplicity. Mass is the result of applying gravity by reducing space. Gravity is not the result of mass. Gravity dismisses space and by doing that the stronger gravity can place more particles fitting into less space occupied where that reducing of the space is bringing about extensive mass increases into the volumetric occupied confining more material into less space. That produces monsters like a Black hole having enormous mass and being without space. If gravity was not about the reduction of space bringing about the increase of mass because more material will fit into less space, the Black hole does not make any sense. Stars reduce their volumetric size as they gain in mass by creating enormous gravity applying to the sphere of influence of such a star. Gravity is the increase of heat in occupied space by the reducing of space in a spherical unit. From the offset of the Big Bang to the process the Universe went through development up to the very point where it is at the present, the process was about converting heat to space and space to heat. I named the expanding of material through the process of overheating antigravity. Where gravity is the contracting we find present in our daily life, the Hubble Constant and expansion through heat applying is the other part of the driving engine applying antigravity with expansion where the heat is transforming to space. Space shifts as heat releases space and converts the Universe in one direction bringing about expanding into more space but less dense space. Remember how the heat came down from 10^{34} to 0 K at present? The density of heat in space surely diminished considerably since then to now. But while the heat was diminishing, space grew as a consequence of the heat diminishing. Gravity on the other hand is exchanging heat through concentration by removing space bringing about space loss with increased density of particles and therefore heat concentration. In the centre of all spheres, which all stars are, it is hot. In fact the heat in the centre of the star is the product of the space it concentrates to form heat and in that we can read the gravity the star is able to produce. The ability to secure heat by reducing space becomes the measure of the star. Momentum is the second form of gravity symbolised by Kepler, as k. The Big Bang is the result of heat expanding into the forming of space. Gravity, on the other hand is about concentrating space back to heat, and take recouped heat through to material, acting out a balance of expanding while contracting. This way gravity is applying the onset of the Big Crunch by destroying space while space is converting heat to material occupying space. The Big Crunch is coming about because the Universe is expanding where the two processes are one principle.

The book **an open letter To Selected Academics ISBN 0-9584410-9-X** is about that process adapted by the Big Bang, which never ended and it is still bringing over, that which is in unoccupied space to material being in occupied space. Occupied space holds matter and unoccupied space is empty of solid materials. There is contraction, which we know by the name we gave as gravity. Then there is expansion, which we gave many names being the Big Bang and the Hubble Constant or better known as simply exploding or forming plasma with all the terminology accompanying that simple idea. This I show is antigravity. Apply heat and space and a balloon lifts, where such lifting is antigravity. There is a balance in the Universe where gravity contracts and reforms space and heat expands becoming space and produces space.

Where the manuscript presented came from an actual letter I sent to specific Universities, my aim with that effort was trying to keep the printing cost as low as possible. The motive with the manuscript was an attempt to assemble the most basics of many of my ideas in my books in an effort to place as much emphasis on explaining the combination of my work while keeping the printing cost as low as possible. In the second effort behind the book, I wish to introduce the theory supporting my work. Where at first the book was a letter I wrote in an attempt to promote my theory at many different Universities outside South Africa, the second attempt is to get it published and let people read it as widely as possible. I realise I have to get the ideas out in the open to as wide an audience reading the work as I can reach and have people read as much of the information as the effort will allow. I have to show others how I came about finding the most basic aspects of my theory in the most basic form and show how that follows a line as growth progressed and developed. It is an attempt to convey the thoughts in formulating the concept on which I based all my other books. But with this book the motto was keeping it so simple children must understand what I am trying to convey about my theory and let the other books carry the slightly more complicated arguments. If I could manage to keep the concept that simple and keep a well established link with what mainstream science knows but without compromising facts, I know I could find the key to acceptance and understanding of what I have to say. With the costs as low as possible it will be possible to find a publisher that will publish and sell the book through Amazon Dot COM and Barns and Noble as well as other major book outlets. Without the academics accepting and underwriting my views it is a chance one takes. Another large factor is that only I believe in my views as being correct because never once were any of the principles vaguely explained whereas I explain, define and underwrite such explaining double and triple.

I do think further information is required about why I have to follow this route. I shall gladly explain. I am not connected to the Mainstream physics in any way or form and through my views and the manner I have to promote in the writing of my work severe criticism is involved from upstanding gentlemen and that I found is proven to be a method how not to make friends with the most educated and honourable members of high standings in Mainstream physics. The promoting letter had very little success because I am not sharing much of any opinion with the Academics of the day and in that respect I have to criticize their work. Such criticizing of Mainstream views is apparently not the correct route to follow when trying to communicate with such gentlemen of influential academic standings and in every case they took the criticizing very personal. Why, I do not know because it is their opinions I dispute, not them personally! But with my definite disagreement with mainstream physics, there are no other options; I have but to disagree with Newtonian disciples and none the lesser so than with Newton in person even after his death. But disagreeing is the trademark of my work and I cannot applaud that which I dispute. However, with the position I have in relation to the academic world, my work landed amongst and between the junk mail the Universities received and was discarded as such (or so I suspect). The important persona gave it to the lesser important persona and they were not in a position of authority and had to discard my work with replying in dismal answers. The arguments coming from those were of such a low mental ability in understanding my work that I would not insult the intellect of others with such dismal arguments. If they did read my work it means they never were able to read my work.

That means I have to find an audience willing to accept my views because they read my work as a fact of interest and are not ordered to do so. That I can only achieve by publishing commercially since the work on its own merit is far off from mainstream science and it seems commercial publishers do not wish to take such a gamble in my more expensive projects. In the event of mainstream science coming out and accepting my work, the outlook of science on science will have to change more than dramatically. Such implications will carry a heavy financial burden on the present academic world because heavily promoted theories will become discarded and the opinions of many influential persons will be nullified bringing about that many a PhD will be proven as being incorrect. That means there are many in top positions that will never admit to my views being correct and acceptable in the near future because they have to secure incorrect approaches because of promoting self-interest. But amongst the many there will be a few that will have little to lose and much to gain by supporting me. My quest is to find that few with much authority and a clear and open vision.

If I wish that any acceptance should come about, I have to introduce my work to as wide an audience in the world of physics as possible. I need only three Academics carrying the approval of my work, the three that has the insight to appreciate what I say, the three that has the position to accept what I say and one of them seeing the truth behind my work and acting as a promoter of my work. From that point my views become established. Sir, you should be a businessman and in that you may consider what such acceptance of the work means financially. From that point of view, the academic ignoring will have to stop because then no longer will academics be able to go around and simply ignore me because they can. From then on they must declare on what grounds they find my wok incorrect and that, they will be unable to do. That is how sure I am about the correctness of my work. If my work is incorrect then Kepler was incorrect and if Kepler was incorrect the cosmos was incorrect about facts in the cosmos. With the cosmos in support of my work, then academics will have little chance to dismiss the credibility of my work as they now so easily can do.

If I am in a position to be able for the first time ever to explain what gravity is, people who are acknowledged as leaders in physics accept my explanations as correct, the selling potential of such a book becomes enormous. For this to happen I need one academic in an influential position to accept such views. Every one interested in science then will have to pay for the book to see how I prove what gravity is and how gravity effects even the breathing of people. But finding the man that will be brave or adventurous enough to underwrite my work, will be a task in need of a person that are prepared to spend time and in order to promote my work this individual must understand my work and for this person to understand my work this person must read my work and while all this requirements go begging it is because everyone thinks of Newton as being flawless. The factuality and truth about my work is above dispute, but changing the world's accepting takes the same effort as redirecting the route the Earth follows around the Sun. Before you get all negative, the redirecting happens every second of every day but no one ever thinks about it. The realising of it brings the accepting of the fact and becomes substantiated proof, as I believe my work will be. By revolving around the sun the route the Earth follows changes every second but nobody sees it that way. Explaining it becomes the issue. The other options or routes I can follow through magazines by publishing articles that only allows me three thousand words at a time. That is far too little to promote such a broad concept as I wish to promote. Even this letter is that number allowed many times over. With the promotion of my view, the complete picture in science has to change totally and not just some aspects. Therefore I need three thousand pages at least to make my ideas logical. But the size and shape the book has now is the minimum required space.

At first I wasn't going to search for a publisher as far as concerning this book Called Newtonian Fraud or Newton's fraud or whatever and I had no such intentions. But then suddenly private needs dictated an urgency to come by more money fast and that gave me the idea to try and find help in this manner by using Carte Blanche as an introduction medium.

When finding a publisher I shall accept the normal publisher/ writer agreement although the information, calculations, formulas and the arguments remain my property and that I use in more conclusive books. The value of this book is that in the event of any person wishing to read any other work I wrote has to read this book first in order to understand how I link my work with that of Mainstream science. With the book in the place it is being the introduction it will become an academic prescribed book in the event of my theory being accepted. It is on the accepting part that a publisher finds some gambling and that I admit. But I have no doubt about the correctness and from that assurance I have no doubt about such acceptance because although my work is unexplored by others, it is factually more correct and more proven to be correct than the basics science now have to rely on and is much more sound in principle. It diverts from science but it is very simple to follow. In the final analyses it comes down to a simple case of "my goodness why didn't I think of it" and because of such simplicity my work has, this book retains a good prospect of finding a commercial market. To receive the acceptance it will only take three academics with insight to read my work and realise I am correct about my thinking. From there it becomes a bed of roses but from where I now stand the locating of the academics is quite impossible without me publishing this book. I can assure you that where my other work might be somewhat more

A Science Conspiracy to Conceal and to Cover Up 168

complex to understand, this book is so simple a child can read it. As far as I could establish during the past twenty-seven years of research my views were never promoted in any way before but in the event of such proof coming about it will make my task a thousand times easier.

Module TEN
You are about to
Meet The

Biggest Crime Syndicate that Ever Operated In the History Of the Earth.

There are so many nagging questions that Newtonian brilliance fail to address. Why did the Moon gather from dust so close to the gathering area of the Earth? Why did the moons of all the planets become the dense structure while achieving this incredible feat with such small mass evoking such poor gravity? Why did the Sun not collect the lot since in the normal formula that applies, the dust had to pass by the Sun on its way to where the location is where that the dust then formed planets on the other side of the circle? What prompted Pluto to become one of almost a double star and not reserve all the accumulation into one unit? Why have the unit, so far from the centre and yet have the dust to form such a firmly compacted unit in two separate structures being so closely together? Why did a giant such as Jupiter allow fragments that we now call moons, go flying about the giant planet in an area so small where the dust assembled as spheres of insignificance while Jupiter collected the overwhelming dust in the area to eventually assemble the planet Jupiter? What made the planets choose the bands that they circle in at the specific points they chose to assemble as planets? There are a million unanswered questions that I do not find answers to because I have too a poor mentality to understand Newton. It seems the Newtonians never got around realising these questions, which go unanswered.

Let us go on to investigate what it is in the Newtonian high priest custodian mythology that I am not intellectually suitable to understand.

$$F = G \frac{M_1 M_2}{r^2}$$

We understand from Newton that gravity is on the increase because

$$\frac{M_1 M}{r^2} G$$ This is the part taxed with establishing gravity the force.

This is the belligerent part enduring the conduct of the above where they are taxed with this facto's destruction. The radius is on the decline. This idea founded physics for centuries in the past and no one ever dared to question the feasibility of the statement.

This meant that the Universe was coming to a conclusion. All that has parted somehow has drawn on the mass that developed and the "gravity" was pulling the cosmos into a "grave". That is what I can conclude from the name but it might be presumptuous of me to conclude Newton's thinking in such a manner. This brought serenity to the cosmos. The cosmos had a purpose, a motive, and a function and even moreover it has a conclusion. At a point the book will close and the curtain will draw. It is where the cosmos will end in a grave and all will conclude into a finality that will end that, which started.

The Sun was pulling the Earth that was pulling the Moon and one day we can rest assure that the Sun will absorb the Earth that would have absorbed by then the Earth that by then would have consumed the Moon in turn. First the inner planets have to give and finally the outer planets have to give to this force of magic being inspired by the mass that Newton saw everything must have because everything had to have mass. The there is a question arising in search of a valid answer...with all the mass available where does the expanding Universe fit into the bigger picture?

Because those who should know these things tell me that do not understand Newton I do not fit into the crows every one I meet is part of because in that group they obviously understands Newton or at least are being part of the crowd that understand Newton. I am seemingly the only person that does not fit into the group that everyone knows are the wise or pretending being wise where that crowd as a group is in total understanding about all of Newton and his forces. To acknowledge my not understanding what there is not to understand I had to go into a personal solo debate with myself commenting about my not understanding to myself on the facts and in order to try and understand what Newton never understood by his own admission, I acted as an appointed narrator and by talking about such and other facts, I bought into the open questions normally never raised in intellectual conversation, such as where is the expanding Universe expanding too? Some then said I went mad because I started talking to myself but I found in my person the only person that understood what I did not understand about what Newton also seemingly never understood. When one does not understand, it is best to go into an argument and clear the matter, but since I am the only one in the world that does not understand Newton I have to argue with myself. Some less sensitive persons think of this method as talking to one-self...but that low I shall never stoop, since I go into a solo debate sharing my intelligence with another of equal intelligence. (That sounds better than saying one is talking to yourself, does it not?)

Newton said everything had to have mass because if everything did not have mass everything did not have gravity and with everything not having gravity everything could not be pulling everything else to the final conclusion that Newton predicted. The most important rule in physics is that although God can be wrong, being wrong is just not possible if you are Newton. This statement I put to all those academics that have on their answering machines the Lord's Prayer recited and verses from the Bible but they would rather face criticism coming to God than allow any criticism about Newton.

The Universe contract by the mass it has with the gravity it produced to find the measurable force that keeps the lot spinning and contracting.

Then came an absolute disaster by the name of Edwin Powell Hubble, the man that will never deserve the Nobel prize because he screwed everything up that Newtonians regard as sacramental and holier than the Bible.

Hubble, Edwin Powell

Hubble, Edwin Powell, 1889–1953, American astronomer, b. Marshfield, Mo. He did research (1914–17) at Yerkes Observatory, and joined (1919) the staff of Mt. Wilson Observatory, Pasadena, California, of which he became director. Building on V. M. Slipher's discovery that galaxies had strong shifts to the red end of their spectra, Hubble used the Cepheids in nearby galaxies to demonstrate that they lie far beyond the Milky Way. Because of an incorrect understanding of the Cepheids, this distance was vastly increased years later.

Hubble's law is the statement in astronomy that galaxies move away from each other, and that the velocity with which they recede is proportional to their distance. It leads to the picture of an expanding Universe and, by extrapolating back in time, to the Big Bang theory.

In 1929 Edwin Hubble first formulated the law that was named after him after he saw an expanding Universe in his telescopes eyepiece. Hubble compared the distances to nearby galaxies to their redshift, found a linear relationship, and interpreted the redshift as caused by the receding velocity. His estimate of the proportionality constant, now known as **Hubble's constant**, was however off by a factor of about 10. Furthermore, if one takes Hubble's original observations and then uses the most accurate distances and velocities currently known, one ends up with a random scatter plot with no discernable relationship between redshift and velocity. Nevertheless the relationship was confirmed by observations after Hubble.

The law can be stated as follows:

$v = H_0 D$ where v is the receding velocity of a galaxy due to the expansion of the Universe (typically measured in km/sec), H_0 is Hubble's constant, and D is the current distance to the galaxy (measured in mega parsec Mpc).

One can derive Hubble's law mathematically if one assumes that the Universe expands (or shrinks) and that the Universe is *homogeneous*, meaning that all points within it are equal.

For most of the second half of the 20th century the value of H_0 was estimated to be between 50 and 90 km/sec/Mpc. The value of the Hubble constant was the topic of a long and rather bitter controversy between Gérard de Vaucouleurs who claimed the value was 100 and Alan Sandage who claimed the value was 50.

Hubble also suggested that the clusters of galaxies are distributed almost uniformly in all directions, although more recent studies show that clusters are combined into huge super clusters of galaxies: at this new level, however, the distribution appears to be even. He was the first to offer observational evidence to support the theory of the expanding Universe, presenting his findings in what is now known as Hubble's law. With Milton Humason, Hubble classified the different types of galaxies including irregular galaxies, three types of spirals and barred spirals, and elliptical galaxies. Included in his writings are A General Study of Diffuse Galactic Nebulas (1926), Extra-Galactic nebulas (1927), Spiral Nebula as a Stellar System (1929), The Realm of the Nebulas (1936), and The Observational Approach to Cosmology (1937).

This put a lion amongst the pigeons and the lion devoured any cat that could enter the pigeon den. The Universe was upside down and this was latterly. What everyone thought was coming was in fact going. What was reducing was expanding and what was to become less became more.

We know Newtonians are not well versed in Horse Trading. If they don't like the rules they bend the rules and if they don't bend the rules they break the rules. They don't take prisoners when it comes to obliging by mathematical principles and they never adhere to science as a coach because now they have the

formula $F = G \dfrac{M_1 M_2}{r^2}$ and now they can coach science.

What would you say were the reaction launched by Newtonians? Newtonians corrupted corruption by enlisting the biggest fraud ever concluded. This outmatched Newton's by a billion to one and Newton's effort in corrupting Kepler's work already surpasses nothing ever cooked up by any gangster ever.

The Academics in charge of official policy never held Newton in content. They never reinvestigated Newton. They never flinched an eyelid at Newton's claims. Not once did one raise an eyebrow against Newton's work. The reaction was that if the Universe did not prove Newton correct then when will the Universe start proving Newton correct. If the Universe was expanding when is the contracting coming in place? They did not say wait a minute that makes Newton a fool. No, they said the foolishness of the Universe must immediately end and start adhering to what Newton predicted. The Universe must come right and it must do so fast. The Universe must attend to Newton's statement and get a grip on its act. The Universe must show its critical density immediately so that Newton should not be ashamed of the actions of the Universe. They never said once that with the Universe expanding that means that Newton actually stands to be corrected. No…they said the Universe is wrong. God and his Universe must be wrong because Newton is correct. Never once was the idea of incorrectness associated with Newton. Never did any person thought of Newton in terms of being disproved. The reaction was condemnation to the cosmos. The mistake the cosmos made has to be corrected. There has to be a density mismanagement committed by the cosmos. The cosmos is inadequate, faulty deficient and impotent. The Universe lacks the mass required to substantiate Newton. There was a critical density to be investigated in which Newton would be vindicated. With the Universe in fault at this stage to prove Newton's absolute trustworthiness then it is up to the Universe to prove when the Universe will become trustworthy. When can one trust that the Universe will correct its miss management of mass and start contributing to Newton's correctness? The issue never went the way of questioning Newton's legality in the statements about contraction that Newton made...the disputing of concept legality went the way of the Universe. If the Universe can dare to show expanding in the face of Newtonian contraction then when will we call the Universes bluff? When will we show the Universe is up to now good and when will the Universe face its flaw and correct its incorrect ways? When will the Universe stop the farce it is busy with and begin to act responsible? The name they came up with to enlist the rehabilitation of the cosmic disgrace was the critical density.

If the cosmos was wrong then when will the cosmos become right. When will the cosmos surrender to what is correct and start to abandon its incorrect approach to the physics of the Universe?

They could not dispute Hubble but they could dishonour his dishonouring Newton by not honouring Hubble with the Nobel Prize. They could disgrace the achievements of the person that showed the tenacity to disgrace the cosmos by indicating the cosmos dared to not enlist Newton's rules. By never honouring Hubble, they could disgrace his disgraceful discovery and at the same time they could reinstate the honour of the cosmos by finding the critical density of the material that will supposedly start to rectify the unbelievable behaviour the cosmos dared to show. The will find out when will the cosmos agree to adhere to Newton's gravity law.

The Academics did what Newton did best. They reinvented the truth to reinstate the Newtonian lies. They conceived more corruption to cover Newton's fraud.

They employed the services of the man that filled the centre of the Universe to calculate the mass of all the material in the Universe as to account for any shortfall there might be to bring a final conclusion in the contraction Newton instated.

The person that did the investigation had to be in a place where such a person could see where all the material in the cosmos was. That person had to view the cosmos from a point that gave the person the advantage to see as far south as possible while also being in a perfect position to see as North as the Universe can go as to find the view of all the mass that the Universe holds between point south and north. At the very same point that person had to have the ability to see east and west as far as the directions can go so that the person can find where every morsel of material is located. The person had to be just as far from the top of the Universe as the person was from the bottom of the Universe to find the view where every crumb of material is hiding. After all the persons was given the task to find and calculate all the mass in the Universe and that was first and foremost then to locate every square inch of material that held mass could have the ability to provide the gravity that will bring about the contraction there has to be to validate Newton and vindicate the cosmic disobedience about Newton.

The person must be in the right position from where the person could glance at the entire Universe. That person must be in the centre of the Universe because only from the centre of the Universe can the person view all the mass there is and that might be hiding in the Universe. The person had to think himself as filling the centre of the Universe which from there he had to be in the academic centre of the Universe which will allow him the vantage that all the light he might use to see the material in the Universe will be heading straight to him. If one light photon pass him he would not be able to bring that particle that the light represents into his calculation and we know that some galactica seems so far distant, they are represented by one photon. Therefore, only a person that fills the centre of the Universe will be able to see all the light that is coming by gravity to the centre of the Universe and being in the centre of the Universe will place the person in the centre of all light moving. That is the most critical position the person has to have and that is the most ardent precondition that the person has to achieve to be able to calculate the mass in the entire Universe.

Fortunately at the time there was one person available that met all the requirements and criteria for the task. He was what was perceived to be the centre of intelligence in the Universe and he was in the country that has the idea amongst it inhabitants of forming the centre place in the Universe and he was part of the people that was vain enough to think them to form the centre of the Universe because he was representing the rights of the Man that formulated the centre of all physics in the Universe, which was Newton. And because of such importance Newton was being the man that brought the entire Universe into disrepute because the entire Universe was at fault in its showing of being in disagreement with the laws Newton (where Newton is the man representing what is correct and what is incorrect) instated. So we have the most brilliant mind being Einstein the intellect at the centre of the Universe forming the physics centre of the Universe, which is American physics in a country that think they are in the centre of the Universe, which is the United States of America, with the people that holds their position as the centre of the Universe which is the American's view about their role in the entire Universe, representing the man in the centre of all physics being Newton, where he placed in physics mass to hold all gravity to form that which preserve the centre of the Universe. What a hell of a lot they were...and now every one knows where the centre of the Universe is...it is where Einstein found his critical density theory in the heart of America in the centre of American physics

They have no other person for this masquerade than Albert Einstein in person. Einstein fitted the profile to a T because Einstein had the image and the flair and the reputation that was required. Remember they were not looking for a result. They were trying to establish a time delay. They did not try and find why the Universe was not doing what Newton said the Universe was doing. They were not placing the incorrectness at the door of Newton. They were putting the Universe at fault. They were in search of when Newton will become correct and not why Newton is incorrect. They tried to establish when the renegade Universe will address its mistake and come into line by correcting the mistake God made in sending the Universe in the wrong direction. Placing the Universe at fault and not Newton gave the opportunity not to address Newton and his miscalculation but find the opportunity to present a Universe that is correcting the mistake God made. In that they placed all the apparent blame on the Universe by disclaiming Newton from any resolve.

At the time the Big Bang concept were as much rejected as Christianity was applauded by science. Newton said everything is going to the grave with gravity and that is where everyone wanted the Universe to go, down an eternal grave hence gravity by name (or so I surmise) where everything was going to meet its Maker except the atheists were convinced there was no Maker. The Moon is on its way to clash with the Earth just before the Earth is going to crash into the Sun and God only knows in what the Sun was suppose to crash.

Every one was sharing the Newtonian vision of a contracting Universe where the lot would one day again come together and Creation will end where Creation started some time ago. The Universe has mass that is pulling mass towards one another and we are in the centre of an ever shrinking Universe. That is what the lot of us can see... we are forming the centre of the ever contracting having cosmos where every Newtonian can vividly see with his or her eyes through any telescope that all Newtonians minded scientists are sharing the centre stage of the ever collapsing Universe. The Universe is about to end where all mass contracts into one huge lump of material. But there was someone out to spoil the party and let rats loose in the dancing parlour!

Then along came a man that had a good look at the Universe. He looked at the sky and came to a conclusion the lot was not shrinking but it was expanding. Any one that would look through his eyepiece could clearly see the lot was not shrinking. The lot was growing apart. In some case he said the lot was racing apart. The Universe was growing by miles and not shrinking into nothing.

This unleashed a problem the world had no name for. Everything known to science was at that point devastatingly unknown to science. The world was expanding and not contracting which made the Universe quite wrong. It is impossible to have any vision about Newton being wrong. Newton could never be wrong because Newton was never wrong yet...so if the Universe is out of step with science, then science will correct such an abnormality by finding a way to defraud science and postpone the correcting that the Universe had to comply with since the Universe owed the Master Newton some apology. Did the Universe not know that he whom never can be wrong is in name Isaac Newton! Decisive action was needed. At this point I cannot believe that the most brilliant minds were so naïve and therefore I must suspect deliberate deception. Hubble was far too prominent to blow away and Newton was found wanting. At that point they put the onus of proof not on Newton but turned the focus away from Newton to what the presented as the guilty party. When will the Universe confirm its incorrectness by affirming Newton's obvious correctness? If they had to admit that Newton was wrong, the most intellectual science then had to admit they had nothing to show for all their minds brilliant work.

Science that was defying the likeliness of a living God stood bare and naked for all to see. They put the onus of proof and converting onto the cosmos. When will the cosmos come clean and prove Newton correct. When will the cosmos admit to a mistake and set its crooked ways straight. When will it meet its diverting from Newton and reach a point where the Universe will finally come to comply with what Newton demands. It is the cosmos that is wrong therefore it is time to find out when the cosmos will correct its manner. To deal with such a task they needed a man with a bigger ego than he had an IQ. They needed a person that thought more of his abilities than his ability to grasp any complex situation. They needed a man that was presented as a genius without ever proving his genius. They had a man that filled the centre of the Universe, which then placed the man in a location from where the man could see the entire Universe. They had just such a man. He went by the name of Albert Einstein. For all the genius Einstein had, Einstein failed to see the most simplistic and tiniest mathematical rule. Einstein failed to realise that if there was insufficient mass at the beginning of the expanding Universe, the growth of the Universe will reduce the influence of such mass as a factor because as the radius grows, such growth will restrict the gravity by rendering the mass progressively more incompetent.

If the Universe is expanding as Hubble indicated, the growth of the radius will reduce the influence value of the mass as every second passes. The mass will become more and more wanting for such a task. Yet with this obvious shortsightedness of the genius Einstein, the genius saw him fit enough to calculate and measure something as overwhelming as the Universe. As in the case of Newton, Einstein as an ego driven maniac that saw his abilities fit to measure and master the Universe while his mind was to simple to recognise the most basic principle of mathematics, the principle of relevancies or ratios. What a mathematical genius that turns out to be. While the radius enlarges, at the same proportion does the influence of the mass factor reduce and the mere fact that the radius increase shows that at no stage further into the future can the mass stem the growth of the radius because the radius overpowered the mass factor already. Unless there is new material entering the Universe at a point, which is impossible, the entire concept is fraud.

The idea was never to admit wrongdoing on the part of Newton and Newtonian science but to post pone, delay and divert attention away from the truth. If there was not enough mass to start with, no dark matter can kick in later on and start secondary mass frenzy that at that stage will then be enough to bring about the required mass potential that will turn around the Universe from expanding to contracting. To establish a scenario that would hide all deception they got the man that has a bigger ego than an IQ, they tell the world this man is a genius while the fool does not no the least of mathematical principles because his Master Newton did not no the least of mathematical principles and they got him to measure the Universe. While they did not even have any device (and will never have such a device) through which anyone would be able to see the entire Universe, they set of a scandalous misconception that this Einstein could calculate all the mass in the Universe.

Off course as can be expected, there was not enough mass and there will never be enough mass because there is no such a thing as mass in the entire Universe. When the deceit played out to the full, they, those I refer to as fraudsters being the paternity of physics elaborated on the delusion by trying to find dark matter that is hidden. If the dark matter did not develop enough contraction at this time, there is no chance in the future to develop enough gravity because the factor of what mass supposedly should have is tarnishing and tarnishing as the Universe expands. The bigger the radius becomes the less would the mass effect be.

The community of astrophysics are trying to frame a picture where they set the stage in the way that if the Universe were stretched to a point the mass would not tolerate any more expanding. The mass will get frustrated in some way and show resistance to the increasingly elastic expanding. The gravity constant (I suppose) must prevent any further expanding. How they ever got to such an argument I never could tell. They surmise that outer space is consistently overall filed with nothing and when this nothing is stretched to the limit, the nothing would resist in growing more nothing or become further nothing and the nothing would stop other nothing to enter outer space in the community represented by nothing. If ever there is a faculty ruled by absolute inconsistency and rubbish as the motto of logic it has to be astrophysics.

Every measured kilometre represents nothing. Every mm is one of nothing. We on Earth are 149×10^6 kilometres holding nothing away from the Sun. Only they can argue that outer space is nothing with material here and there. If that is the case then which has more nothing between the Sun and Pluto or the Sun and Mercury. The distance between the Sun and Pluto is more, therefore that which outer space is made of is more than in the case of Mercury and the Sun. Therefore Pluto has more nothing between the Sun and the planet than Mercury has between the planet and the Sun. Only astrophysics and all the geniuses guarding the principal of astrophysics can put a calculated value by measure on nothing. In fact Mercury has hundred times less nothing between the planet and the Sun than is the case with Pluto. Since my days at school I was always under the impression that a hundred times the value of outer space being nothing is numerically expressed as (zero = 0 x 100 = 0), but where the genius that is such a prevailing part of astrophysics take the stage we find that Pluto can have 100 times more nothing than the amount or distance measuring nothing than Mercury has. The figure containing nothing that puts Pluto at the edge of the solar system is one hundred times more nothing than what Mercury has where Mercury becomes the first planet in the solar system. That is astrophysics. The brilliant minds of the mathematicians hold no rules apart from what they can calculate. Astrophysics is the only department throughout the Universe where normal rules don't apply since because with mathematics they can bend all laws as they wish…in fact Newton started the trend with his deceit.

Only the guardians of astrophysics policy can know why the undetected dark matter will start producing gravity to change the expanding to contraction. Would the fact that it is detected, change the influence it established? Or is it merely to extend the cover up and allow the deceit to linger until the following generation. There is no mass and any one that says there is mass, let such a fraudster then explain why all the planets irrespective of size or density, spin around the Sun at the same sped as all the others. Let them prove that the Universe acknowledge big and small and let them show how Jupiter can move at the same pace as does Mercury and Pluto while Jupiter is so many times more massive than the other two mentioned. More condemning evidence is yet to come because the astrophysics tricksters did not leave the corrupting of evidence just at that.

The fatherhood of physics never once diverted from acknowledging that Newton's contraction is the prevailing thesis on which the cosmos is built because they accepted that Newton used unlawful arguments and to cover up Newton's fraud which they still use to this day, they then proceeded with further criminality when producing the bluff they established with Einstein just to fool everyone in the normal public. Without ever recalling Newton's contraction theory that is obviously not working or admitting doubt about Newton's testimony to the effect, physics accepted the Big Bang Theory. The Big Bang theory opposes what ever Newton might have implied. The physics paternity however finds it wise to still advocate Newton while admitting to the Big Bang event. Newton said the lot is contracting. Go on and marry that with the Big Bang that says everything is expanding. You can't promote both except is you can define why we would see the two merge.

The Universe comes from a point the size of a Neutron. That makes the radius parting the Universe infinitely small. It just about removes the radius as a factor. At the very same implication it takes the pulling of the mass (if there are pulling forces converted by mass) to a level it will never again have. As soon as the distance between the objects holding mass started to grow, the power and influence of the mass factor started to diminish in the same ratio. If the mass were incapable of contracting the Universe then, it will forever remain contracting the Universe. Then you may ask what the story is? Read on and you will learn how far Mainstream Physics stray from the truth and how big a cover up the paternity is protecting.

According to science the Universe started with singularity. Quoted directly from the Oxford dictionary of Astronomy the following:

The definition of singularity is as follows:

Singularity: a mathematical point at which certain physical quantities reach infinite values for example, according to the general relativity the curvature of space-time becomes infinite in a black hole. In the Big Bang theory the Universe was born from singularity in which the density and temperature of matter were infinite. The average daily temperature was "$10^{\alpha\beta}$ to 10^{34} K".

Then the second "day" the daily average temperature came down to 10^{34}K and 10^4K. That is fine, but if the temperature was in Kelvin, then what was 0^0K. In order to make sense of the scale used there must be a minimum to secure a maximum otherwise the maximum can just as well be the minimum and is only advocated to impress humans applying earthly standards.

By using a scale as $10^{\alpha\beta}$ to 10^{34} K, it places the lower temperature at a modern 0^0K to make sense of standards. If that was the temperature the standards were lowered, compromising something to gain something, because something had to grow larger for heat to reduce. We know space grew larger bringing heat down to reduce.

Being the onlooker the viewer has to maintain one position. From that position some particles would be circling a centre point, as the particles would be coming towards the onlooker. The other matter would be circling the centre point while rushing away from the onlooker.
At the very end the single dimension may come into the dynamics but where the single dimension comes in the factor of zero is removed.
If there is space, there is a flow of light and a flow of light has to produce lines in relation to angles forming space between them. Something must be present to confirm space because there is an absolute difference between being in space and no space to be found. If there was a line that formed nothing that one line that forms nothing would completely destroy the other lines' chances of ever forming a triangle, let alone having all lines and they then have a total being zero. As shown in the example no line can form zero and therefore no mathematical equation as far as it extends to cosmology can ever bring about zero as a number. While there is space present there has to be three dimensions relating to each other by time and in three dimensions there has to be three lines in relevancy to each other by angles formed holding space in (at least) six opposing sides. Removing one line must bring about a flat Universe and that then will constitute nothing.

Cosmology is about light flowing by means of lines indicating space obeying the rules enforced by time in motion and light flowing dictates crossing space and across space light is using lines. The book: **_An open letter Announcing Gravity's Recipe_** is dealing with the subject finding singularity by removing the concept of nothing from outer space. By diminishing nothing one uncover singularity and the effort brings in a new perspective not yet introduced.

For your benefit I will shortly give a summary by which I hope to interest you in reading the manuscript: Compressing space produces heat Releasing heat will bring expansion of space bringing about space. **We call such a release of heat an explosion. In other words heat translates to space and space concentrates back to heat. The one is a product of the other where space forms expanded heat.**

They are quick to show the time that was applying at the time being some thousandth of a second or the heat that was present being numbers we have no name for. The other side of the story they ignore. They ignore the other side of the story because in that respect it puts their promoting of Newton down to madness. If you reduce the radius applying at the present back to what it was at the time of the initiating of the Big Bang, you must also increase the influence gravity and mass had at that moment by the same number you are decreasing the radius. That is pure mathematics and the most basic physics of all concepts.

The shrinking radius will increase the effectiveness of the influence of the gravity that the mass can produce by the margin of the shrinking of the radius. If the Radius was infinite at that point,

then that means the gravity was eternal. With the entire Universe being as big as a Neutron, the Universe was the size of an atom. If the Universe were the size of an atom and the mass within that Universal atom could not prevent the Universe exploding into immeasurable atoms, then it would not be able to retract all the atoms into one unit again. If there was not enough mass to start the contraction, there can be no contraction of mass that is producing the gravity at this stage. If the gravity is of such a nature that it allows a continuous growth of the radius, then the radius firstly cannot be zero as Newton suggested and the extending of the radius proves there is no contraction in the way Newton had everyone to believe. If Newton's mass contracting mass is true, then on the other hand it must have resulted in an implosion as that which can never repeat again. With Newton's formula of $F = G\dfrac{M_1 M_2}{r^2}$ **forming gravity, then the Big Bang is just not possible because from that formula the Big Crunch must respond.**

The critical density is the biggest and most elaborate case of fraud ever perpetuated by any group of people or persons. They did not say Newton was incorrect. They said they are on a mission to see when the Universe will correct itself and prove Newton correct. Newton remained correct while they gave the Universe the chance to mend its ways. If there was not enough mass to start with, why will there be enough mass in the future? If the Universe is growing the mass in the Universe is not pulling the Universe into contraction. Only an over bearing egomaniac with an ego outweighing his common sense by a margin of many times to one, will take the opportunity to calculate the mass in the Universe. Step outside tonight and see what there is to see outside in terms of mass. They put the exercise to a formula in order to calculate what no man can even presume. If you wish you can read the following example I offer to prove what elaborate criminal scheme the entire venture called the critical density is and to what fraud it amounts. Here follows a part taken from a web site that tries to prove fraud by elaborating on the fraud by implying more corrupt fraud. You are welcome to read the following frauds if you wish, but to my mind it is not worth the paper I allocated to its print. Nevertheless, here it is:

A simple, non-relativistic, derivation of the critical density can be performed as follows

> *Recall that kinetic energy of a body of mass m moving with velocity v is $\frac{1}{2}mv^2$, and that gravitational potential energy of a body of mass, m, at a distance, r, from a second mass, M, in a radial field is GMm/r.*

In order to find the escape velocity of a spacecraft from a planet we can work out that the spacecraft will escape if it has enough kinetic energy that gravity will be unable to slow it to a standstill. Using the two expressions above we can write

> $\frac{1}{2}mv^2 >= GMm/r$

As the kinetic energy of the craft is converted to gravitational potential energy while it escapes. If the two are equal it would stop at a very great distance (infinity), if the kinetic energy were bigger, then it still has velocity when at infinity.
We can do the same calculation to see if a galaxy is able to escape from the attraction of all other galaxies or whether it will be stopped, turned around, and caused to fall back. If this latter situation is the case then the future of the Universe is finite. The galaxies, which are at present travelling away from each other, will stop, turn around, and head back towards some common centre where everything will be compressed, perhaps out of existence, 'The Big Crunch'. If, however, there is not enough material in the Universe to stop the galaxies then perhaps the Universe will go on forever.
Hubble's law shows that the velocity of a galaxy is proportional to distance,

> $V = H_0 r$

Where H_0 is about 5×10^4 m s^{-1} Mpc^{-1}, 80 km s^{-1} Mpc^{-1} or between 1.587×10^{-18} and 1.62×10^{-18} s^{-1} depending upon whose figure you accept!
So the kinetic energy of a galaxy can be written

$$\tfrac{1}{2}m\,(H_0 r)^2$$

The mass of all the material inside a sphere of radius r is given by

$$M = \tfrac{4}{3}\,Pi\,r^3\,Rho$$

Where Pi is the average density of the Universe.
Substituting these two expressions for the first equation above gives

$$\tfrac{1}{2}m(H_0 r)^2 = Gm(\tfrac{4}{3}\,Pi\,r^3\,Rho\,)/r$$

This can be simplified to

$$\cancel{m}(H_0 \cancel{r})^2 = G\cancel{m}(\tfrac{8}{3}\,Pi\,\cancel{r}^2\,Rho)$$

and rearranged to

$$Rho = 3H_0^2/8\,Pi\,G$$

The above my friends are hogwash. It is what represents the utter most thoughtless and mindless arrogant mismanagement of all facts. If the mass was not up to the grade to pull the Universe into contracting, it is incorrect to presume the formula will apply. By finding that the Universe is growing in size, one should question the authenticity of the formula $F = G\,\dfrac{M_1 M_2}{r^2}$. This formula can only apply to a shrinking or contracting Universe. To try and integrate the formula as pronounceable in the case of the expanding Universe, is committing fraud.

$F = G\,\dfrac{M_1 M_2}{r^2}$ In the using of the formula such as Newton recommends the radius determine the influence of the mass.

$$\frac{M_1\,M}{r^2}\,G$$

In mathematics to put any number in relation to the value of another number is to agree that the top part f the formula will find a suitable and applying measure by the measure and the size of the bottom part.

$$\frac{M_1\,M\,G}{r^2}$$

If the bottom part is bigger than the lower part, the bottom part will be a fraction of the top part. If the top part is smaller than the top part, the bottom part will be a fraction of the top part. The bottom part and the top part will always and without reservation have one of the two smaller or then a fraction of the other, except in cases where the two are equal. In such cases there will be a unifying number coming about as one. Putting two factors into a ratio or a relevancy, places the one value in charge of the other value while the other value is dependent on the first value. Not one of the two then can be dismissed as nothing or having no value because the other value holds the measure of the first value.

$$\frac{M_1 M}{r^2} G$$ Appreciating that the top is sizably more than the bottom value will put the top part in a numerical superiority and the answer will not be in any fraction, which includes a mathematical fraction of less than one.

$$r^2$$ Having the radius growing, means the top part of the formula that represents the mass is shrinking in influence.

Edwin Hubble proved that $F = G \dfrac{M_1 M_2}{r^2}$ is not applying because the cosmos is not shrinking. To further use an already falsified formula is cooking the books to a point where the books they cook starts to burn.

By enlisting $F = G \dfrac{M_1 M_2}{r^2}$ you are placing the mass and the mass as a multiplied unit with the gravitational constant in relation to the radius value that applies. That means if the radius does not shrink but it grows, the mass does not reduce the radius but it promotes the radius into growing. By incorporating the incorrectness into what one then tries to establish, is committing fraud by falsifying the facts derived from fraud even further. There simply is no formula as $F = G \dfrac{M_1 M_2}{r^2}$ so then there can be no integrating the formula to whatever form $\frac{1}{2}mv^2 >= GMm/r$ is required. The cosmos proved that $F = G \dfrac{M_1 M_2}{r^2}$ is a farce and Newton should be investigated. The Newtonians elaborated on Newton's fraud by investigating the Universe of fraud. There just is no mass found in the entirety of the Universe.

Read my book and you will find that the Universe is not space but the Universe is time. There is no space in the Universe, for through expanding the Universe placed time in space. As time develops it changes density into space. The Big bang is about substituting the development of time into forming space. Time is space because space came about as time progressed. You cannot calculate the Universe in terms of $4/3 (\frac{4}{3} \Pi r^3 / 3)$.

Guess how surprising was the result that Einstein produced. Einstein shockingly and most surprisingly confirmed a constant defiant Universe. Einstein produced results confirming not only Hubble but also that Newton will have to wait for longer to be proven correct. A new plan was devised to vindicate Newton and his correctness. If the cosmos did not play ball then the Brainy Bunch will cook up a real witch's brew that not even the cosmos can dare to defy.

A search went out to find dark matter. Dark matter will result in correcting Newton if nothing else can. If one can't see dark matter then no one can prove dark matter and while no one can prove dark matter, then it also becomes true that disproving dark matter is impossible. It worked with the instigating of the original fraud that Newton invented. Not ever proving Newton and his mass that creates gravity by attraction between the mass factors also brought about that not one person ever thought of disproving the mass deception. With the mass being impossible to prove, it also makes the mass of dark matter impossible to disprove. Then on what grounds do I lay my charges of the biggest hoax ever created to defraud tax payers out of their livelihood by stealing tax money and diverting the proceeding of the tax collector to establish a criminal inspired corruption never yet before experienced by man?

Kepler introduced space a^3 growing by the time T^2k that allows the space a^3 to move. That is time T^2k allowing space a^3 to be. For space a^3 to move about it will use the motion that time provide in the rotation of a^3 as well as the displace net of a^3. In space there is no mass because in space there is no proof of mass.

$$F = G\ \frac{M_1 M_2}{r^2}$$ If there was not enough mass to redirect the expanding Universe into contraction, then there can never be enough mass to redirect the flow of the cosmos. What will produce more mass if there is not sufficient mass at this point? Why would the dark matter be gravity dormant and what will enlighten the dark matter into activating gravity. If it is mass and the mass is suppose to establish gravity, the mass then has to establish gravity whether the mass is dark or light. What will make matter, that is visible more active in creating gravity than matter not seen by man? Why will dark matter come into gravity later on, as it at present holds the gravity dormant? Why will the dark matter at present not form sufficient gravity by mass but will later on become energised and jump start the cosmos into contracting. If the mass is there, the mass should charge gravity if mass does charge gravity. Why would dark matter play hide and seek and hide their potential gravitational abilities for a later date?

$$F = G\ \frac{M_1 M_2}{r^2}$$ The suggested formula indicates that it is mass that produces the gravity by which the force pulls structures and thereby it reduces the distance between the objects. The mass is proportional charging gravity. The mass is responsible for the force of gravity by the measure that the mass has. A lot of mass will charge a lot of gravity and a little mass will charge less gravity. The mass is there in relation to the establishing of the force and forcefulness of the force proportionate to the amount of gravity applying by the production in relation to the mass. If the mass is there, the gravity is there. If the gravity is absent the mass is not present. If the mass cannot reduce the gravity, what then is the point of trying to establish when the gravity will come about, since the mass is obvious lacking as a quantifiable amount to charge enough gravity to contract the expanding Universe? If the gravity is not available, the mass is not sufficient and therefore, if the mass is not sufficient in the first place how the hell will the mass become sufficient later on. The mass is not an elastic band that allows expanding up to a point where after it will reduce as the elastic energises.

$$F = G\ \frac{M_1 M_2}{r^2}$$ The formula does not compensate for any such suggestion. The notion alone is one huge farce inspired by criminal minds to cover up flaws in the top banner bearers of the science world of physics. The whole issue goes against the grain of physics and mostly against all mathematical principles. If they were incompetent novices trying to address a school project that has a mathematical inscribed theme that goes way past their abilities and they are in far too deep water to tread, then yes I can find some degree of honest miscalculating and incorrect judgement of mathematical founded facts. They hold the top notch of all mathematical insight. They father the laws of mathematical principles. Therefore, in that light they woefully and deliberately avoided the truth and rendered their honesty to distrust. The fact that they skipped inspecting Newton and created a farce to mislead any thought the public may have about the mistake, shows their deliberate action to avoid the truth about the matter and carry on in criminal intent. I challenge any one bearing the title of professor in physics to prove otherwise. Let any academic in physics show where I overstep my boundaries when I charge them of deliberate acting with the motive to corrupt the truth and avoid the true impact of Hubble's findings.

$$\frac{M_1\ M}{r^2}\ G$$ If the mass was prominent from the start in relation to the distance separating the two objects, then the contracting of the object will reduce the radius parting the objects to become a factor of one that has no influence on the formula.

$$\frac{M_i M_j G}{r^2}$$

If the mass grows and the mass keeps growing, the mass will reduce the input of the mass on the gravity. The further the mass goes apart from the mass, it shares gravity with the less the influence will be of the mass exerting gravity. If the gravity was insufficient from the beginning then it will remain insufficient as the expanding progresses. The gravity does not grow as the radius increases. It is the other way around and as mathematicians of notoriety they knew it. That makes their criminal intent even more appalling than what it would have been if the actions of criminality came from the midst of others being persons in a less trustworthy disposition. However, Kepler gives all the answers they tried to criminally cover up.

This we dealt with here is only the tip of the iceberg. When one starts to dissect the inner workings of the Critical Density Theory and find their way of thinking about how the operation will carry out its purpose then it is clear how much the theory proves how the madness can grip the idiot's mind and unbalanced thoughts go out of control. If the mass stops the molecules in the process leading to turn about of direction of Universal flow, it will stop the less massive first and this will have the more massive molecules plough into the slowing as well as stopping smaller molecules.

The more massive molecules would look like trains colliding with cars and bicycles and removing the bicycles from the face of the Universe. At that point a second Big Bang will come about that totally destroys everything that was not destroyed by the first Big Bang.

This is what happens when a lot of criminals can go about, as they are allowed to work unchecked. As they are blindly trusted by all those unsuspecting, honest persons in the general public holding no criminal intentions they never think that these Academics are never being controlled, which means the Academics can steal billions from the coffers of the tax paying public. That they then also do in finding funding for the most bizarre research that is so fruitless it falls in the category of the insane. One such a venture that comes to mind is the monumental fraud of research into alien life coming from somewhere n the outer space. This is only one of so many that they use purposely on many levels. In other instances the fraudsters apply the same modus operandi by inventing some bizarre mathematical formula that must supposedly serve some purpose and then they unleash this fraud onto the public at large albeit to try and cover up Newton's fraud or to defraud the public in other ways.

The latest book that is supposedly written by Steven Hawking is a perfect example of the degree of mad corrupt and idiotic methods they employ to commit such fraud. They invent a formula. The formula in theory is meaningless because the practicality of the formula can never be tested and must therefore be accepted. The reasoning behind such a formula is as unrealistic as the critical density crime venture, which I just explained and has as little practical function as does the critical density theory. As I show they go about to bend and cheat mathematical principle and logic at will to corrupt the truth as to find proof to their meaningless venture. By applying even the least of logic they prove callow. But since they deal with honourable people they take the public to the cleaners

The people forming the public are too scared to ask questions about the sanity behind the reasoning because the people feel incompetent in reading the mathematics. That uncertainty these evil gangsters use most shrewdly to their financial benefit. Behind this fear of feeling incompetent that the general public has is a rational of not asking questions because they do not wish to feel stupid and this the academics in astrophysics exploit for their criminality while the gangsters continue with their evil exploits.

The criminals calling themselves astrophysics academics then realising this flaw that the public have and then go on to commit gang rape on the unsuspecting public, which are also those they see as witless beings they can manipulate and control at will. I challenge those fraudsters to sell me that lame idea that the latest Hawking book explores and see how I plough their fields to uncover pure bullshit in front of a TV audience.

I call them criminals because notwithstanding the position any person has, when such a person pretend to be what the person is not, or to have what the person does not have, or to launch a project that the person intending to benefit from the results and the research of such a project but in launching the project the person intentionally has too spread untruths and corrupt lies will filly, then this makes that that person intends criminality. By telling not the truth one tells a lie. By spreading lies one commit fraud and deception. There is no small medium and large way of deceiving a person because untruths are lies. In doing so they intentionally divert to truth to sustain a lie and such actions is criminal. Even not admitting to the fraud is a way f committing intentional diverting of the truth, which is the sustaining of criminal behaviour. Any behaviour not being in line with the truth is committing criminal activity and that is the narrow and the broad of the lot.

Module Three

Brainwashing and Mind Control is an Everyday Practice in Physics

If you are a student in physics then you should read the following information. One could think of another name for physics and that would be Newton's mythology. It is about the subject of gravity and is most important. The "Newton's mythology" comes from the fact that students have to learn what never was proven. Do you realise that it is an accepted practise that all students that are studying physics on all levels are subjected to the most intense brainwashing and thought control found any where on Earth? This must be some sort of a joke you may think but thinking that way in disbelief is just what those practising the mind control wish you to think!

I came upon a mistake concerning physics.

This mistake is about the cosmic phenomena called gravity. Detecting the mistake is simple because it is uncomplicated to understand. Academics in Science say that a feather will fall with the same speed as what a large rock would fall.

That is according to Galileo and that is accepted as a principle in physics. For the first time ever since the time Newton introduced gravity I seem to be the person that questions this interpretation.

Has anyone ever explained how the idea of a feather falling as fast as a hammer fits into the idea that mass pulls mass and how the falling by the gravity forms power that is exerted by mass as a hammer has much more mass than even what a large feather has. How on Earth do these two concepts of a feather falling equal to a hammer proved by Galileo fit into this interpretation they use of mass causing objects to fall?

How can a large mass pull as equal as a small mass pulls to travel equal at the same speed over the same distance and still be driven by the power of mass creating gravity. Have you given this idea a good thought? By me scrutinising this concept I disagree and by me disagreeing I am silenced by those in power. When any person disagrees with any academic in any lecture hall about mass not forming a picture as being the factor responsible for pulling gravity and you come to a conclusion that you doubt the mass part that they bring into the picture as being responsible for establishing gravity the academics wipe you from the table with a swipe because then they contemplate that you are so stupid you fail to see facts and you are to stupid too understand physics. They even in some cases go on to say physics is not for stupid people!

I have been at odds with academics for years and only because of the superior positions they hold in office are they able to bully me into silence. Academically I am not from their league and neither am I from their ranks and with me not being part of their ranks they are of the opinion that it disqualifies me to have any opinion. Being what they are gives them the rite that they may regard or disregard all opinions when they do not fancy the opinion. They may silence whatever I may say notwithstanding my correctness and validity. Absolute power corrupts and they are the living example of that.

Due to the important positions Academic hold in the huge academic institutions such castles of power gives them free sanctuary from where they can hide their criminal ploy of deceit. They do not need to explain anything but to themselves amongst themselves and their deeds go totally unchecked. That makes them be the untouchable and unapproachable powerful from where they rule with absolute authority. This unquestionable authority gives them the locations erect a cover and give them the opportunity to hide behind that wall of absolute superiority and suppress little persons such as I into silence and submission notwithstanding...Whatever I have to say can never go past their scrutiny and can never pass their sanctions.

What Newton saw as gravity can't withstand even the slightest test of proof and I showed that it is not possible to use Newton's formula as Newton suggested it applies to mathematically calculate gravity. I come back to this issue later on. I have tested Newton's thinking and the book I offer to you for investigation serves as the testimony to all the testing I did on Newton. This any body who can see, will see when reading this book, I tested Newton from all the angles to see if he possibly could be correct but found his thinking wanting every time. The truth about Sir Isaac Newton's concepts I came to conclude, was that the reality is that it is not in any way overstated to declare that Newton conspired to defraud science and moreover that he committed blatant mathematical corruption in trying to prove the concept he had about what he thought forms gravity. There is no backing for Newton's ideas and even the ideas which are in use are not in the form that Newton said it applies where physics in daily use serves as the best discredit to Newton bringing no proof about any of the claims that Newton made on matters concerning science in cosmic gravity.

I show that every thought Newton introduced that later proved useful and was correct, was what he stole from another far better cosmologist called Johannes Kepler. Not one of his laws are directly relating to any concept Newton ever introduced at any stage but is the result of academic theft he committed against a much larger figure that preceded him by almost a century. But he stole, he lied and he raped the work of a predecessor in order to defraud the world of science in his time. Newton brought no original input into science except that he gave a concept the name "gravity" and even that is inappropriate. Newton made suggestions that break every mathematical principle he could think of. That, Newton did in his attempt to win over the prevailing academic thinking of the day in his time as to lay some sort of groundwork to form backing for his ideas on physics and to attempt to explain gravity or what he thought gravity is. If this is shocking and sounds outrageous, then a lot more shocking detail awaits the reader in this book.

Newton's claims about the principles he declared as being responsible for guiding physics carry no proof and after I realised that, I was able to start forming another line of thought on gravity. After formulating my concept about how gravity was truly formed, I had to introduce my ideas to academics in physics. In my quest to find the method how gravity formed I used the four phenomena and the principles of these phenomena as well as determining in which way each phenomenon applied. Then I placed each one in the way that were known how they work and then implicated that specific formula's function mathematically in forming gravity in the cosmos. This was no easy task but I did it and by formula shows that my argument is logic and the mathematics prove that it works well.

They proclaim to understand what flows out from what they understand but such concepts become meaningless because of many inconsistencies. To name but one such an example is the explanation they put forward in the Tunguska event. To claim that a mini Black Hole went through the Earth is demeaning just going on the basis that they claim there can be such a thing as a mini Black Hole. Such statements are beyond the ridiculous and to achieve some degree of believability from the public they create scenarios, which use arguments that are entangled with deception, such as what is obvious in the case I mentioned. What they declare as unwavering facts can't even be supported in the least form when tested. Even the least degree of verification of correctness is absent and Newton lacks all evidence of authentication in any investigation of even the simplest terms. It is as if they never read with interest that which they explain and they never scrutinise that which they advocate. They give values that are senseless and make that which they say meaningless. In all this they use billions of tax dollars to prove what they have no idea of. They try to commit matter to fusion while they have no idea why matter would fuse at all!

Now I am taking my case to the members of the public so that the truth must be brought into the open. I have had the tour they give and then more came my way. I never got around swallowing the mass creating gravity part where science is of the opinion that mass pulls as gravity is… Academics condemned my work and therefore me and for six years where I could not get a publisher to come around and bother to read my work let alone seriously proposing a publishing contract. I had to finally go private with the publishing as all doors shut in my face as soon as the academics read the content of my work because from the nature of my work I take Mainstream science head on and am confrontational on most aspects of astronomy. There does not seem to be any publisher that wants to go head bashing with the establishment of science on official science principles, which I have to do to convey my message in a no uncertain language. If you also have doubts about the academic's indisputable correctness please read on and confront either them or me on everything you read here.

After reading this letter you will have to take sides because you will know the truth.

Then you either become partner in the crime as you cover the truth up or you will be part of the truth and help me confront them to acknowledge the truth.

Should you think this page is some sort of a prank then answer the following simple question to yourself in utter honesty? If there is a Big Bang with everything moving apart, how does that support Newton's contraction? Tests results received after the Moon landing show the Moon and Earth are moving apart! Yet students learn about mass pulling mass and that puling by mass forces togetherness by contraction.

The entirety of physics rests on this one formula $F = G \dfrac{M_1 M_2}{r^2}$ The questions concerning that which you are

studying and that touches every aspect you are academically concerned with, is that if everything is moving apart, how does that support Newton's idea that everything is coming together...and please don't let them fool you with Einstein's Critical Density idea! If there was mass seen or unseen in the Universe and mass generated gravity and gravity does the pulling then why is the mass not at this moment doing the pulling. What is all that mass of so many supposed stars doing at present while waiting to get to work where it will only later, much later form a force of gravity that then will bring about this pulling of the Universe? What makes the mass slumber in darkness to one day form a pulling force? What has the "darkness" or the fact that we don't see the mass got to do with the idea that the mass at present is not forming gravity that is forming a pulling force? You are taught that gravity pulls objects to the centre and obviously gravity then has to ultimately pull everything to the centre of the Universe. That is what the

Critical density research that Einstein initiated wishes to establish. The idea is that $F = G \dfrac{M_1 M_2}{r^2}$ makes the mass create a force that will destroy the radius and ensure everything is going to come together eventually at one point where the radius then will be no more. If that is the case, then where is that point? If everything is destroying the radius, then it must end at one specific point.

In the classes you attend a physics lecture, has any one confirmed a location where one might find the centre of the Universe to confirm the ultimate destination of $F = G \dfrac{M_1 M_2}{r^2}$? If you wish to apply a Gravitational constant as a calculated factor then it is apparent that one must know to where such gravity is pulling since it then is the gravity that is predominantly keeping everything apart. Then the gravitational constant is what is resisting the collapse of the Universe. If there is a force, then where is the force taking the pulling...if it is a gravitational constant applying through out outer space then where is it having a centre base?

I wrote a book in which I found a means to define gravity. This feat I accomplish and by my effort it was done this for the first time ever. For the first time ever runs further back than since the time Newton introduced gravity. Before I achieved that discovery, I firstly had to find the centre of the Universe because it is there that I could locate gravity. I now am able to show how gravity forms because I have detected the centre of the Universe. But by my effort in finding the location I disrupted everything Academics in physics hold holy and for that I am most unwanted in the presence of the Academics charged with guarding the ethics of physics. In short, I clash head on with Newtonian principles. During my research I discovered abnormalities and inconsistencies about mistakes the Arch fathers in physics must be aware of but is hiding with all their considerable influence. I will come to some of the inconsistencies later on but the discovery also introduces a much better vision about many new aspects that I discovered but in reality was never before realised in science. But these discoveries discard and blacken the Newton reputation totally and therefore the academics dispute my work totally in order to save their Newtonian reputation. The road I took in my search for truth concerning physics was never smooth and the resistance I came across coming from the academic sector was almost unbearable. Academics guarding physics will never allow an outsider to enter their domain and dislodge Newton from being god that is without the intruder paying a heavy price for trying to do so and in this matter I was and still I am seen as being in the role reserved for such an intruder. It is not about my work they detest but it is my rebutting of Newtonian thoughts that they reject! However such intruding allowed me to find so much that I was not supposed to find, which was reserved to all that studied physics the insider information that is available but because of that it was only allotted to the most inner circle and the insider information that I share with you. By finding the centre of the Universe enabled me to find a point the Universe is controlled from. In achieving the locating of the centre of the Universe I had to step on some very important toes, which made me very unpopular. With my unpopularity rating this high as it does, I never qualified for help and those that would help found my ideas intolerable whereby I only found rejection instead of help as I tagged along. Because of this insider rejection I had to resort to private publishing because from the nature of my work I take Mainstream science head on and am confrontational on most aspects of astronomy. This is the only road to go if one wishes to lay axe to the root of the insider corruption they are guilty of. In that sense there does not seem to be any publisher that wants to go head bashing with the Physics Custodian establishment of science on official science principles, which I have to do to convey my message in no uncertain language. I argue that if it is the

correct practise to use $F = G \dfrac{M_1 M_2}{r^2}$ to calculate gravity then the radius holding the gravitational constant must lead one to the centre of the Universe. With nobody willing to publish my work as I confront science dogma and principles all the way, I had to go the road alone and fight the battle by my private effort.

This is only one of many points that I make on this one issue and there are so many other issues one may think of those in terms of counting in numbers in many hundreds or even in thousands. If the Sun for instance has mass that is apart from the Earth and the Earth also has mass and there is a gravitational constant in between the Sun's mass and the Earth's mass we have the radius in that location. It then must be the gravitational constant that fills the space that the radius holds. It is rather obvious that while the radius is filling the vacant space between the Sun and the Earth it is the only place left where the gravitational constant can hide. To find the centre of the Universe I had only to find the gravitational constant that holds the centre. Through my venture I discovered one person that knows what gravity is! Newtonians went and filled that space reserved for the gravitational constant having a measured value with nothing! How can nothing have a value of 6.67 X 10 $^{-11}$ while also being filled with nothing as it is nothing filling the nothing of outer space?

If you think scientists know what gravity is do not be duped that easily because no one in science remotely knows what gravity is...not even Newton knew what gravity is except Kepler... and because of what Kepler introduced I now know I can prove what gravity is. Gravity is precisely what Kepler said gravity is and only Kepler new where to find the centre of the Universe because only Kepler knew what gravity is all about.

Try to get an answer from any academic person in physics about where the centre of the Universe is, is like trying to touch the moon.

I dispute Newton and so do all students learning physics because Newton's arguments are an onslaught on human intellect. Think of the resentment that students have towards Newton under normal conditions when they have to cope with understanding the Newton principles Mainstream science says are applying and how that confusion of what is possible and what Newton suggests is possible clashes with their intellect which makes them feel stupid. Students hate Newton because they don't understand Newton and for that they are accused of not having the intellectual capacity to follow Newton. Every student from the past going into the present and even including those forming a future generation of students will purchase a book that is showing that Newton's legitimacy is cracking up when exposed to some vivid scrutiny. This fact gives the book a selling potential like no other book in the past could do. Yet I am unable to find a publisher because publishers need academics to assure the correctness of the information in the book and academics would cover up Newton's errors at all costs. There is a total denial about the truth and as long as those academics have the opportunity to brainwash students in to accepting Newton's unproven and ridiculous concepts and as long as their misconduct of mind control by fact manipulation goes unchallenged, the process will go from generation to generation as it has been going for the past three hundred and fifty years.

In short I will now explain what I explain throughout the book you are about to receive and which is named

Newton's Fraud or whatever it will be named as. The Newtonian formula $F = G \dfrac{M_1 M_2}{r^2}$ is the formula used by science to explain and define gravity. It says the that the ($\underline{\textbf{M}_1} \times \underline{\textbf{M}_2}$) mass of one object pulls the mass of another object and this process in relation with a gravitational constant ($\underline{\textbf{G}}$) (a supposed force keeping the Universe attached) and the pulling subsequently destroys the radius ($\underline{\textbf{r}^2}$) being between the objects. That says that objects **ALWAYS MOVE CLOSER _BY FORCE_** in relation to _**MASS**_. Newton submitted the suggestion that objects fall as MASS provides the force that will cause the falling by the inducing of a force he named gravity which he subsequently only proposed was the acting suppositious force. I disprove this formula in so many ways in this book and I show that this formula and the ideas Newton introduced just don't stand up to even the smallest tests. Then, if Newton's idea on gravity has validity and mass is responsible for objects falling, then all objects that are in a process of falling must be subject to mass and in that idea rests differentiation and discrimination in size and compactness producing speed variations. If any and all falling is subject to the variation mass introduces and the influences coming about is the result of mass interfering in the gravity force being generated, this then must bring different speeds to cause substantial variation in the falling of different objects holding different mass factors. There can't be conformity in the falling of all objects while such falling is the result of the discrepancy that mass has to inflict due to variations that result in mass differentiations. This is a vital issue that science eludes and has all clever ways to avoid direct questioning. This part science just run around and never addresses and avoids confronting the issue. This avoidance of confronting the issue whish will disprove the validity of Newton is done with such cunning as you will not believe. The fact that objects fall due to conformity in the falling, science accepts but portrays a picture of deceit that mass brings falling distinction and therefore equal falling doesn't happen, while they at the same time admit to Galileo's presentation that falling of all objects are equal in tempo, irrespective of size or any form of differentiation. While they promote the obscurity that Newton and Galileo is in harmony the truth about their deceit is that the two can never have the same issues. That I prove is a fact and also I show how big a part this is in the overall covering up of Newton's initial fraud.

I have written several books in which I challenge the thought process of Mainstream physics and especially Sir Isaac Newton's arguments about physics. I am of the opinion that even though everyone thinks of Sir Isaac Newton as the genius who established every aspect that is used in modern physics today, but in spite of every other person hailing Newton, I remain of the opinion that the man did not have a foggy clue about any of the principles driving the concept that he named as gravity, or what brought about gravity according to his explaining of what forms gravity. I am able to explain gravity but it doesn't even vaguely resemble Newton's version of gravity. I can explain gravity by proving my explaining with the use of simple mathematics. I use Johannes Kepler's formula to back up my statements. By using Johannes Kepler's formula I found a way to prove there are four phenomena found in the cosmos. There are the four phenomena applying in tandem that together forms gravity. They are: The Titius Bode law; The Roche Limit; The Lagrangian Point System and; The Coanda effect. As the phenomena don't support Newton's vision on cosmology, the phenomena has no support amongst Mainstream science although they did apply it with a positive results in locating the missing planets at the time of their discovery. When they located unknown and undetected planets in the past, the existing of the phenomena was never disputed but when the argument of proving them comes to mind, then they are dismissed as some coincidental abnormality occurring. But since it holds no similarity to Newton's view on science, Mainstream science rather disclaimed the validity of the phenomena than they would find fault with Newton's ideas. In the mind of science the cosmos can be wrong and God can be wrong but Newton can never be wrong. In using the four correct principles correctly, which I back up with the correct mathematical interpretation thereof in support of the function that each phenomena has in forming gravity, I did a far better job than what Sir Isaac Newton did and what I achieved is of a far more acceptable level as well as being mathematically far more correct than what Sir Isaac Newton did achieve with his guessing about issues he couldn't explain. To be successful in my quest to find an explanation for gravity, I had to redirect all my concepts I previously had and also alter all the otherwise normally accepted thinking on physics. I had to find the phenomena and I had to dissect the function of each phenomenon as well as mathematically valuate the phenomena. In this process I realised that to come to realise what gravity is, I had to realise that gravity is not what Newton foresaw. Planets have no mass and neither has the Sun got mass except the mass Newtonians wish to credit planets with.

Bigger planets don't move faster because they have more mass and smaller planets are not further from the Sun because they have lesser mass. All planets big and small spin at the same speed around the Sun and in relation to the Sun and all planets are scattered going around the Sun while being big and small where all sizes are well mixed. This is because planets have no mass except in the imagination of Newton and his devoted followers. The mass of the Earth never plays a role in physics and the mass of planets do not draw any of the planets closer to the Sun and let one physics professor bring proof that the planets do draw nearer to the Sun!

They just can't because planets do not have mass that can produce a pulling gravity! If and when the mass of the Earth do not feature as a factor in any formula that is used in physics, then the mass of the Earth is no factor playing part in gravity. This then can only indicate that the Earth has no mass. If there is an absence of mass as a factor that influences physics, this can only be as the result that the Earth mass has no gravitational presence in any physics formula. Gravity does have the value of $g = 9.81 \text{ Nm/s}^2$ but that I explain and the value $g = 9.81 \text{ Nm/s}^2$ I prove as well. With that evidence being that clear, then the mass that the Earth should supposedly have, does not produce gravity as Newton suggested. Prove me wrong by getting gravity at $g = 9.81 \text{ Nm/s}^2$ from using either any

of Newton's formulas being $F = G\dfrac{M_1 M_2}{r^2}$ or $F \, \alpha \, \dfrac{M_1 M_2}{r_2}$ and $F = \dfrac{r^2}{M_1 M_2}$. Let me see Newtonians do that

and I will become a believer in Newton! The Earth has no mass because physics can't show the Earth's mass playing part in calculating formulas and if there is no mass that plays a part that should produce gravity, and then mass can't be responsible for the producing of gravity as Newton declared. That makes Newton's suppositions total rubbish and that makes Newton responsible for a crime of defrauding and falsifying the science of physics. If you, the reader is able to get academics in physics as far as even reading this argument I make, then you are more influential than I can ever be. They plainly dismiss all these arguments with arrogance by discrediting my credentials!

The scientific presumption is that gravity is established when one object holding mass is pulling another object having mass and forces the two abject to move toward each other. The entire basis of all physics rests on this formula where init it is believed that mass produces all gravity by distinction of differentiation in density as well as size and if physics is anything to go by, then what ever is proven, such proof must stem from and be in support of

well as being supported by this formula $F = G\,\dfrac{M_1 M_2}{r^2}$. It is the formula that keeps the entire Universe in place

and all of Newton's accuracy solely depends on $F = G\,\dfrac{M_1 M_2}{r^2}$ as a formula that has to be truthful and

unquestionably accurate. The mass is the crucial factor because the mass is in a position where the mass destroys the distance of the radius from both ends equally. The mass generates a force and the force produces the gravity and the gravity produces the pulling and the pulling is what the time depends on that we have left to enjoy a Universe. We have to appreciate Newton's finding of mass until Kingdom comes because if not for mass, Kingdom is coming either tomorrow or never. Then we also accept the formula $F = G\dfrac{M_1M_2}{r^2}$ has been tested and proven so many times by science that there is no other formula on Earth that has endured the testing that Newton's gravitational formula $F = G\dfrac{M_1M_2}{r^2}$ has under gone. The force of gravity has the mass that would generate the gravity whereby the pulling of the other object orbiting and in also generating by mass the other object will also force gravity onto the first object $F = G\dfrac{M_1M_2}{r^2}$. What goes up must come down. We fight our mass because we fight gravity the entire time during one life span we live through. When I jump the force of my mass that generates the gravity by which I pull the Earth and by which the Earth pulls me back and the pulling is the result of the mass of the Earth that pulls me down again while at that moment I am pulling the earth up again, thus the square value coming about in the radius factor. When I fall my mass kicks into action and by mass I hit the ground at a rate my mass will determine. Newton is a genius because Newton realised all these wonderful happenings. Newton saw that a planet pulls another planet by the gravity that the mass of the planet charges. But consider that if mass is what brings about falling, it then implies that objects just cannot fall equal but have to fall differently and according to their mass. If mass has nothing to do with the falling then objects must fall equal and in equanimity through out the entire distance of travel while in the process of falling and that fact goes without argument. Mass brings variation and conformity is the result of mass not applying! The Universe is in a state of contracting $F = G\dfrac{M_1M_2}{r^2}$ as Newton's formula must indicate. The objects are drawing closer to each other all the time.

Now marry that thought with the ever expanding Big Bang beginning and the Newton's concept of a Universe shrinking which totally contradicts the reality that Hubble found to be true and that there is a Universe out there of which we are part of that is exploding in expansion. To the world they declare openly that Newton's contracting Universe and Hubble's expanding is the same thing and we must wait for the Universe to admit being incorrect and start to employ Newton's contracting. They gave this blaming of the Universe going the wrong way on the Universe being the incorrect party because they are looking for mistakes in the Universe and wait to find out when the Universe will start to comply with Newton and start shrinking because the Universe has to stop this ridiculous expanding since Newton said the Universe is contracting. Since Newton just can't be wrong, therefore the blame of such silly contradicting of Newton has to be found at the door of the Universe. This blame game and detecting how far and why the Universe went wrong in disobeying Newton they named the Critical Density Theory and is the biggest scam and covering of fraud ever invented by any group of persons any time during the history of man. If the Big Bang is true (and it is true), then Newton just doesn't fit! In my book I show how this led to the biggest criminal cover up man has ever devised and was initiated by a person called Albert Einstein. The entire philosophy behind the Critical Density Theory is a scam and is even as ridiculous as what the rest of Newton is. You are about to read how far Newtonians will employ criminal cover up to form a blanket of deception!

The Newtonian formula $F = G\dfrac{M_1M_2}{r^2}$ explains the comet arriving at the Sun, drawn by the mass of the Sun, pulling the mass of the comet as the comet comes closer to the Sun, but then if Newton's $F = G\dfrac{M_1M_2}{r^2}$ has any validity the comet has to crash into the Sun after arriving. If gravity by mass was pulling the comet towards the Sun in the manner as Newton insisted in the Newtonian formula $F = G\dfrac{M_1M_2}{r^2}$, then try and get any academic to explain why and how the comet moves away from the Sun and into the black yonder. After reaching the comet, the comet avoids colliding with the Sun as the formula $F = G\dfrac{M_1M_2}{r^2}$ would suggest and head into the darkness of outer space. The comet then is moving directly in the opposite direction of what Newton's formula $F = G\dfrac{M_1M_2}{r^2}$ would have us believe as the comet is not suppose to be pulling away because it is the mass pulling that was in place when the comet was drawn by mass as Newton stated. Does mass then start pushing mass to get the comet floating away from the Sun? Mass establishing gravity by pulling of a force is a gimmick Newton suggested but is unproven and it is nothing less than foolhardy to believe that mass does the pulling of the comet. Try and get those academics in physics to sensibly admit this reality and then in the explaining be sensible by using their Newton formula $F = G \dfrac{M_1 M_2}{r^2}$ as Newton's formula presents the law to show how this going away happens when mass is doing all the pulling at first. Try and get any Newtonian academic to explain this escaping of the comet from the mass of the Sun in the face of mass pulling mass. Some try to use the idea that the momentum drags the comet around the Sun but the mass will pull the comet into the Sun if $F = G \dfrac{M_1 M_2}{r^2}$ applies. Newton never created a detour as the mass pulling mass forms a linking straight line running from the centre of the Sun to the centre of the comet. Newtonians always bring more deceit to cover up Newton's fraud. These questions I address are otherwise never asked by students because students are brainwashed to accept and not think about asking questions. In presenting my work I can and I do answer the questions raised above but my answers do not fit the Newtonian visions of mass doing the pulling and because it contradicts Newton, I am ignored. In my following describing Newtonians is not to moan and grumble but it is to show the means and the manners they use to fight and when using such utter arrogance, despicable high and mighty autocracy with plain bullying tactics and megalomania. They have this attitude that only they are wise enough to think and the rest is mindless dehumanised animals walking on hind legs. If they fought fair and used intelligence it would not be that bad but to use dirty tactics when confronting me by just dismissing my views from a position of having authority is coward ness. By bullying me from holding a position of being able to ignore me and I can do nothing about it doesn't frighten me, it angers me!

If you are a student in the science of physics, then ask your Educated Masters to please explain the following abnormalities you are about to read in this book and insist on a clear explanation about the inconsistencies they promote while tutoring physics as if the physics they present are the most flawless and accurate institution there has ever been. Ask those academics supporting Newton about the following flaws that no one, except me, ever mention. Get them to explain the inconsistencies they never talk about. Wise up and confront those charged with tutoring physics and see who should you believe. Then get informed instead of brainwashed.

One very simple example, which I mention now at this point but I do not elaborate on this matter any other place in the book since in this book I wish to limit space, used, is mentioning the gravitational constant. If any one wished to bring in an explanation by employing the gravitational constant also introduced in the Newtonian formula $F = G \dfrac{M_1 M_2}{r^2}$ then using this gravitational constant is one of the ultimate bogus ploys Academics use to confuse the public.

Newton first envisaged the idea that it is mass standing in relation to mass that is destroying the radius found between the two objects forming gravity as presented by the formula $F = \dfrac{r^2}{M_1 M_2}$ but subsequently the notion as well as the formula used changed to $F \propto \dfrac{M_1 M_2}{r^2}$. To get Newton's miscalculation $F = G \dfrac{M_1 M_2}{r^2}$ to work with some dignifying crookedness' they devised a constant of sorts going by the title as the gravitational constant and is this constant holding the symbol **G** in $F = G \dfrac{M_1 M_2}{r^2}$ It is put in place as being the same as all the gravity but is apparently that gravity that fills the space between the Earth and the Moon. Now comes the Newtonian part... This same space filling ingredient called the gravitational constant and holds a measured value of 6.67×10^{-11} where it is using this value while it is playing its part in filling all the space we find between the Sun and the Earth as well as the Sun and Pluto and everywhere there is space in outer the gravitational constant is the space-filler to have in that space being filled. If you think of space then we have such space filled with a gravitational constant at a value of 6.67×10^{-11}. This was the case in the days when it was accepted that ether was filling the space the gravitational constant filled and therefore ether might have had the value of 6.67×10^{-11}. Then after finding no evidence of ether, the ether that was not filling the gravitational constant was miraculously and by a stroke of Newtonian magic removed and replaced with...nothing...yes, nothing is now filling the space ether filled before they realised ether was not filling the space but the marvellous part is that nothing that the replaced ether took from the ether that is not there the value given to the gravitational constant and now while space is filled with nothing it still holds the measured value of 6.67 X 10⁻¹¹.

Newtonians are most adamant the Universe outside material is filled with nothing and nothing form outer space. Let's quickly ponder on this for a second or two and find out how much of this concept is palatable. They are then saying there is a long line of nothing standing the one after the other where one nothing is following the nothing in front while the nothing in front is leading the nothing behind. The nothing is lines forming rings here every ring ends at a point that the nothing that form the line of nothing connect linking in a chain of nothing from the Sun all the way to Pluto and even far beyond, in fact as far as the mind can take nothing and then nothing links in a line even further.

This line of nothing linking in a chain has a measured value that consists on the gravitational constant and each one of any of the nothing we mention has the value of 6.67 X 10⁻¹¹. Well even Red Riding Hood is going to sound more believable than does the most intellectual minds in mathematics…and from where does this first thought originate that the Universe comprises of nothing…from Newton of course when Newton said that the spin of an object cancels the space in which the object spins or in mathematical terms

Newton's formula holding and explaining the Universe portrayed as being $F = G \dfrac{M_1 M_2}{r^2}$ is completely wrong.

If everything is in contraction then by now some places should already contracted large areas of space leaving gaps at other places where cosmic holes should by now be in place. If contraction brings about the gain of space in some parts of the cosmos as Newtonians say, then there has to be a part of the Universe that is losing space and by losing space a shortage of space must bring holes in space to appear where space is reducing! The gain of one part must bring about the loss of another part. In all of this that I have mentioned thus far in this letter, it does not even form a drop in the ocean compared to the incorrectness I present on science in my books about science. One should think that a book that challenges the dogma of Mainstream science and bring a new view or if only then just another view on science has to be commercially viable and should have some sort of selling potential. One should imagine that there has to be some publisher that recognises the potential and would have the courage if not the business sense to put such a book in the book shops. But I found that thinking that way is pure daydreaming. A book such as the book I have to offer that takes on Newton and starts to strangle Newtonians has more scope to become controversy than does the Da Vinci code because it affects a wider audience.

Since 1977 I was convinced that there was something amiss in the approach science took on the matter of gravity. In the work I presented I based my theory on the discipline Johannes Kepler introduced. But the more I pursued my

goal of forming another gravity concept other than supported by Newton's view that $F = G \dfrac{M_1 M_2}{r^2}$, the more I

had to confront the thinking of Newtonian inspired culture when approaching academics. I was never in doubt that Newton was ultimately wrong and that the Newtonian dogma of gravity was incorrect, but never wished to directly attack Newton as a scientist. But in the end I had to change my approach of being polite because it was clear to me that academic culture would not change and see the logical arguments. In September / October 2005 I wrote letters to nine academics heads of departments at Universities in South Africa that was involved in cosmology. Each one I wrote a letter to was heading a cosmology department at a South African University and in the letter I informed them that I intended to show why Newton was incorrect and therefore what mind games they as physics academics and tutors were playing by protecting Newton's dogma and Newtonian religiosity, I was going to uncover and make public the fraud Newton committed and what the extend of the fraud was that they were intentionally covering up. I knew before I wrote the letters that no one is going to take me serious but then there is a price to pay for every mistake one makes and they were blind to what I said when I said I was going to expose them. Even at this point they do not see that they are intentionally covering up Newton's criminal behaviour and hat they are committing intentional fraud by covering up Newtonian misconceptions. They are so involved in a culture of crime it is not possible for them to separate criminal fiction from factual reality. The covering up syndrome and culture prevailing in physics will prevent them from having any negative work on Newton or about Newton to be published.

By merely putting gravity in the Universe that is acting as a mysterious FORCE that is pulling towards a common point in an allocated general centre is rather avoiding the question with simplicity because the question about how and why remains unanswered. Not knowing the answer will leave you empty and unfulfilled because of being a student and not knowing is the same as suicide n a mental level. Ask yourself the following: If gravity pulls towards a centre and gravity holds the Universe attached the question arising from that simplistic answer is then … where is the centre of the universe?

Should and if you decide read my letter addressed to students it will bring along a new perception about Kepler. Science sees to it that Kepler stays the least appreciated Cosmologist where as in truth Kepler proved gravity, proved singularity, proved space-time, proved the Big Bang, proved every dynamic most of the wise persons afterwards thought about. Yet no one gave Kepler any recognition up to now because science denies Kepler his limelight.

By not confronting the establishment, you give the establishment grounds to allure you into being sheepish. Because they see you, as just another stupid senseless student they have the opinion that they can brainwash you into accepting these fallacies that I am about to tell you. They will literally brainwash and condition your mind to accept what they never yet were able to prove. If you feel I come across far too strong then put correct values in

place of the symbols in $F = \dfrac{r^2}{M_1 M_2}$ or in $F \: \alpha \: \dfrac{M_1 M_2}{r_2}$ and see according to your own opinion how totally

ridiculous Newton was and physics at present is.

They are of the opinion you will swallow any rubbish they throw your way just because every generation before you, were mind controlled in the way they are about to put their control over you. You may think this is big words but read on and see after you come to know all the facts if I exaggerate even in the least. They see you as slow-witted and mindless because they think they are the academics being superior making you the inferior. If you are not aware of the facts beforehand they know you will follow their teaching without asking questions.

They think that your naivety makes you mindlessness will incapacitate you into their control. They don't want you to ask nosy questions about contradictions existing and they refuse to answer. This process of brainwashing and mind controlling in physics has been in progress for hundreds of years. Just answer how a feather and a large hammer fall equally while mass drive gravity as a force. If you can't...well they can't either! Their task is not to explain but to mislead since they think you can't think while they think they know how to control you.

The following letter addressed to students does not aim to represent the full entirety of the copy of the original letter addressed to students but is reduced considerably to aid any possible potential reader in the examining what the purpose is of the information this letter addressed to students wish to announce. Anybody and everybody are aware that all objects fall at an equal rate. If an object such as a car weighing one ton falls at the same pace as a person weighing fifty kg how does mass come into the picture by committing a force to do the pulling? Mass has to pull because according to their teaching it is mass that establishes gravity. However, mass is a factor that produces differentiation whereas all objects show equality during their fall. If it is mass that is establishing the force gravity, all objects must fall at different speeds. That they do not do as they all fall equal. That means physics is wrong from the start because mass cannot have any input in objects falling.

Physics students, it is your duty to pull the plug on the powers of the ALL-POWERFULL Academics in Physics and stop their dishonesty. It is your task as the as the next physics generation to stop the criminals that are filling the corridors and the lecture halls of physics departments throughout the world. Stop their teachings by forcing them to stop their criminal fraud. Force them to explain the deception called THE CRITICAL DENSITY, which is a conspiracy to commit fraud. Let them explain how an expanding Universe can suddenly and abruptly turn in direction of developing and start to contract as Newton stated it is doing at present, and when facing all other concluding evidence showings it was expanding since time began. Tell them to prove that the cosmos will begin to contract doing its turn about by using other proof than merely Newton's say-so. Tell them to bring proof with evidence that the cosmos is contracting as Newton said. Then force them to admit to the fraud they are precipitating in, which is THE CRITICAL DENSITY conspiracy... In THE CRITICAL DENSITY conspiracy all they say is that they are waiting to see when the cosmos would stop its criminally insane behaviour and start to listen to the laws of Sir Isaac Newton.

With The Critical Density shambles the modern Newtonian set out to defraud the world in the same manner as their Master Sir Isaac Newton has done centuries ago. Newton said the cosmos is contracting. When Hubble proved the cosmos is not contracting, Newtonians looked where the cosmos went wrong by not following Newton guide lines he so clearly set the cosmos to follow. It has to contract and not expand

To all those that feel disgusted by me accusing the greatest name in science that ever lived being Sir Isaac Newton of fraud, please go on and prove me wrong!

$F = \dfrac{r^2}{M_1 M_2}$ This is the formula Newton used with which Newton proved gravity. Now prove gravity by using this

formula. Do the following to prove me wrong.

To find the force of gravity one has to multiply the mass of the Earth (M_1) with your personal mass (M_2) and then divide the distance there is between you and the Earth (r^2). Using these factors by multiplying (M_1) and (M_2) and dividing with (r^2) should present gravity coming from mass. But science uses a fixes value to calculate gravity.

Now, convince your mind about my correctness. Do the simple calculations.
Take the mass of the Earth (M_1).

Multiply the Earth mass by your personal mass that any scale should indicate (M_2).
After multiplying the two mass factors, then proceed to the following step by dividing the multiplied mass factors with the square of the radius there is between your feet and the Earth (r^2), which should not amount to more than a few billionth of a millimetre.
If the answer in front of you is not 9.81 Nm/s^2 then there is something very wrong.
The incorrectness has to be one either of two possibilities presented:
The measured value of gravity is not 9.81 Nm/s^2 as science uses it, or

Sir Isaac Newton's formula suggested as $F = \dfrac{r^2}{M_1 M_2}$ is complete fraud...

Now which is it...you can decide...
The force of gravity that the world of physics uses to do measurements is 9.81 Nm/s^2. If your answer you have in calculating your force of gravity is not 9.81 Nm/s^2, then it is either this measuring value of gravity that is wrong or it is Newton's $F = \dfrac{r^2}{M_1 M_2}$ that is wrong because by the calculation you did, the calculated answer you got could not possibly have deliver a measured value of 9.81 Nm/s^2. After all, science maintains it is the pulling of mass that delivers the force of gravity! If by using the factors of mass and the radius does not accumulate to 9.81 Nm/s^2, then how can mass deliver gravity.

To teach students that $F = \dfrac{r^2}{M_1 M_2}$ are the measuring formula in determining gravity, while knowing very well it is not totalling gravity at 9.81 Nm/s^2, then doing that to students while enforcing a thinking pattern in the minds of a student is committing brainwashing because by forcing examinations on students, expecting them to confirm the falsified statements used that the tutors present as correct is brainwashing, a way of enforcing mind control and it is manipulating the thinking process of students.

If you can't prove that my manner of thinking is incorrect and you keep surmising that science is correct then recalculate the formula or start reading the rest of their fraud.

Gravity is a constant of 9.81 Nm/s^2. This is used in all cases of scientific calculations.

Please use your intellect to explain the following mathematical expressions as Newton suggested that the principles of the different formulas applied.

Let any one of them prove this. Let any one of them explain this as a mathematical principle. This is all mathematical formulated expressions and has to be proven accordingly. This is not linguistic suggestions but is used in terms of mathematical accountability! This is palpably false

Mass is an individual factor that is different on anything on which it is applied as a measuring factor. How can something as different as mass that is never constant even on Earth form a constant such as the force of gravity and still be the same in all cases?

Sir Isaac Newton's says that $a^3 = T^2$. I have to believe Sir Isaac Newton when it is said that three dimensions are equal to two dimensions or in mathematical terms that $a^3 = T^2$ on no more grounds than that Sir Isaac Newton said so and without having any other proof to back the statement. Remember, Kepler never said $a^3 = T^2$, that is the part coming from the fantasy of Sir Isaac Newton. Kepler said $a^3 = kT^2$ which places three dimensions on one side holding three dimensions equal on the other side of the equation. There is a^3 on the one side of $=$ and then there is kT^2, which is $k^1 \times T^2$ which is $k \times T^2(^{1+2=3})$ and that makes $a^3 = kT^2$ having three dimensions on the one side being equal to three dimensions on the other side. There is no way in heaven or hell that one can have the third power being equal to the second power or have a cube that is equal to a square, even if you are Sir Isaac Newton. There is no one on Earth that will tell me that $10^3 = 10^2$. There is a case that $10^3 = 10^2 \times 10$ or that $2^3 = 2^2 \times 2$ but never can it be that $2^3 = 2^2$. Not even when Sir Isaac Newton is doing the saying so. If one says that in the event where $a^3 = kT^2$ one may assume that $a^3 = a \times a^2$ or $k^3 = k \times k^2$ or even that using $T^3 = T \times T^2$ will also bring equality but never can $a^3 = T^2$...and then there are academics that try to convince me that $a^3 = T^2$ because Sir Isaac Newton was of the opinion that $a^3 = T^2$ and furthermore they expect me to also believe that it is true that Sir Isaac Newton has never been wrong on any suggestion and because no one could ever find Sir Isaac Newton to be wrong, I have to accept that $a^3 = T^2$ and take it as the absolute truth without questioning this abnormality!

 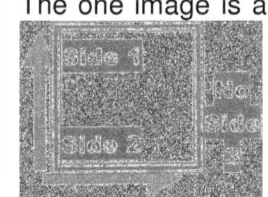 The one image is a cube with three sides. The other totally different image is a square having two sides. **Sir Isaac Newton** said the two are equal while they can never be equal since they are one dimension apart **Sir Isaac Newton** convinced so many generations of idiots considered as being the wise amongst the wise and fooled those to the point where these stooges are willing to believe they are wise enough to believe that a cube is equal to a square and only on the ground that **Sir Isaac Newton** said so.

Sir Isaac Newton proposed and moreover convinced the world of science, and this includes every one and all members that should be the most intellectual bunch living on Earth in human form, that they and the entire world should accept that the inexplicable $a^3 = T^2$ is correct and that the biggest trick in fraud can be played on a bunch of fools all willing to be stupid enough to pretend they are clever enough to see that $a^3 = T^2$ and they are so stupid they pretend to be so clever that they will accept that $a^3 = T^2$ which when translated in words means that two dimensions are equal to three dimensions. This is the same as stating that a person's reflection coming back from the mirror is the same as the person filling reality while standing and looking at his image in the mirror. In this group hosting the most advanced minds man can produce there are a big enough bunch of zombies pretending to be mentally superior while being big enough idiots that are foolish enough not to think and not to ask questions but be small minded to the point that they will accept that a cube is equal to a square $a^3 = T^2$ just simply going on the say so of **Sir Isaac Newton's**

When **Sir Isaac Newton** says $a^3 = T^2$ that does not prove that $a^3 = T^2$. It only proves **Sir Isaac Newton** was the worlds biggest and best silver tongue devil and cheated an entire Earth load of scientists for almost four hundred years. He fooled the supposedly wisest humans we all think there can be to pretend to be wise so that they can hide their stupidity while they only focus on their stupidity by not questioning the validity of $a^3 = T^2$. You bring me one other con artist and fraudster that can manage that. It takes some doing to fool so many people for so long and leave all those fooled feeling good about themselves in that they are fooled. **Sir Isaac Newton** was the biggest con artist ever to live and never again will the world experience an equal to **Sir Isaac Newton**. It is no small wonder that science is infested with atheism because science upholds disdainful lies based on mediocre understanding about truth applying as a reality and crooked science! Newton is all lies and shambles and reading this book will prove that.

If science cannot prove God's existence, it is not God that does not exist, but it is science failing and therefore it is then that specific view about science that should be re-examined since it is the view on science that is proving as being incorrect. This fact is what the so very brilliant and intellectually mindful Newtonian atheist should remember when they fail in their science altogether. That their science fails altogether and that failing it does in all its splendour, is facts I am delighted to prove! The fact is Newton's views were never tested and that the Newtonian views on science were never challenged before and because of that Newton principles never withstood diligent scrutiny before. When **Sir Isaac Newton** is investigated even in the flimsiest of manners, well accepted facts seems to become very suspect, to say the least. This becomes evident when concluding all the facts this book presents. Now, in this book, for the first time Newton is tested and such testing is the proof you gain by reading that which I uncover. What I bring into the open is unseen facts, which I present you with as I take you on a tour through an avenue of facts I introduce in this work. The lack there is in sensibility concerning **Sir Isaac Newton's** principles this book proves. The theories of **Sir Isaac Newton** require proof, which was never given while God never needs proof and that is what science constantly seeks. When science perpetually ignored my concerned calling on and ignored my calling on them because (I suppose) they were finding my concerns wanting, in my final letter to them I promised them never to contact them personally again by any and by all means. I also promised them a fight. This is the fight I promised. I was not worth noticing so I was ignored. I now am calling on the public, as I am ignoring their reputations.

All prospective, intending and otherwise possible potential readers of anyone of the two books called **An Open Letter On Gravity** are hereby seriously advised to read Part 1 of **An Open Letter On Gravity** first before advancing onto Part 2 of **An Open Letter On Gravity**, and only then afterwards and after completed reading Part 1 of **An Open Letter On Gravity** then the reader should advance by reading Part 2 of **An Open Letter On Gravity** because by first reading Part 2 of **An Open Letter On Gravity** the answers might seem to remain questionable but when reading Part 1 of **An Open Letter On Gravity** first the questions will become answerable. If it is said by using easier language to explain the expression: then by not knowing the questions, the answers are not well defined but knowing the questions, the answers become rather simplified as it is self explained and much better understood in explaining.

Still my challenge is and remains there where I **challenge everyone and all persons notwithstanding title or position to show me how they can maintain with clear and lily white conscience that Sir Isaac Newton did not corrupt science in all aspects by committing fraud in mathematics and do so after reading the two titles An Open Letter On Gravity Part 1 and Part 2.**

AN OPEN LETTER ON GRAVITY Part 1 disputes the correctness of the formula $F = \dfrac{r^2}{M_1 M_2}$

Using the formula above as Newton did does not imply a suggestion or carry an idea across as a thought but must be seen to be acting as confirmation about a fact because one cannot suggest anything mathematically, one can only confirm a fact mathematically. There is no mere suggesting of any possible movement in a specific direction of any suspected behaviour by an object moving from and to a point as suggested but this is saying that the gravity of the Earth measured in mass at it's totality is colluding with the falling body's measured in mass as the two factor's diminish the radius from both ends. This is used to back up a fact!

This Is a Book That Is Not Afraid To Show How The Paternity of Newtonian Science in Physics Openly Cheats To Cover Their Oversight In an All Out Effort to Hide Newton's Misjudgement

Newtonians say the force F of gravity is 9.81Nm/s^2
Newtonians also say the force F of gravity is

$$F = \frac{r^2}{M_1 M_2}$$

Then in terns of mathematical principles

Newtonians say $F = \dfrac{r^2}{M_1 M_2}$ **is 9.81Nm/s^2 where both are equal to the force F of gravity**

$F = \dfrac{r^2}{M_1 M_2}$ **= 9.81Nm/s^2. That is the way that proving with mathematics is done and what does it prove? SIR ISAAC NEWTON'SFRAUD**

Science teaches that a feather and a hammer have different mass while they fall equal in time through an equal distance travelled. All things fall equal in time and distance when subject to the same environment. If gravity was mass related, then this was not possible, because then objects must fall according to mass. Falling objects bears no evidence of mass playing any part in falling. Any two objects holding different mass fall equal in time and in distance when sharing similar conditions, which suspends mass altogether as an influencing factor. Galileo proved different mass fall equally under similar conditions. That fact about Galileo, science does embrace, although this strongly contradicts Newton's impressions about mass inflicting gravity. Acknowledging Galileo must make the work of Newton incorrect and also corrupt. On TV we see how all objects, such as cars, humans and bags fall at the same pace, which sets a standard totally against Newtonian mass principles that produce the falling, and proves Newton wrong because mass then does not underwrite gravity in any way or form at all. The formula

$F = \dfrac{M_1 M_2}{r^2}$ would suggest mass taking all the responsibility for such falling that takes place. Newtonians declare gravity as the force of gravity F, that is = equal to gravitational constant G, when it is multiplied by the mass M$_1$ and the mass M$_2$ after which then the product of the three factors influencing gravity is divided by the square r^2 distance between mass pulling the mass that destroys the distance between the two objects. If mass pulls mass as Newton said, the Big Bang is not possible but the Universe is notwithstanding Newton's claims, expanding (growing apart). If the mass destroys the radius separating the objects, then the comet has to collide into the Sun, but it doesn't. If mass forms gravity, every planet must orbit at a different pace, which they do not, as all planets orbit at the same pace around the Sun. Planets don't give the slightest hint that they obey Newton's suggested cosmic laws by implementing mass. The truth is that mass is the resistance of any independent material to deform and to acquire mass the individual object relinquishes independent motion. Mass is the reluctance to deform and integrate into a larger structure and becoming a unit of the larger or holding structure. Mass comes about when the falling of any object stops the motion of the falling. Mass prevents further falling, it does not sustain further falling. Gravity is the moving of the object to the centre of the Earth while falling.

I say gravity is movement while mass is obstructing independent movement, which is what gravity is. Mass is not forming the factor responsible for gravity or movement, but prevents further movement. A body falls by gravity. Mass obstructs further falling, while gravity remains present as a factor that brings the tendency or inclination to move or the attempt to continue moving. Mass hinders movement and therefore mass can't enhance or produce movement or gravity. Mass prevents or blocks gravity. Gravity is the motion that defines the individual identity of any object's structural form by rendering motion while reserving independence in granting free space from other manipulating objects. By saying this, I am awarded the cloak of death by Academics ignoring my correspondence as if I never addressed their mailbox.

This is not the only untruth that the Paternity called Mainstream Science is keeping concealed as a cover up that is wrapped under an airtight blanket of deception. If you sit in class and listen while also experiencing the sinking feeling that the facts you hear are not adding to a total you are comfortable with while you disagree with what is said then you better read on because this letter addressed to students has it at task to show all that will read this document how much discrepancies academics lay on unsuspecting students that trust Academics with their future and their life. Do you as students realize the inconsistencies that physic Academics present you with when portraying that what they teach you as being the solemn truth.

Students tell your Professors to stop deceiving and stop trying to control your minds with their fraud. Those Academics tutoring you are telling facts about gravity that has never been proven.
That is mind control.

They wish for you to accept facts on gravity that they hold as the truth. They claim those truths are beyond questioning yet with the least examining those truths they stand by then proves to be totally void of substance because it was never corroborated by one single experiment.

Should you question that mass produce gravity they will expel you from University by letting you fail your examinations and it was never proven. They will expel you and have you fail tests should you question their authority on the matter of gravity while at the same time they can't for one second bring evidence in support of what they wish you to accept as the unquestionable truth.

That's brainwashing by mind control because if you don't accept their baseless fact as God given truths they dismiss your academic career.

It is either put up and shut up or be gone. Academics do put mind control to work on unsuspecting students by forcing students never to question the legality of statements they offer as being sound and correct.
What they present as correct I prove in this very letter addressed to students are openly laughably totally incorrect and by just reading my evidence you will see how feebly easy it is to rubbish it. Take the evidence I am about to share with you and confront them with the fabrication of facts that they present. Go on and challenge those teaching you with the falsified facts as I challenge any one to prove me wrong.
What they maintain is gravity is total incompetent nonsense and can't be corroborated at all but what they can't corroborate because they don't understand I prove to be that which the Universe employs to form gravity. There are four phenomena they dismiss because they have no idea what they are. I studied each one and formed an explaining by implementing Kepler's formula as the Universe gave it to Kepler.

The by understanding the formula and implementing the content into the four phenomena I am able now to prove what forms the motion we think is gravity and when reading it then the Universe makes sense. All the questions in these books I managed to answer while they can't … and in the books I answer a lot more questions than what I ask here in the rest of the web site while Science fail to answer any...

2   +.........!......4......U......2......?

What the Christians find hard to admit, is that they destroyed thousands of years of knowledge through their insanity, and if it were not for the conservative efforts of Islam followers, like AL-BATTÄNÏ, MUHANNAD IBN JÄBÏR (858 – 929) an Arab astronomer, things would have been so much different today. He was born in modern Turkey and is known by the Latinized name Albetegnius. He was one of the first Arab astronomers to grasp the importance of accurate observations. He produced a set of tables, including a catalogue of star positions more accurate than in Ptolemy's Almagest that was to influence medieval European astronomers. Al-Battänï refined the values of the precession of the equinoxes, the obliquity of the ecliptic, and the length of the tropical year, and found that over the course of the year the Earth-Sun distance varies.

APOLLONIUS OF PERGE (c.262-c.190BC) who was a Greek mathematician was presumably born in the region, which now is the modern Turkey. He showed that the ellipse, parabola, and hyperbola are all curves formed by a plane intersecting a cone in different ways, i.e. that they are conic sections. The orbital path of an unperturbed body moving in a gravitational field follows one of these three curves, as would come to be appreciated by later astronomers such as E. Halley, who translated Apollonius's book Comics. Apollonius also originated the mathematical concept of motion based on epicycles and deferent's, later taken up by Hipparchus and Ptolemy to explain planetary motion.

Some sources say the BIG BANG THEORY is the most widely accepted theory of the origin and evolution of the Universe, but it seems as if this theory nowadays is the only excepted theory. According to the Big Bang theory, the Universe originated from an initial state of high temperature and density and has been expanding ever since. The theory of general relativity predicts the existence of a singularity at the very beginning, where the temperature and density were infinite. Most cosmologists interpret this singularity as meaning that general relativity breaks down at the Planck era under the extreme physical conditions of the very early Universe, and that the very beginning must be addressed using a theory of quantum cosmology. With our present knowledge of hinge-energy particle physics, we can run the clock back through the lepton era and hydron era to about a millionth of a second after the Big Bang, when the temperature was 10^{13}K. By adopting a more speculative theory, cosmologists have tried to push the model to within 10^{-35} of the singularity, when the temperature was 10^{28}K.

The Big Bang theory accounts for the expansion of the Universe; the existence of the cosmic background radiation; and the abundance of light nuclei such as helium, helium-3, deuterium, and lithium-7, which are predicted to have been formed about 1 second after the Big Bang when the temperature was 10^{10} K. The cosmic background radiation provides the most direct evidence that the Universe went through a hot, dense phase. In the Big Bang theory, the background radiation is accounted for by the fact that, for the firs million years or so (i.e. before the decoupling of matter and radiation), the Universe was filled with plasma that was opaque to radiation and therefore in thermal equilibrium with it. This phase is usually called the primordial fireball. When the Universe expanded and cooled to about 3 00 K it became transparent to radiation. The discovery of the microwave background in 1965 resolved a long-standing battle between the Big Bang and its then rival, the steady-state theory, which cannot explain the blackbody form of the microwave background. Ironically, the term Big Bang was initially intended, to be derogatory and was coined by F. Hoyle, one of the strongest advocates of the steady state.

BIG BANG CHRONOLOGY

Era	Time after Big Bang	Temperature
Planck era	0 to 10^{43} s	7 to 10^{34} K
Radiation era[a]	10^{-43} s to 30 000 years	10^{34} to 10^4 K
matter era[b]	30 000 years to present	10^4 to 3 K

[a]   The time from about 10^{-6} or 10^{-5} s to about 1 s or so is subdivided into the hydron   and lepton eras.

[b]   Includes the recombination epoch, which took place about 300 000 years after the Big Bang, at a temperature of about 3 000 K.

The BLUESHIFT and the
A Doppler shift of light towards the blue end of the spectrum, caused when the emitting source is approaching us.

The German mathematician and astronomer BODE, JOHANN ELERT (1747 – 1826), published a formula in 1772, now known as Bode's law, which yielded the approximate distances of the six known planets, from which he predicted the existence between Mars and Jupiter of an undiscovered planet. His major publication was Uranographia (1801), a comprehensive atlas of the entire sky showing over 17 000 stars and nebulae. For fifty years, he oversaw the publication of astronomical data in the Berlin Academy's yearbook. BODE'S LAW
A numerical sequence announced by J.E. Bode in 1772, which matches the distances from the Sun of the six planets then known. It is also known as the Titus-Bode law, as it was first pointed out by the German mathematician Johann Daniel Titus (1729-96) in 1766. It is formed from the sequence 0, 3, 6, 12, 24, 48, 96, and 192 by adding 4 to each number. The planets were seen to fit this sequence quite well – as did Uranus, discovered in 1781. However, Neptune and Pluto do not conform to the 'law'. Bode's law stimulated the search for a planet orbiting between Mars and Jupiter that led to the discovery of the first asteroids. It is often said that the law has no theoretical basis, but it does show how orbital resonance can lead to commensurability.

BODE'S LAW								
Planet	Mercury	Venus	Earth	Mars	Ceres	Jupiter	Saturn	Uranus
Bode's law distance	4	7	10	16	28	52	100	196
Actual distance (10^{-1} AU)	3.9	7.2	10	15.2	28	52	95	192

BOK, BARTHOLOMEUS ('BART') JAN (1906 – 83)
Dutch-American astronomer. His lifelong work was the study of the Milky Way, much of it in collaboration with his wife, Priscilla Book, née Fairfield (1896-1975) . In particular, he investigated its structure, its distribution of stars, interstellar matter, and star-forming regions. In the 1930's, he discovered the objects now called Bok globules, and demonstrated that stellar associations are made up of young stars. In the early 1950s, with J.H. Oort and others, he pioneered the mapping of the Galaxy at radio wavelengths.

BOLTZMANN CONSTANT
(Symbol k or k_B) A constant that relates the kinetic energy of a particle in a gas to the temperature of that gas. It has the value $1.380\ 658 \times 10^{-23}$ joules per Kelvin. The particles can be molecules, atoms, ions, or electrons. The Boltzmann constant relates pressure, p, and temperature, T, by the equation $p = nkT$, where n is the number of particles per unit volume. In astrophysics, this equation is of importance in understanding the interiors and surface layers of stars, and the atmosphere of planets. The constant is named after the Austrian physicist Ludwig Edward Boltzmann (1844 – 1906).

BRAHE, TYCHO (1546 – 1601) was the man who Johannes Kepler was associated with, in their joint study about the orbiting routes of the planets This Danish astronomer died before he could accomplish his life task. Nevertheless, Kepler completed the task. He was the most accomplished observer of the pre-telescope era, expert in constructing instruments for making accurate naked-eye positional measurements. He first gained fame through his report (De nova stella, 1573) of the 1572 supernova in Cassiopeia. In 1576, he constructed Uraniborg, an observatory on the island of Hven in the Baltic (a second observatory, Stjerneborg, was built in about 1584). He calculated that the comet seen in 1577 had a highly elongated orbit, which would pass through several of the 'spheres' on which the planets were supposedly carried, and this led him to doubt the reality of Aristotle's planetary model. However, he rejected the heliocentric system proposed by Copernicus. In the Tychonic system, although the planets orbit the Sun, the Sun itself (and the Moon) revolves around a stationary Earth. Tycho made major contributions to the study of the Moon's orbit. In 1597 he moved to Prague, and employed J. Kepler as his assistant. Kepler later made use of Tycho' s observations when deriving his laws of planetary motion.

BRANS-DICKE THEORY was the rival theory to Einstein's work; publish to serve as an alternative to Einstein's theory of general relativity. I will later in this article come to this, the most widely excepted theories of all, which attempts to incorporate Mach's principle. Among many other things, this theory predicts that the value of Newton's presumption of the so-called, gravitational constant, G, should change

with time. The two persons responsible for introducing this theory were the American physicists Carl Henry Brans (1935 -) and R.H. Dicke.

BUTTERFLY DIAGRAM
A graph on which the latitudes of sunspots are plotted against time. It shows how spots migrate from higher latitudes (30 – 40° north or south) towards the equator (latitude 5° or so) throughout each sunspot cycle, in accordance with Spörer's law. The shape of the distributions, when plotted for both northern and southern hemispheres, resembles the wings of a butterfly.

CALLIPPUS (c. 370-c.300 BC.) was yet another Greek astronomer and mathematician who lived in the golden Greek Roman era. He modified the theory, which was held, by an earlier scientist by the name of Eudoxus. Adding to the existing Earth-centreed spheres, extra ones for the Sun, Moon, and some of the planets changed Eudoxus's principle. This meant that there was a total to 34 different spheres. Aristotle further refined Callippus's model. Callipus calculated accurate lengths for the seasons, as measured between solstice and equinox.

CHANDRASEKHAR, SUBRAHMANYAN (1910 – 95)
Indian-American astrophysicist, born in modern Pakistan. He was the first to identify whit dwarf stars as end-products of stellar evolution, showing how a star collapses when there is no longer any radiation pressure to counteract its own gravity, producing degenerate matter. He calculated an upper limit (the Chandrasekhar limit) beyond which a star would enter a more dramatic final phase, presaging the existence of neutron stars. He studied how stars transfer energy by radiation in their atmospheres, publishing his findings in *Radiative Transfer* (1950). Chandrasekhar shared the 1983 Nobel Prize for physics with W.A. Fowler.

CHANDRASEKHAR LIMIT
The maximum possible mass of a degenerate star, above which it will be unable to support itself against the inward pull of its own gravity. For a star with no hydrogen content the limit is 1.44 solar masses, which is thus the maximum possible mass for a white dwarf. A degenerate star with a mass greater than this limit would collapse under gravity to become either a neutron star or a black hole. It is named after S. Chandrasekhar.

CHANDRASEKHAR-SCHÖNBERG LIMIT
The maximum mass of a star's helium core that can support the outer parts of the star against gravitational collapse, once the hydrogen as its centre has been exhausted. The limit is about 10 – 15% of the total mass of the star. If the mass of helium in the core exceeds this limit, the central parts collapse while the outer part expands rapidly to become a red giant. Calculations suggest that this happens only in massive stars. The limit is named after S. Chandrasekhar and the Brazilian astrophysicist Mario Shonberg (1916 -)

COPERNICUS, NICOLAUS (1473-1543) was according to the Anglo Americans a Polish churchman and astronomer although this is just another political inspired propaganda because his parents were both German (in Polish, Mikolaj Kopernigk). While he was completing his studies, he had realized that the Earth revolves around the Sun and not vice versa. Such a view was in that time, held to be heretical. As I pointed out in the first few articles, the Church regarded the geocentric world-view of Ptolemy as consistent with its doctrines. Copernicus set down his basic ideas around 1510 in the Commentariolus, which he circulated anonymously, because of the Islam link. In 1512-- 29 he conducted his study and concluded the observations that he needed to support his theory, while carrying out ecclesiastic and local administrative duties. In this time, he had to defend his mother in court on charges of witchcraft. In 1539, the Austrian astronomer and mathematician Georg Joachim von Lauchen (1514-74), known as Rheticus, became a pupil of Copernicus and began to spread his ideas. The published work was openly spread as the Copernican system, in spite of the life-threatening dangers connected with such a "crime", in 1543 in the book De revolutionibus orbium coelestium. However, the reality of a heliocentric Solar System was only commonly accepted, after the work of Galileo and J. Kepler.

CURVATURE OF SPACE-TIME
A property of space-time in which the familiar laws of geometry no longer apply in regions where gravitational fields are strong. In general, relativity the geometry of space-time is intimately connected with the distribution of matter. In a space of only two dimensions, such as a flat rubber sheet, Euclidean geometry applies so that the sum of the internal angles of a triangle on the sheet is 180°. If a massive object is placed on the rubber sheet, the sheet will distort and the paths of objects moving on the sheet will become curved. This is, in essence, what happens in general relativity.

In the simplest cosmological models, based on the Friedmann universe, the space-time curvature is simply related to the mean density of matter, and is described by a mathematical function called the Robertson-Walker metric. If a Universe has a density greater than the critical density, it is said to have positive curvature, meaning that space-time is curved in on itself, like the surface of a sphere; the sum of the angles of a triangle drawn on a sphere is then greater than 180°. Such a Universe has finite size and also finite lifetime; it is a closed universe. A surface of a saddle, on which the sum of the angles of a triangle is less than 180°. Such a Universe would be infinite and would expand forever; it is an open universe. As Einstein-de Sitter Universe has critical density and so is spatially flat (Euclidean) and infinite in both space and time.

CURVE OF GROWTH

A method for determining the temperature and chemical abundances of stellar atmospheres. From the different profiles of weak and strong absorption lines of a given element, a diagram can be constructed showing how the equivalent line widths increase (or 'grow') from weak to strong. The shape of this diagram is related to the total abundance of the element. Computer modeling of stellar atmospheres has now largely superseded the technique

DICKE, ROBERT HENRY (1916-97)

American physicist and astronomer. In 1961, he suggested that the gravitational constant varies with time. In 1964, with the Canadian-born American physicist Phillip James Edwin Peebles (1935 -) and others, he began to develop a hot Big Bang theory, independently of G. Gamow. The theory predicted the existence of the cosmic background radiation, discovered shortly after by A.A. Penzias and R.W. Wilson. He also invented the Dicke radiometer and Dicke switch, and in 1957 set out what has become known as the weak anthropic principle.

DE SITTER, WILLEM (1872-1934) a Dutch mathematician and astronomer, was an early supporter of the theory of relativity, assessing its implications for astronomy. From the relativity of time and space theory, he derived, what is now, the de Sitter universe, the first theoretical model of an expanding Universe. In other publications, he refined the orbits and masses of Jupiter's Galilean satellites, and showed the rotation of the Earth to be gradually slowing. The DE SITTER UNIVERSE is a model of an expanding Universe in which there is no matter or radiation but a cosmological constant drives the expansion. W. De Sitter published this proposed theory in 1917 when he was strongly influence by the work and views of Albert Einstein. This model is physically unrealistic. The vantage this theory holds is that it introduces the idea that the real Universe is expanding at a certain precise rate. An expansion phase, which is very similar to that in the De Sitter model, also plays an important role in modern theories of the inflationary universe.

At the time the DURAC COSMOLOGY principal was firstly introduce it was receive with great skepticism by the scientific community of the day. Today more and more scientists are less skeptical about his views His is a cosmological theory built around the so-called large numbers hypothesis, which relates the fundamental constants of subatomic physics to large-scale properties of the Universe such as its age and mean density. It is due to the British mathematical physicist Paul Adrien Maurice Durac (1902-84). Durac's theory is not widely accepted, but it introduced ideas related to the anthropoid principle .My personal opinion is that Durac's ideas is dismissed rather on ground that does not suit the fashion trend of those in influence, rather than on grounds of in acceptance.

ALBERT EINSTEIN, (1879-1955) is worldwide viewed as the Motsart and Beethoven of the world of physics. This German-Swiss-American theoretical physicist's theories on the relativity still is widely not yet fully understood, by ALL members of science. They may think it helped to shape 20th century science but if implemented correctly, the profound implications that it will have on astronomy, would change all current views and wish full thinking. The special theory of relativity (published in 1905) arose out of the failure to detect the ether, and built on the work of the Dutch physicist Hendrik Antoon Lorenta (1853-1928) and the Irish physicist George Francis Fitzgerald (1851-1901). It yields the relation $E = mc^2$ between mass and energy, which was the key to understanding energy generation in stars. The general theory of relativity (announced 1915, published in expanded form 1916), which encompasses gravitation, assumes great importance in very large-scale systems, and rapidly had an impact on cosmology. Astronomy has furnished observational evidence to support these theories. Einstein produced no subsequent work of great significance, searching unsuccessfully for a theory that would link relativity with electromagnetic forces (a so-called grand unified theory).

Then the two wizards, mentioned above, joined forces and shared opinions to produce the EINSTEIN-DE SITTER UNIVERSE. They suggest a type of Universe in which the mean density of matter matches precisely the critical density of matter through out the universe. Such a model will not actually collapse, but will expand forever with a continually decreasing expansion rate. This model lies on the dividing line between a closed Friedmann Universe (which collapses) and an open Friedmann Universe (which does not). This model has the mathematical virtue of simplicity, in that it is spatially flat. It names after A. Einstein and W. de Sitter.

ETHER

A hypothetical medium once thought to permeate all space, through which electromagnetic radiation supposedly traveled; formerly spelt aether. Based on this supposition, the Earth should move with respect to the ether, and it was predicted that the speed of light would vary when measured in different directions. Experiments in the 19th century (e.g. the Michelson-Morley experiment) failed to detect any such variation in speed. The ether is now regarded as unnecessary, since it is recognized that electromagnetic radiation can propagate through empty space.

EXPANDING UNIVERSE
Any model Universe in which the space between widely separated objects is expanding. In the real Universe, neighboring objects such as close pairs of galaxies do not move apart because their mutual gravitational attraction exceeds the effect of the cosmological expansion. However, the distance between two widely separated galaxies, or clusters of galaxies, will increase as the Universe expands.

EUDOXUS OF CNIDUS   (c.400-c.350BC) was another product was of the golden pre Christian era which, of course then made him a Greek mathematician and astronomer, as he was born in modern Turkey. He developed a model of planetary motion in which the Sun, Moon, and planets travels, each to a specific route around the Earth on a series of 27 Earth-centreed spheres, with axes at different angles and rotating at different speeds. This theory of heavenly spheres, as modified first by Callipus and ultimately by Ptolemy, remained the orthodox view of the Universe for two thousand years. Eudoxus is also reputed to have introduced the constellation system from Egypt.

Far from me being in a position that I am, to criticize past Masters, however I do get the distinct feeling that Einstein came foreword with an opinion, and Hubble brought foreword the undeniable proof of galactica shift and the marriage of these concepts, resulted in the EXPANDING UNIVERSE theory. This theory envisaged a model of a Universe in which the space between widely separated objects is expanding. In the real Universe, neighboring objects such as close pairs of galaxies do not move apart because their mutual gravitational attraction exceeds the effect of the cosmological expansion. However, the distance between two widely separated galaxies, or clusters of galaxies, will increase as the Universe expands.

Again, I as an outsider, get the distinct impression that in the FRIEDMANN-UNIVERSE there is such an enormous strive to pleas the opinions of as many persons as possible, whilst bringing it all under one umbrella. The, all inclusive, all pleasing model, portrays an expanding Universe containing matter and radiation, but without a cosmological constant. Such a Universe is both homogeneous and isotropic. There is, in fact, a family of such universes including those, which expand forever (open universe), those that eventually collapse (closed universe), and the particular example of the Einstein-de Sitter universe, which has a critical density of matter. The geometric of space-time in these universes is described by the Robertson-Walker metric and is, in the preceding examples, negatively curved, positively curved, and flat respectively. The Friedmann models, originated by the Russian mathematician Alexander Alexandrovich Friedmann (1888-1925), form the bases of the standard Big Bang theory.

The GENERAL THEORY OF RELATIVITY, is by far, the most widely excepted, best known and least understood universal theory of them all. I do realize that I have a very good chance of landing myself before the modern day version of the inquest for saying this, but I do sometimes wonder if Einstein himself understood his concept. At a certain place in this book, I mention the fact that I am the only one that grasps Einstein's theory. This I say, not because I am opinionated by myself, but to the contrary, I find it unacceptable that the modern day scientists go out of their way in ignoring the factuality of this theory. If I am wrong then why does scientist, go on and on about achieving the speed of light, when this fact contradicts the whole meaning of every thing Einstein ever tried to explain. In this, well known theory, A. Einstein introduced to the world in 1915 a first ever view that describes how the gravitational fields of matter affect space and time. The theory predicts that gravitational fields change the geometry of space and time, causing it to curve. This curvature is apparent in a number of ways. First, light is bent in a gravitational field, a prediction that was confirmed by photographic measurements of the positions of stars near the limb of the Sun made during a total solar eclipse in 1919. The same effect manifests itself in a delay in radio signals from distant space probes as the signals pass the limb of the Sun. The curvature of space near the Sun also causes the perihelion point of Mercury's orbit to move forward, by 43" per century more than predicted by Newton in his view on the orbit of gravity. In the orbits of pulsars in binary systems, the advance of perihelion can amount to several degrees per year.

Another effect predicted by general relativity is the red shift of light caused by gravity. This later proved correct in the demonstration of the red shift of lines, which is present in the spectra of the Sun and, more noticeably, white dwarfs. Other predictions of the general theory include the gravitational lens effect; gravitational waves; singularities; and the invariance of the universal gravitational constant, G. General relativity was developed from the principle of equivalence between gravitational and inertial forces.

The **GÖDEL UNIVERSE** is the most bizarre concept ever thought up by any one. It has Hollywood written all over it. This is a most outrageous and unusual cosmological model, which represents a rotating universe. This model possesses a number of strange mathematical features, including the fact that it allows time-travel to occur within it. It is due to the Austrian-American mathematician Kurt Gödel (1906 – 78).

GOULD, BENJAMIN APTHORP (1824 – 96) was an American astronomer, which, in 1849 he founded the Astronomical Journal. In 1870, he moved to Argentina, founding the National Observatory at Córdoba, where he initiated the Córdoba Durchmusterung, a southern equivalent of the catalogue produced by F.W.A. Argelander for the northern stars. The name Gould's Belt is in his honor.

GOULD'S BELT
A band of hot, bright stars (types O and B) forming a circle around the sky. It represents a local structure of young stars and interstellar material tilted at about 16° to the galactic plane. Among the most prominent components of the belt are the bright stars in Orion, Canis Major, Puppis, Carina, Centaurus, and Scorpius, including the Sco-Cen Association. The belt has a diameter of about 3 000 l.y. (about one-tenth the radius of the Galaxy), and the Sun lies within it. Viewed from Earth, Gould's Belt projects below the plane of our Galaxy from the lower edge of the Orion Arm, and above the plane in the opposite direction. The belt is estimated to be about 50 million years old, but its origin is unknown. It is named after B.A. Gould, who established its existence in 1879.

To curb the confusion that presented it in the accumulation of the various theories a new theory was officially accepted and named the GRAND UNIFIED THEORY (GUT). So many new, information, became apparent, because of the development in technology, an attempt to describe the weak and strong nuclear forces and electromagnetism in a single mathematical theory. The unification of the weak force with electromagnetism was achieved in the electro weak theory. Before about 10^{-12} seconds after the Big Bang, by which time the Universe had cooled to about 10^{15} K, the electromagnetic and weak interactions acted as a single physical force; in the cooler temperatures since then, they have been distinct. Attempts to unify the electro weak force with the strong nuclear force have been only partially successful. It is thought that the temperature for their unification is of the order of 10^{27} K, which occurs only 10^{-36} s after the Big Bang. Particles surviving to the present day from this phase are possible candidates for non-baryonic dark matter. Unification of the GUT interaction with gravity may take place at higher energies still, but there is no satisfactory theory, which unifies all four physical forces. Such a theory would be called a theory of everything (TOE).

GRAVITATIONAL COLLAPSE
The collapse of a body that is unable to support itself against its own gravity. Gaseous bodies undergo such collapse if they are not hot enough for their gas pressure to balance gravity. This can happen in the early stages of star formation, or when nuclear burning ceases in a star's core. The time taken for such collapse decreases rapidly with increasing density, varying from about 100 000 years for the birth of a new star to less than a second for the formation of a neutron star. Star clusters may undergo a similar collapse if the random motions of their constituent stars are insufficient to offset gravitational effects, either during their formation or at an advanced stage of their evolution.

GRAVITATIONAL REDSHIFT
The redshift of light or other electromagnetic radiation caused by a strong gravitational field; also known as the Einstein shift. It arises because radiation loses energy as it passes out of the gravitational field of the emitting body. Therefore, the frequency of the radiation decreases and its wavelength is shifted to the red end of the spectrum. The redshift at wavelength λ is given by $Gm\lambda/c^2r$, where m is the mass of the body, r is the distance of the emitting region from the centre of mass, c is the speed of light, and G is the universal gravitational constant. A gravitational redshift has been observed in the light from some white dwarfs, and would result in the rapid fading out of a black hole in the process of formation as seen from outside.

GRAVITATIONAL WAVE
A wave-like motion in a gravitational field, produced when a mass is accelerated or otherwise disturbed. Gravitational waves travel through space-time at the speed of light, and their amplitude is proportional to the rate of acceleration of the body producing them. The strongest sources are those with the strongest gravitational fields although the waves, like the force of gravity itself, would be very weak. Gravitational waves have not yet been observed directly. However, the decay in the orbital period of the binary pulsar PSR 1913 + 16 is attributed to loss of energy through gravitational waves.

GRAVITON

A hypothetical particle or quantum of gravitational energy, predicted by the general theory of relativity. Gravitons have not been observed but are predicted to travel at the speed of light and to have zero rest mass and charge. A graviton is the gravitational equivalent of a photon.

GRAVITY ASSIST
The technique of using the gravitational field and orbital velocity of a planet to alter a spacecraft's trajectory and velocity; also known as a gravitational slingshot. As the spacecraft makes a close fly-by of a planet, its direction of travel is altered and it picks up additional speed from the planet's orbital velocity. The technique was first used by Mariner 10, which flew past Venus on its way to Mercury in 1974. The two Voyager probes made fly-bys of Jupiter, considerably shortening the time they took to reach Saturn. Voyager 2 subsequently used gravity assists from Saturn and Uranus to take it to Neptune. Other probes to use gravity assists were Giotto, Galileo and Ulysses.

GRAVITY GRADIENT
The direction of a gravitational field at a point within the field. Near a massive body such as a star or planet, the gravity gradient points to the body's centre. An elongated object in orbit about the body will revolve with its long axis pointing towards the body's centre.

In example, the Moon's longest axis lies along a line towards the body's centre. Artificial satellites can be oriented in orbit by making use of the gravity gradient.

H_2O maser
A maser source in which the water (H_2O) molecule is excited to maser action. They are the most widely distributed of all the cosmic masers. There are many different H_2O maser lines. The first to be discovered, in 1969, was the powerful line at 22.2 GHz (13.5 mm) in the Kleinmann-Low Nebula in Orion. Other H_2O lines at higher frequencies are difficult to observe with ground-based radio telescopes because of strong absorption by water vapor in the Earth's atmosphere. Water masers are found in star-forming regions, circumstellar envelopes, comets, and in the nuclei of some active galaxies in the form of mega masers.

Another link in the long chain of the period that blossomed during early Roman, late Greek period was HIPPARCHUS OF NICAEA (c.190-c.120BC) an astronomer, geographer, and mathematician, born in modern Turkey. He put Greek astronomy on a more scientific footing, introducing arithmetic and early trigonometric methods Without any aid of lenses his many accurate astronomical observations resulted in, as far as I know, a catalogue of 850 stars, which included their coordinates and dividing them into six magnitude classes. Ptolemy incorporated the catalogue and other findings by Hipparchus in the Almagest. Hipparchus made surprisingly accurate measurements of the precession of the equinoxes, the length of the year, and (from observations of eclipses) the Moon's distance. He may have been the inventor of the astrolabe.

The HOT BIG BANG is a logic conclusion derived from the BIG BANG THEORY and shares the same principles. The BIG BANG THEORY was the brainchild of Father Lemaitre's a Catholic priest who saw the Universe growing the same way as an egg does.

Father LE MAÎTRE, GEORGE ÉDOUARD (1894-1966) was a Belgian priest and cosmologist who was the first person to embrace the fact that the Universe expanded from an infant stage. His model of an expanding Universe (1927) was superior to that of W. de Sitter in that it took into account mass, gravitation and the curvature of space. Similar models had been proposed in the early 1920s by the Russian mathematician Alexander Alexandrovich Friedmann (1888-1925). Lemaître argued further (1931) that the quantum theory supported an origin in the explosion of a 'primeval atom' or 'cosmic egg' into which was originally concentrated all mass and energy. As modified by A.S. Eddington, Lemaître's model provided the springboard for G. Gamow's Big Bang theory.

LE MAÎTRE'S UNIVERSE
A model of the Universe containing a cosmological constant term, named after G.E. Lemaître. In this model, space has a positive curvature but expands forever. The Lemaître Universe is both homogeneous and isotropic. The most interesting aspect of such a Universe is that it undergoes a so-called coasting phase in which the cosmic scale factor is roughly constant with time.

This theory was an evolutionary process to act as an alternative term for the standard Big Bang theory. The word 'hot' was initially used to distinguish it from a rival theory, which had a cold initial phase. The existence of the cosmic background radiation requires that the Universe must have been hot in the past if the Big Bang picture is correct.

HOYLE, FRED (1915 -)

English astrophysicist and cosmologist. In 1948, with H. Bondi and T. Gold, he proposed the steady-state theory of the Universe in which matter is continuously created. Subsequently abandoned by most astronomers in favor of the Big Bang (so named from a dismissive remark by Hoyle), the steady-state theory nevertheless stimulated much important astrophysical research. Particularly significant was the work by Hoyle, W.A. Fowler, and G.R. and E.M. Bridge on nucleosynthesis in stars. As well as his suggesting for example those viruses and perhaps other life forms have been brought to Earth by comets. He has also been a popularizer of astronomy.

The man that (to my humble opinion) took cosmology into a new dimension was HUBBLE, EDWIN POWELL (1899-1953), the American astronomer. He first studied nebulae, concluding in 1917 that the spiral-shaped ones (which we now know as galaxies) were different in nature from diffuse nebulae, which he found to be gas clouds illuminated by stars. From 1923, using the 100-inch (2.5-m) telescope at Mount Wilson Observatory, he resolved the outer regions of the spiral nebulae M31 and M33 into star, identifying over 30 Cepheid variables in them. This proved that such 'nebulae' were truly independent star systems like our own – other galaxies. In 1925, he devised the so-called tuning-fork diagram of galaxies, dividing them into ellipticals, spirals, and barred spirals, which he believed to indicate an evolutionary sequence. By 1929, Hubble had good distance measurements for over twenty galaxies, including members of the Virgo Cluster. By comparing distances with their velocities, as revealed by the redshifts in their spectra, he concluded that galaxies were receding with speeds that increased with their distance, a relationship known as the Hubble law. This was powerful evidence that the Universe is expanding. The dynamics of his work was so far reaching everybody (including Einstein had to revise their theories to accommodate his findings. His findings are the most disputed, undisputed observations in all of history. The HUBBLE CLASSIFICATION is a widely used system for classifying galaxies according to their visual appearance, illustrated on the tuning-fork diagram. The sequence is based on three criteria: the relative sizes of the central bulge of stars and the flattened disk; the existence and character of spiral arms; and the resolution of the spiral arms and / or disk into stars and H II regions. The system was originated by E.P. Hubble.

The sequence starts with round elliptical galaxies (EO) showing no disks. Increasing flattening of a galaxy is indicated by a number which is calculated from $10(a - b)/a$, where, a, and b, are the major a minor axes as measured on the sky. No elliptical is known that is flatter than E7. Beyond this, a clear disk is apparent in the ventricular or SO galaxies. The classification then splits into two parallel sequences of disk galaxies showing spiral structure: ordinary spirals, S, and barred spirals, SB. The spiral types are subdivided into Sa, Sb, Sc, Sd (Sba, SBb, SBc, SBd for barred spirals). With each successive subdivision, the arms become less tightly wound (but more easily resolvable into stars and H II regions), and the central bulge becomes less dominant. Two types of irregular galaxy are defined. Irr I galaxies show rather amorphous, irregular structure with perhaps a hint of a spiral arm or bar, and can be placed at the far end of the spiral sequence. Irr II galaxies are sufficiently unusual to defy assignment to any of the other types, although this category encompasses only about 2% of bright or moderately bright galaxies in the nearby Universe. The original, erroneous idea that the sequence might be an evolutionary one led to the ellipticals refers to, as early-type galaxies, and the spirals and Irr I irregulars as late-type galaxies. Color and amount of interstellar material vary systematically along the Hubble sequence: ellipticals are red and contain little interstellar gas or dust, whereas late spirals and Irr I galaxies are blue, with significant amounts of interstellar material. The relatively faint dwarf spheroidal galaxies were not recognized as a separate type in the Hubble classification. Some variants of the Hubble classification use plus and minus signs to subdivide classes, so that Sa^+ is later than Sa, but earlier than Sb^-. The importance of the HUBBLE CONSTANT is still to this day, underestimated. This "constant" is well explained, for the first time, I might add, in this book. The Symbol H_o is the figure that relates the speed of an object's recession in the expanding Universe to its distance in the Hubble law. It represents the current rate of expansion of the Universe. This important cosmological parameter is usually measured in units of kilometers per second per megaparsec. In the Big Bang theory, H_o varies with time and it is therefore more properly known as the Hubble parameter. Its value is not accurately known but is thought to lie between 50 and 100 km/s/Mpc, recent research tending to favor values towards the lower end of this range. In the HUBBLE DIAGRAM, a graph plots either the redshift, or velocity of recession of galaxies against their apparent magnitude or distance from us. The Hubble law appears in the form of a straight line on such a plot. The original diagram, presented by E.P. Hubble in 1929, was the first indication that the Universe is expanding. The Hubble diagram is now mainly used to test the geometry of the Universe, since at large distances any departures from the simple linear form of the Hubble law should show up as a curve. The HUBBLE FLOW is the general outward motion of galaxies resulting from the uniform expansion of the Universe. All motions lie in a radial direction from the observer, and the velocities are proportional to the distance of the galaxies. The real pattern of galaxy motions is not exactly of this form, particularly close to us, because of the mutual gravitational interaction between galaxies; some nearby galaxies are even moving towards the Milky Way. At large distances, however, the discrepancies are small compared with the Hubble flow. All these findings are incorporated in the HUBBLE LAW, which is the mathematical equation of the principle law that governs the expansion of the Universe. According to the law, the apparent recession velocity of galaxies is proportional to their distance from the observer. In mathematical terms, $v = H_o r$, where v is the velocity, r the distance, and H_o the Hubble constant. The law was put forward in 1929 by E. P. Hubble.

HUBBLE RADIUS
A distance defined as the ratio of the velocity of light, c, to the value of the Hubble constant, H_o, This gives the distance from the observer at which the recession velocity of a galaxy would equal the speed of light. Roughly speaking, the Hubble radius is the radius of the observable Universe. Depending on the precise value of the Hubble constant, the Hubble radius lies between 9 and 18 billion l.y. This data is the basis on which the age of the Universe depends and is the HUBBLE TIME. The time required for the Universe to expand to its present size, assuming that the Hubble constant has remained unchanged since the Big Bang. It is defined as the reciprocal of the Hubble constant, $1/H_o$. Depending on the precise value of the Hubble constant, the Hubble time is between 9 and 18 billion years. In the standard Big Bang theory, the actual age of the Universe is always less than the Hubble time, because the expansion was faster in the past.

KANT, IMMANUEL (1724 – 1804) was the German philosopher, which proposed a cosmogony, published in 1755, in which the Solar System forms, via a disk, which condensed out of primordial material. The Solar System was part of a larger system (what we would now call a galaxy), and many of the nebulae seen by astronomers were in fact other galaxies, which he termed island universes. Kant was influenced by I. Newton's theories, which he termed island universes. Kant was, as everyone ells up to now, influenced by I. Newton's theories and by the English philosopher Thomas Wright of Durham (1711-86).

The Scottish physicist, Lord W. C KELVIN (1824-1907) was born in Ireland. He originated the thermodynamic temperature scale, and considered the consequences of energy dissipation in the Universe. Kelvin made one of the first scientific attempts at estimating the Earth's age, based on known cooling rates of materials, although his result (20 – 400 million years) was far too low. He also calculated the solar constant. He produced the KELVIN SCALE as a temperature scale in which the zero point is defined to be equal to $-273.16°$ Celsius. This zero point is also known as absolute zero. The thermodynamic temperature is expressed in Kelvin, symbol K.

The German mathematician and astronomer KEPLER, JOHANNES (1571-1630)
German mathematical and astronomer became Tycho Brahe's assistant in Prague in 1600 A. D. where he undertook to complete the tables of planetary motion Tycho had begun. Kepler first calculated the orbit of Mars. He spent much time trying to reconcile Tycho' s accurate observations of the planet with a circular orbit, but concluded (in Astronomia nova, published in 1609) that Mars moved instead in an elliptical orbit. Thus, he established the first of his laws of planetary motion. A theory that the Sun controlled the planets by a magnetic force led him to the second and third of his laws, which were published as part of his treatise on theoretical astronomy, Epitome astronomiae Coernicanae (1618-21). The Rudolphine Tables (named after Tycho's patron, the Holy Roman Emperor Rudolph II) of planetary motion appeared in 1627 and were still in use in the 18th century. Kepler also wrote De Stella nova, on the supernova of 1604 and Diptirce on optics and the theory of the telescope. The overall view followed in this book **Matter's Time in Space** places the true significance of his work in true contents. In KEPLER'S EQUATION is the equation that relates the eccentric anomaly of a body in an elliptical orbit to its mean anomaly. The equation is $E - e \sin E = M$., where E is the eccentric anomaly, M the mean anomaly, and e the eccentricity of the orbit. It is important as one of the mathematical relations enabling the position of a planet about the Sun, or a satellite about is planet, to be calculated from the orbital elements for any time. However this only relates to the solar system, and KEPLER'S LAWS only apply in the contents of the solar system. The three laws governing the orbital motions of the planets, discovered by J. Kepler is as follows: The first law states that the orbit of a planet is an ellipse with the Sun at one focus of the ellipse. The second law states that the radius vector joining planet to Sun sweeps out equal areas in equal times. The third law states that the square of the orbital period of each planet in years is proportional to the cube of the semi major axis of the planet's orbit. The first law gives the shape of the planet's orbit; the second describes how the planet must continuously vary its speed as it follows its orbit, moving fastest at perihelion and slowest at aphelion. The third law gives the relationship between the planets' average distances from the Sun and their periods of revolution. Instead of placing, the true value to Kepler's laws I. Newton placed his own interpretation to Kepler's laws, and in doing this, he willfully destroyed the principle working of the Creation. Through Newton's tunnel vision, he applied his own miss interpretations to the correct presumptions of Kepler. Newton reduced the implication that Kepler findings hold by introducing to the law of gravitation. he then went about and changed it to three laws of motion. I. Newton generalized Kepler's first law, verified the second law, and showed that the third law should be amended to the form; $4 \pi^2 a^3 / T^2 = G (m + m_p)$. In this, the value of T and a are the period of revolution and semi major axis of the orbit of a planet of mass m_p about the Sun of mass m, and G is the gravitational constant. The major aim of this book is to correct these misgivings of Newton.

KERR'S view of the BLACK HOLE is one of a rotating black hole, as distinct from a non-rotating Schwartzschild's black hole. Black holes are expected to rotate rapidly, since the stars that formed them would have been rotating; hence, they will be Kerr black holes. Several consequences arise from the addition of rotation to a black hole. First, the event horizon becomes elliptical, and its surface area, become less than that for a static black hole of the same mass. If the black hole were rotating sufficiently quickly, the area of the event horizon would reduce to zero, leaving the central singularity visible from outside (a naked singularity). Second, there is a region around a rotating black hole, the ergosphere, in which objects are forced to spin around the black hole. The outer edge of the ergosphere is the static limit. Third, a new, inner event horizon forms, and it becomes possible to travel through the black hole, and emerge into a new Universe or perhaps another part of our own Universe, through this second event horizon. Rotating black holes are named after the New Zealand mathematician Roy Patrick Kerr (1934 -), who first described their properties in 1963. Rotating black holes with electric charge are called Kerr-Newman black holes, but in practice, black holes are unlikely to have any significant electric charge.

KINETIC ENERGY
The energy possessed by a body by virtue of its motion in space. It is equivalent to the work that would be done if the moving body were brought to rest. Kinetic energy is equal to $\frac{1}{2}mv^2$, where m is the mass of the body and v is its velocity. A rotating body has kinetic energy $\frac{1}{2}I\varpi^2$, where I is its momentum of inertia and ϖ is its angular velocity.

KIRCHOFF, GUSTAV ROBERT (1824-87)
German physicist. With the chemist Robert Wilhelm Bunsen (1811-99) he established the principles of spectral analysis. In 1859 he reasoned that the Fraunhofer lines in the solar spectrum indicated that light from the photosphere was being absorbed at those wavelengths by the Sun's atmosphere. Furthermore, he realized that the Fraunhofer D lines were produced by sodium in the Sun's atmosphere, and other Fraunhofer lines would therefore reveal which other elements were present in the Sun. From then on, astronomical spectroscopy developed rapidly in the hands of others such as P.A. Secchi in Italy and W. Huggins in England.

KIRCHOFF'S LAWS
Three laws concerning spectra, stated in 1859 by the German physicist G.R. Kirchoff:
1.   A solid, liquid, or gas under high pressure, when heated to incandescence, produces a continuous spectrum.
2.   A gas under low pressure, but at a sufficiently high temperature, produces a spectrum of bright emission lines.
3.   A gas at low pressure (and low temperature) , lying between a hot continuum source and the observer, produces an absorption line spectrum, i.e. a number of dark lines superimposed on a continuous spectrum.

KUIPER, GERARD PETER (1905-73)
Dutch-American astronomer. In a search for new planetary satellites he discovered Miranda (in1948, orbiting Uranus) and Nereid (in 1949, orbiting Neptune). His spectroscopic studies revealed methane in the atmospheres of Uranus and Neptune. He also found methane bands in Titan's spectrum, demonstrating that the satellite has an atmosphere. He suggested the existence of the Kuiper Belt as the source of short-period comets. Kuiper was advisor on many American lunar and planetary missions, and proposed the idea of flying infrared telescopes on board high-altitude aircraft, which led to the Kuiper Airborne Observatory.

KUIPER BELT
A region of the outer Solar System containing an estimated 10^7-10^9 icy planetesimals, or comet nuclei. The Kuiper Belt is an inner, flattened extension of the Oort Cloud. It lies more or less in the same plane as the planets and extends outwards from around 30 AU (the orbit of Neptune) to perhaps 1 000 AU. Such a vast reservoir of comets beyond Neptune was first proposed in 1951 by G.P. Kuiper. In 1992, the British-born American astronomer David Clifford Jewitt (1958 -) and the Vietnamese-born American astronomer Jane Luu (1963 -) discovered the first Kuiper Belt object, 1992 QB_1. This has a diameter of about 200km, semi major axis 44.0 AU, orbital period about 296 years, perihelion 40.9 AU, aphelion 47.1 AU, and inclination 2°.2. Since than, dozens more have been found. The Kuiper Belt could be the source of most periodic comets. Unusual objects such as Chiron and Pholus may have originated in the Kuiper Belt.

LAGRANGE (-TOURNIER), JOSEPH LOUIS DE (1736-1813)
French mathematician, born in Italy. In celestial mechanics he studied perturbations and stability in the Solar System. He examined the three-body problem for the Earth, Moon and Sun (1764) and the motion of Jupiter's satellites (1766). In 1772 he found the particular solutions to the problem that give rise to the equilibrium positions now called Lagrangian points. Lagrange also studied the Moon's liberation. LAGRANGIAN POINT One of five points at which small bodies can remain the orbital plane of two massive bodies; also known as liberation points. Three of the points lie on the line joining the two massive bodies: L_1 lies between them, while L_2 and L_3 have the two bodies between them. These three points are unstable, slight displacements of a body from then resulting in its rapid departure. the fourth and fifth points (L_4 and L_5) each form an equilateral triangle with the two massive bodies, 60° ahead of and behind the smaller body in its orbit around the larger one. A well-known example of bodies flying at the L_4 and L_5 Lagrangian points are the Trojan asteroids in Jupiter's orbit. Among Saturn's

satellites, Telesto and Calypso lie at the L_4 and L_5 Lagrangian points in the orbit of the much larger Tethys. In similar fashion, tiny Helene precedes Saturn's satellite Dione, keeping 60° ahead of Dione. The Lagrangian points are named after the French mathematician J.L. de Lagrange, who first calculated their existence.

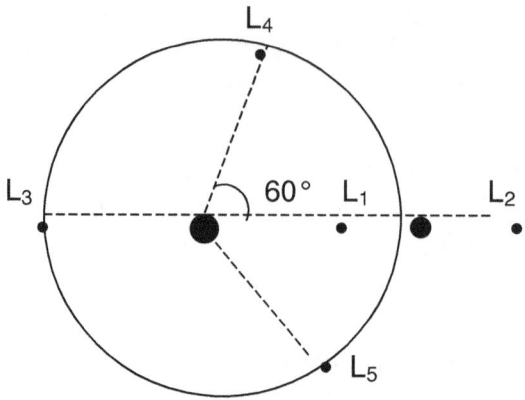

LAGRANGIAN POINT: *The Lagrangian points are five equilibrium points in the orbit of one body around another, such as a planet around the Sun.*

MAGELLANIC CLOUDS
The two irregular galaxies that are satellites of our own Galaxy, easily seen with the naked eye in the southern hemisphere like detached portions of the Milky Way. They are named after the Portuguese explorer Ferdinand Magellan (1480 –1521), who described them during his voyage around the world. Both Clouds are believed to orbit our Galaxy in a plane nearly perpendicular to its disk, and may eventually spiral into the Galaxy.

MATTER ERA
In the Big Bang theory, the era that started when the gravitational effect of matter began to dominate the effect of radiation pressure. Although radiation is massless, it has a gravitational effect which increases with the intensity of the radiation. Moreover, at high energies, matter itself behaves like electromagnetic radiation because it is moving at a speed close to that of light. In the very early Universe, the expansion rate was dominated by the gravitational effect of radiation pressure but, as the Universe cooled, this effect became less important than the gravitational effect of matter. Matter is thought to have become predominant at a temperature of around 10^4K, roughly 30 000 years after the Big Bang. This marked the start of the matter era.

MEAN DENSITY OF MATTER
The density of material that would be obtained if all the matter contained in galaxies were smoothed out across the universe. Although stars and planets have densities greater than the density of water (about 1 g/cm^3), the cosmological mean density is extremely low (less than 10^{-29} g/cm^3), or 10^{-5} atoms/cm^3) because the Universe consists mostly of virtually empty space between galaxies. The mean density of matter determines whether the Universe will continue to expand.

NEWTON, ISAAC (1642-1727)
English physicist and mathematician. He developed his principal theories of gravitation, optics and mathematics in 1665 and 1666. In 1668 he made the first working reflecting telescope. Most of his work remained unpublished for long periods, partly because of criticisms by c. Huygens and the English scientist Robert Hooke (1635-1703) of his early work on the corpuscular theory of light. However, in 1684 E. Halley persuaded him to organize his work on the celestial mechanics of the Solar System, which was published as the Principia. Newton's other major work, Opticks, was not published until 1704. It contains his corpuscular theory of light, and the theory of the telescope. His greatest mathematical achievement was his invention of calculus, independently of the German mathematician Gottfried Wilhelm Leibniz (1646-1716). His profound influence on physics and astronomy is reflected in the phrase 'Newtonian revolution'.

NEWTON'S LAWS OF MOTION
Three laws published in 1687 by I. Newton concerning the motion of bodies.
1.   A body continues in a state of uniform rest of motion unless acted upon by an external
        force.
2.   The acceleration produced when a force acts is directly proportional to the force and
        takes place in the direction in which the force acts.
3.   To every action there is an equal and opposite reaction.

OORT CLOUD
A roughly spherical halo of comet nuclei surrounding the Sun out to perhaps 100 000 AU (over one-third of the distance to the nearest star). Its existence was proposed in 1950 by J.H. Oort to account for the fact that new comets approach the Sun on highly elliptical orbits at all inclinations. The Oort Cloud remains a theoretical concept, since we cannot currently detect inert comets at such great distances. The cloud is estimated to contain some 10^{12} comets remaining from the formation of the Solar System. The most distant members are loosely bound by the Sun's gravity. There may be a greater concentration of comets relatively close to the ecliptic, at 10 000 – 20 000 AU from the Sun, extending inwards to join the Kuiper Belt. Oort Cloud comets are affected by the gravitational influence of passing stars, occasionally being perturbed into orbits, which take them through the inner Solar System.

OPEN UNIVERSE
A universe, which expands forever and has an infinite lifetime. A Friedmann Universe with a density less than the critical density is an example. It is not yet known whether our Universe is of this type.

OPPENHEIMER-VOLKOFF LIMIT
The maximum mass that a neutron star can have without it being overwhelmed by its own gravity. Calculations put this between 1.6 and 2 solar masses, although the exact figure is uncertain. A neutron star with a mass greater than this is expected to collapse further into a black hole. The limit is named after the American physicist (Julius) Robert Oppenheimer (1904-67) and the Russian-born Canadian George Michael Volkoff (1914 -).

PENROSE PROCESS
A process for extracting energy from a rotating black hole (i.e. a Kerr black hole). The process requires sending a mass into a trajectory in the ergosphere around the black hole, against the direction of rotation of the black hole. While inside the ergosphere the mass splits into two parts, one of which enters the black hole while the other escapes. Given a suitable trajectory, the emerging fragment may possess a total energy (i.e. rest mass plus kinetic energy) greater than the total energy of the mass that went in. The extra energy has been extracted from the rotational energy of the black hole, which must therefore slow down slightly. The process is named after the English mathematician Roger Penrose (1931 -), who discovered it in 1969.

PLANCK CONSTANT
(Symbol h) A constant that relates the energy of a photon to its frequency. It has the value 6.62076×10^{-34} Js. It is named after the German physicist Max Karl Ernst Ludwig Planck (1858 – 1947)

PLANCK ERA
In the Big Bang theory, the fleeting period between the Big Bang itself and the so-called Planck time when the Universe was 10^{-43} s old and the temperature was 10^{34}K. In this period, quantum gravitational effects are thought to have dominated. Theoretical understanding of this phase is virtually non-existent. It is named after Max Planck (1858-1947)

PLANCK'S LAW
A mathematical description of the energy radiated at different wavelengths by a black body: $E = hf$, where E is the energy of a photon and f its frequency. It was formulated in 1900 by Max Planck (1858-1947), who realized that energy is radiated in discrete packets, which he called quanta, and it formed the basis of quantum theory. The quantum of light is a photon, the energy of which depends on its wavelength.

PTOLEMAIC SYSTEM
The ancient Greek geocentric model of the Solar System, as described by Ptolemy . It may be traced back through the work of, for example, Hipparchus, Apollonius, Callippus and Eudoxus. The Earth is placed at the centre of the Universe, and around it revolve the Moon, Mercury, Venus, the Sun, Mars, Jupiter and Saturn. Beyond Saturn is the sphere of the fixed stars. In the basic model, each body moves along the circumference of a small circle, the epicycle, whose centre in turn follows the circumference of a larger circle, the deferent, centred on the Earth. In later refinements, Ptolemy introduced two points equally spaced on either side of the Earth: the eccentric and the equant. The centre of the epicycle revolved around the eccentric, not the Earth, and the orbiting body moved uniformly with respect to the equant. As a computational device the Ptolemaic system predicted planetary movements, including their retrograde motion, tolerably well, and survived with minor amendments until displaced by the Copernican system in the 16th century.

PTOLEMY (CLAUDIUS PTOLEMAEUS) (2nd century AD)
Egyptian astronomer and geographer. He produced the Almagest , a compendium of contemporary astronomical knowledge, drawing on writers, such as Plato and Hipparchus, whose works were kept in the great library at Alexandria. His Ptolemaic system was a geocentric model of the Universe. Highly contrived as it now appears, it

successfully accounted for the observed apparent motions of the planets, and remained largely unquestioned until the 16th century, when it was challenged by N. Copernicus. Ptolemy's Geography enjoyed a similar period of dominance (it convinced Columbus that he could sail westwards to India); his Tetrabiblos was an astrological treatise.

ROBERTSON-WALKER METRIC
A mathematical function describing the geometry of space-time in a model which incorporates the cosmological principle. In general, a metric relates physical distances or intervals between events separated in space and / or time to the coordinates used to describe their position. General relativity deals with four-dimensional space-time in which the separation between space and time coordinates is not obvious. In a homogeneous and isotropic cosmology, however, it is possible to define a unique time coordinate, called cosmic time, and three spatial coordinates. The Robertson-Walker metric is the most general possible four-dimensional metric function compatible with homogeneity and isotropy. In general, it describes a curved space which is either expanding or contracting with cosmic time. It is named after the American mathematician and cosmologist Howard Percy Robertson (1903 – 61) and the English mathematician Arthur Geoffrey Walker (1909 -)

The Roche limit is:
The region surrounding each star in a binary system, within which any material is gravitationally bound to that particular star. The boundary of the Roche lobes is an equipotential surface, and the lobes touch at the inner Lagrangian point, L_1, through which mass transfer may occur if one of the components expands to fill its lobe. It names after the French mathematician Edouard Albert Roche (1820-83).

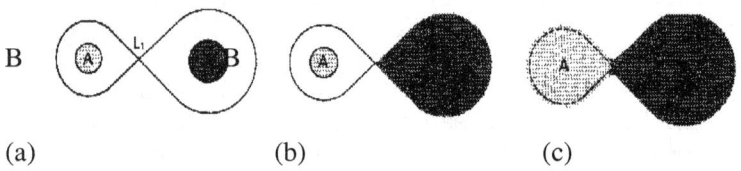

(a)   (b)   (c)

THE ROCHE LOBE: In a binary system, the Roche lobes of components A and B meet at the L_1 Lagrangian point. (a) In a detached system, neither star fills its Roche lobe. (b) In a semidetached system, one massive component, B, fills its Roche lobe. (c) In a contact binary, both components overfill their Roche lobes and share a common envelope.

SCHWARTZSCHILD BLACK HOLE
A non-rotating black hole with no electrical charge. This is the simplest case of a black hole, predicted by K. Schwartzschild in 1916 but is unlikely to be found in reality. The most likely form for black holes is the rotating Kerr black hole.

SCHWARTZSCHILD KARL (1873 – 1916)
German astronomer. He established a method for determining stars' brightness from photographs, comparing their visual and photographic magnitudes to obtain their colour index. In 1905 he obtained ultraviolet photographs of a solar eclipse, and went on to study energy transfer in the Sun, deducing that its outer regions had a layered structure. In1916 he showed that, in the general theory of relativity, a sphere of material (approximating to a star) collapsing under its own gravitational field past its Schwartzschild radius would cease to radiate energy (i.e. it would become a black hole). His son, Martin Schwartzschild (1912 – 97), became a naturalized American and studied stellar evolution.

SCHWARTZSCHILD RADIUS
The radius of the event horizon of a black hole. At the Schwartzschild radius the escape velocity becomes equal to the speed of light. The more massive the black hole, the larger the Schwartzschild radius. For a body of mass M the Schwartzschild radius is $2GM?c^2$, where G is the gravitational constant and c is the speed of light. It was first calculated in 1916 by K. Schwartzschild.

SPECIAL THEORY OF RELATIVITY
A theory proposed by A. Einstein in 1905, based on the proposition that the speed of light in a vacuum is constant throughout the Universe, and is independent of the motion pef the observer and the emitting body. A consequence of this proposition is that three things happen as an object's velocity approaches the speed of light: its mass goes up, its length shortens in the direction of motion, and time slows down. Hence, according to special relativity, no object can ever reach the speed of light because its mass would then become infinite, its length would become zero, and time would stand still. In addition, Einstein concluded that the mass of a body is a measure of its energy

content, according to the famous equation $E = mc^2$, where c is the speed of light. This equation describes the conversion of mass into energy in nuclear reactions within stars.

STEFAN-BOLTZMANN CONSTANT
(Symbol σ) A constant (appearing in the Stefan-Boltzmann law) that relates the luminosity of a black body to its thermodynamic temperature in Kelvin. It has the value 5.67051×10^{-8} W/m^2/K^4. Also called the Stefan constant.

STEFAN-BOLTZMANN LAW
A law relating the energy emitted by a black body (such as a star) to its temperature; also known as Stefan's law. According to the law, the total energy radiated in watts per square meter is proportional to the fourth power of its thermodynamic temperature in Kelvin; hence a doubling of temperature leads to a sixteen-fold increase in energy output. Expressed mathematically , $E = \sigma T^4$, where σ is the Stefan-Boltzmann constant. The total power per square meter can vary from 3 μW for the microwave background radiation, to75 MW for the sun, and thousands of gigawatts for hot stars such as whit dwarfs. The law was discovered by Joseph Stefan (1853-93) and derived theoretically by Ludwig Edward Boltzmann (1844-1906).

THERMAL EQUILIBRIUM
1.   A state in which two objects, or an object and its surroundings, have the same temperature so that there is no exchange of heat energy between them. For example, a telescope mirror should ideally be in thermal equilibrium with its supports and with the atmosphere to prevent distortion of the mirror or the creation of air currents within the telescope's tube.

2.   A state in which the available energy of an object is distributed uniformly among all the possible forms of energy; also known as thermodynamic equilibrium. for example, deep inside a star the radiation field, the kinetic energy, the excitation energy, and the ionization levels will all have equal amounts of energy. Furthermore, all processes are in balance so that, for example, there will be as many ionizations of helium per second as there are combinations of free electrons and helium ions. The condition of local thermodynamic equilibrium is often taken as an approximation when modeling stellar atmospheres.

UNIVERSAL TIME (UT)
A world-wide standard time-scale , the same as Greenwich Mean Time. Universal Time is the mean solar time on the meridian of Greenwich. It is defined as the Greenwich hour angle of the mean sun plus 12 hours, so that the day begins at midnight rather than noon. It is closely linked to Greenwich Mean Sidereal Tim (GMST), since the mean sidereal day is a precisely known fraction of the mean solar day. In practice, UT is determined by a formula from GMST, which in turn is derived directly from such observations of the meridian transits of stars. The version of UT derived directly form such observations is designated UTO, which is slightly dependent on the observing site. When UTO is corrected for the variation in longitude due to the Chandler wobble, a version of Universal Time, UT1, is derived which has genuine world-wide application. When UT1 is compared with International Atomic Time (TAI), it is found to be losing approximately a second a year against TAI. Broadcast time signals use the time-scale known as Coordinated Universal time (UTC). This is TAI with an offset of a whole number of seconds. The offset is adjusted when necessary by the introduction of a leap second, and UTC is always kept within 0.9 s of UT1.

UNIVERSE
Everything that exists, including space, time, and matter. The study of the Universe is known as cosmology. Cosmologists distinguish between the Universe with a capital 'U', meaning the cosmos and all its contents, and Universe with a small 'u' which is usually a mathematical model derived from some physical theory. The real Universe consists mostly of apparently empty space, with matter concentrated into galaxies consisting of stars and gas. The Universe is expanding, so the space between galaxies is gradually stretching, causing a cosmological redshift in the light from distant objects. There is growing evidence that space may be filled with unseen dark matter that may have many times the total mass of the visible galaxies. The most favoured concept of the origin of the Universe is the Big Bang theory, according to which the Universe came into being in a hot, dense fireball about 10-20 billion years ago.

VIRIAL THEOREM
A way of estimating the total mass of an object such as a galaxy or a cluster of galaxies from the movement of its individual members. The theorem states that the average gravitational potential energy of the constituent objects is twice their average kinetic energy. Calculations with the virial theorem show that galaxies and clusters contain up to ten times as much mass as can be seen telescopically, providing strong evidence for this existence of large quantities of dark matter. A modified version of this theorem, called the cosmic virial theorem, applies on cosmological scales. It relates the statistics of galaxy motions an the correlation function (which describes the way galaxies cluster in space) to the average density of the Universe. Since the first two quantities are measurable, the

density parameter can thus be estimated. The usual result obtained is around 0.2, indicating that there is dark matter on cosmological scales, but not enough to reach the critical density.

VOGT-RUSSEL THEOREM

The theorem, found valid except in rare circumstances that there is only one internal structure possible for a star of given mass and chemical composition. The calculation of that structure depends on knowing how quantities such as pressure, rate of energy production, and opacity depend on local gas properties such as temperature and chemical composition. The mass-radius and mass-luminosity relations in main sequence stars are among the theorem's consequences. It is named after the German astronomer Heinrich Vogt (1890-1968) and H.N. Russell.

WEIZÄCKER, CARL FRIEDRICH VON (1912 -)

German theoretical physicist and astrophysicist. In 1938, independently of H.A. Bethe, he proposed that stars generate their energy via the carbon-nitrogen cycle, converting hydrogen into helium by nuclear fusion. In 1944 he set out a modern version of the nebular hypothesis proposed in the 189th century, first by I. Kant and later by P.S. de Laplace, to account for the origin of the Solar System.